D1006389

# Bully
# for
# Brontosaurus

Ontogeny and Phylogeny
Ever Since Darwin
The Panda's Thumb
The Mismeasure of Man
Hen's Teeth and Horse's Toes
The Flamingo's Smile
An Urchin in the Storm
Time's Arrow, Time's Cycle
Illuminations (with R. W. Purcell)
Wonderful Life

# Bully for Brontosaurus

*Reflections in Natural History*

Stephen Jay Gould

W·W·NORTON & COMPANY
NEW YORK    LONDON

Jacket design by Mike McIver

Jacket painting by C.R. Knight, *Apatosaurus* (formerly called *Brontosaurus*),
Courtesy Department of Library Services, American Museum of Natural
History. Neg. trans. no. 2417 (3)

The text of this book is composed in 10/12 Baskerville, with the display set in
Baskerville. Composition and manufacturing by the Haddon Craftsmen, Inc.


Library of Congress Cataloging-in-Publication Data
Gould, Stephen Jay.
  Bully for brontosaurus : reflections in natural history / Stephen
Jay Gould.
    p.   cm.
  Includes index.
  ISBN 0-393-02961-1
  1. Natural history—Popular works.   2. Evolution—Popular works.
I. Title.
QH45.5.G68   1991
508—dc20                                              91-6916

ISBN 0-393-02961-1

W.W Norton & Company, Inc., 500 Fifth Avenue, New York, N.Y. 10110
W.W. Norton & Company, Ltd., 10 Coptic Street, London WCIA IPU

*Pleni sunt coeli*
*et terra*
*gloria eius.*

*Hosanna in excelsis.*

# Contents

*Prologue  11*

## 1 | HISTORY IN EVOLUTION

1   George Canning's Left Buttock and the Origin of Species  21

2   Grimm's Greatest Tale  32

3   The Creation Myths of Cooperstown  42

4   The Panda's Thumb of Technology  59

## 2 | DINOMANIA

5   Bully for Brontosaurus  79

6   The Dinosaur Rip-off  94

## 3 | ADAPTATION

7   Of Kiwi Eggs and the Liberty Bell  109

8   Male Nipples and Clitoral Ripples  124

9   Not Necessarily a Wing  139

## 4 | FADS AND FALLACIES

10   The Case of the Creeping Fox Terrier Clone  155

11   Life's Little Joke  168

12   The Chain of Reason versus the Chain of Thumbs  182

## 5 | ART AND SCIENCE

13    Madame Jeanette    201
14    Red Wings in the Sunset    209
15    Petrus Camper's Angle    229
16    Literary Bias on the Slippery Slope    241

## 6 | DOWN UNDER

17    Glow, Big Glowworm    255
18    To Be a Platypus    269
19    Bligh's Bounty    281
20    Here Goes Nothing    294

## 7 | INTELLECTUAL BIOGRAPHY

*Biologists*
21    In a Jumbled Drawer    309
22    Kropotkin Was No Crackpot    325
23    Fleeming Jenkin Revisited    340

*Physical Scientists*
24    The Passion of Antoine Lavoisier    354
25    The Godfather of Disaster    367

## 8 | EVOLUTION AND CREATION

*The World of T. H. Huxley*
26    Knight Takes Bishop?    385
27    Genesis and Geology    402

*Scopes to Scalia*
28    William Jennings Bryan's Last Campaign    416
29    An Essay on a Pig Roast    432
30    Justice Scalia's Misunderstanding    448

## 9 | NUMBERS AND PROBABILITY

31    The Streak of Streaks    463
32    The Median Isn't the Message    473
33    The Ant and the Plant    479

# 10 | PLANETS AS PERSONS

34      The Face of Miranda    489
35      The Horn of Triton    499

*Bibliography    513*
*Index    525*

# Prologue

IN FRANCE, they call this genre *vulgarisation*—but the implications are entirely positive. In America, we call it "popular (or pop) writing" and its practitioners are dubbed "science writers" even if, like me, they are working scientists who love to share the power and beauty of their field with people in other professions.

In France (and throughout Europe), *vulgarisation* ranks within the highest traditions of humanism, and also enjoys an ancient pedigree—from St. Francis communing with animals to Galileo choosing to write his two great works in Italian, as dialogues between professor and students, and not in the formal Latin of churches and universities. In America, for reasons that I do not understand (and that are truly perverse), such writing for nonscientists lies immured in deprecations—"adulteration," "simplification," "distortion for effect," "grandstanding," "whizbang." I do not deny that many American works deserve these designations—but poor and self-serving items, even in vast majority, do not invalidate a genre. "Romance" fiction has not banished love as a subject for great novelists.

I deeply deplore the equation of popular writing with pap and distortion for two main reasons. First, such a designation imposes a crushing professional burden on scientists (particularly young scientists without tenure) who might like to try their hand at this expansive style. Second, it denigrates the intelligence of millions of Americans eager for intellectual stimulation without patronization. If we writers assume a crushing mean of mediocrity and incomprehension, then not only do we have contempt

11

for our neighbors, but we also extinguish the light of excellence. The "perceptive and intelligent" layperson is no myth. They exist in millions—a low percentage of Americans perhaps, but a high absolute number with influence beyond their proportion in the population. I know this in the most direct possible way—by thousands of letters received from nonprofessionals during my twenty years of writing these essays, and particularly from the large number written by people in their eighties and nineties, and still striving, as intensely as ever, to grasp nature's richness and add to a lifetime of understanding.

We must all pledge ourselves to recovering accessible science as an honorable intellectual tradition. The rules are simple: no compromises with conceptual richness; no bypassing of ambiguity or ignorance; removal of jargon, of course, but no dumbing down of ideas (any conceptual complexity can be conveyed in ordinary English). Several of us are pursuing this style of writing in America today. And we enjoy success if we do it well. Thus, our primary task lies in public relations: We must be vigorous in identifying what we are and are not, uncompromising in our claims to the humanistic lineages of St. Francis and Galileo, not to the sound bites and photo ops in current ideologies of persuasion— the ultimate in another grand old American tradition (the dark side of anti-intellectualism, and not without a whiff of appeal to the unthinking emotionalism that can be a harbinger of fascism).

Humanistic natural history comes in two basic lineages. I call them Franciscan and Galilean in the light of my earlier discussion. Franciscan writing is nature poetry—an exaltation of organic beauty by corresponding choice of words and phrase. Its lineage runs from St. Francis to Thoreau on Walden Pond, W. H. Hudson on the English downs, to Loren Eiseley in our generation. Galilean composition delights in nature's intellectual puzzles and our quest for explanation and understanding. Galileans do not deny the visceral beauty, but take greater delight in the joy of causal comprehension and its powerful theme of unification. The Galilean (or rationalist) lineage has roots more ancient than its eponym—from Aristotle dissecting squid to Galileo reversing the heavens, to T. H. Huxley inverting our natural place, to P. B. Medawar dissecting the follies of our generation.

I love good Franciscan writing but regard myself as a fervent, unrepentant, pure Galilean—and for two major reasons. First, I would be an embarrassing flop in the Franciscan trade. Poetic writing is the most dangerous of all genres because failures are so conspicuous, usually as the most ludicrous form of purple prose (see James Joyce's parody, cited in Chapter 17). Cobblers should stick to their lasts and rationalists to their measured style. Second, Wordsworth was right. The child is father to the man. My youthful "splendor in the grass" was the bustle and buildings of New York. My adult joys have been walks in cities, amidst stunning human diversity of behavior and architecture—from the Quirinal to the Piazza Navona at dusk, from the Georgian New Town to the medieval Old Town of Edinburgh at dawn—more than excursions in the woods. I am not insensible to natural beauty, but my emotional joys center on the improbable yet sometimes wondrous works of that tiny and accidental evolutionary twig called *Homo sapiens*. And I find, among these works, nothing more noble than the history of our struggle to understand nature—a majestic entity of such vast spatial and temporal scope that she cannot care much for a little mammalian afterthought with a curious evolutionary invention, even if that invention has, for the first time in some four billion years of life on earth, produced recursion as a creature reflects back upon its own production and evolution. Thus, I love nature primarily for the puzzles and intellectual delights that she offers to the first organ capable of such curious contemplation.

Franciscans may seek a poetic oneness with nature, but we Galilean rationalists have a program of unification as well—nature made mind and mind now returns the favor by trying to comprehend the source of production.

This is the fifth volume of collected essays from my monthly series, "This View of Life," now approaching two hundred items over eighteen years in *Natural History* magazine (the others, in order, are *Ever Since Darwin, The Panda's Thumb, Hen's Teeth and Horse's Toes,* and *The Flamingo's Smile*). The themes may be familiar (with a good dollop of novelty, I trust), but the items are mostly new (and God has never left his dwelling place in the details).

Against a potential charge of redundancy, may I advance the immodest assertion that this volume is the best of the five. I think that I have become a better writer by monthly practice (I sometimes wish that all copies of *Ever Since Darwin* would self-destruct), and I have given myself more latitude of selection and choice in this volume. (The previous four volumes discarded only a turkey or two and then published all available items in three years of essays. This volume, covering six years of writing, presents the best, or rather the most integrated, thirty-five pieces from more than sixty choices.)

These essays, while centered on the enduring themes of evolution and the innumerable, instructive oddities of nature (frogs that use their stomachs as brood pouches, the gigantic eggs of Kiwis, an ant with a single chromosome), also record the specific passage of six years since the fourth volume. I have marked the successful completion of a sixty-year battle against creationism (since the Scopes trial of 1925) in our resounding Supreme Court victory of 1987 (see essays under "Scopes to Scalia"), the bicentennial of the French revolution (in an essay on Lavoisier, most prominent scientific victim of the Reign of Terror), and the magnificent completion of our greatest technical triumph in *Voyager*'s fly-by and photography of Uranus and Neptune (Essays 34 and 35). I also record, as I must, our current distresses and failures—the sorry state of science education (approached, as is my wont, not tendentiously, abstractly, and head-on, but through byways that sneak up on generality—fox terriers and textbook copying, or subversion of dinomania for intellectual benefit), and a sad epilogue on the extinction, between first writing and this republication, of the stomach-brooding frog.

Yet I confess that my personal favorites usually treat less immediate, even obscure, subjects—especially when correction of the errors that confined them to ridicule or obscurity retells their stories as relevant and instructive today. Thus, I write about Abbot Thayer's theory that flamingos are red to hide them from predators in the sunset, Petrus Camper's real intent (criteria for art) in establishing a measure later used by scientific racists, the admirable side of William Jennings Bryan and the racist nonsense in the text that John Scopes used to teach evolution, the actual (and much more interesting) story behind the heroic, cardboard version of the Huxley-Wilberforce debate of 1860.

For what it's worth, my own favorite is Essay 21 on N. S. Shaler and William James (I won't reveal my vote for the worst essays—especially since they have been shredded in my mental refuse bin and will not be included in these volumes). At least Essay 21 best illustrates my favorite method of beginning with something small and curious and then working outward and onward by a network of lateral connections. I found the fearful letter of Shaler to Agassiz in a drawer almost twenty years ago. I always knew that I would find a use for it someday—but I had no inkling of the proper context. A new biography of Shaler led me to explore his relationship with Agassiz. I then discovered the extent of Shaler's uncritical (and lifelong) fealty by reading his technical papers. At this point, luck intervened. One of my undergraduate advisees told me that William James, as a Harvard undergraduate, had sailed with Agassiz to Brazil on the master's penultimate voyage. I knew that Shaler and James had been friendly colleagues and intellectual adversaries—and now I had full connectivity in their shared link to Agassiz. But would anything interesting emerge from all these ties? Again, good fortune smiled. James had been critical of Agassiz right from the start—and in the very intellectual arena (contingency versus design in the history of life) that would host their later disagreements as distinguished senior professors. I then found a truly amazing letter from James to Shaler offering the most concise and insightful rebuttal I have ever read to the common misconception—as current today as when James and Shaler argued—that the improbability of our evolution indicates divine intent in our origin. James's document—also a brilliant statement on the general nature of probability—provided a climax of modern relevance for a story that began with an obscure note lying undiscovered in a drawer for more than a hundred years. Moreover, James's argument allowed me to resolve the dilemma of the museum janitor, Mr. Eli Grant, potential victim of Shaler's cowardly note—so the essay ends by using James's great generality to solve the little mystery of its beginning, a more satisfactory closure (I think) than the disembodied abstraction of James's brilliance.

Finally, and now thrice lucky, I received two years later a fascinating letter from Jimmy Carter presenting a theological alternative to the view of contingency and improbability in human evolution advanced in my last book, *Wonderful Life.* Carter's argument, though more subtle and cogent than Shaler's, follows the same logic—and James's rebuttal has never been bettered or more apropos. And so, by presidential proclamation, I had an epilogue that proved the modern relevance of Shaler's traditionalism versus James's probing.

Some people have seen me as a polymath, but I insist that I am a tradesman. I admit to a broad range of explicit detail, but all are chosen to illustrate the common subjects of evolutionary change and the nature of history. And I trust that this restricted focus grants coherence and integration to an overtly disparate range of topics. The bullet that hit George Canning in the ass really is a vehicle for discussing the same historical contingency that rules evolution. My sweet little story about nostalgia at the thirtieth reunion of my All-City high school chorus is meant to be a general statement (bittersweet in its failure to resolve a cardinal dichotomy) about the nature of excellence. The essay on Joe DiMaggio's hitting streak is a disquisition on probability and pattern in historical sequences; another on the beginnings of baseball explores creation versus evolution as primal stories for the origin of any object or institution. And Essay 32, the only bit I have ever been moved to write about my bout with cancer, is not a confessional in the personal mode, but a general statistical argument about the nature of variation in populations—the central topic of all evolutionary biology.

A final thought on Franciscans and Galileans in the light of our environmental concerns as a tattered planet approaches the millennium (by human reckoning—as nature, dealing in billions, can only chuckle). Franciscans engage the glory of nature by direct communion. Yet nature is so massively indifferent to us and our suffering. Perhaps this indifference, this majesty of years in uncaring billions (before we made a belated appearance), marks her true glory. Omar Khayyám's old quatrain grasped this fundamental truth (though he should have described his Eastern hotel, his metaphor for the earth, as grand rather than battered):

> Think, in this battered caravanserai
> Whose portals are alternate night and day,
> How sultan after sultan with his pomp
> Abode his destined hour, and went his way.

The true beauty of nature is her amplitude; she exists neither for nor because of us, and possesses a staying power that all our nuclear arsenals cannot threaten (much as we can easily destroy our puny selves).

The hubris that got us into trouble in the first place, and that environmentalists seek to avoid as the very definition of their (I should say our) movement, often creeps back in an unsuspected (and therefore potentially dangerous) form in two tenets frequently advanced by "green" movements: (1) that we live on a fragile planet subject to permanent ruin by human malfeasance; (2) that humans must act as stewards of this fragility in order to save our planet.

We should be so powerful! (Read this sentence with my New York accent as a derisive statement about our false sense of might, not as a literal statement of desire.) For all our mental and technological wizardry, I doubt that we can do much to derail the earth's history in any permanent sense by the proper planetary time scale of millions of years. Nothing within our power can come close to conditions and catastrophes that the earth has often passed through and beyond. The worst scenario of global warming under greenhouse models yields an earth substantially cooler than many happy and prosperous times of a prehuman past. The megatonnage of the extraterrestrial impact that probably triggered the late Cretaceous mass extinction has been estimated at 10,000 times greater than all the nuclear bombs now stockpiled on earth. And this extinction, wiping out some 50 percent of marine species, was paltry compared to the granddaddy of all—the Permian event some 225 million years ago that might have dispatched up to 95 percent of species. Yet the earth recovered from these superhuman shocks, and produced some interesting evolutionary novelties as a result (consider the potential for mammalian domination, including human emergence, following the removal of dinosaurs).

But recovery and restabilization occur at planetary, not human, time scales—that is, millions of years after the disturbing event. At this scale, we are powerless to harm; the planet will take care of itself, our puny foolishnesses notwithstanding. But this time scale, though natural for planetary history, is not appropriate in our legitimately parochial concern for our own species, and the current planetary configurations that now support us. For these planetary instants—our millennia—we do hold power to impose immense suffering (I suspect that the Permian catastrophe was decidedly unpleasant for the nineteen of twenty species that didn't survive).

We certainly cannot wipe out bacteria (they have been the modal organisms on earth right from the start, and probably shall be until the sun explodes); I doubt that we can wreak much permanent havoc upon insects as a whole (whatever our power to destroy local populations and species). But we can surely eliminate our fragile selves—and our well-buffered earth might then breathe a metaphorical sigh of relief at the ultimate failure of an interesting but dangerous experiment in consciousness. Global warming is worrisome because it will flood our cities (built so often at sea level as ports and harbors), and alter our agricultural patterns to the severe detriment of millions. Nuclear war is an ultimate calamity for the pain and death of billions, and the genetic maiming of millions in future generations.

Our planet is not fragile at its own time scale, and we, pitiful latecomers in the last microsecond of our planetary year, are stewards of nothing in the long run. Yet no political movement is more vital and timely than modern environmentalism—because we must save ourselves (and our neighbor species) from our own immediate folly. We hear so much talk about an environmental ethic. Many proposals embody the abstract majesty of a Kantian categorical imperative. Yet I think that we need something far more grubby and practical. We need a version of the most useful and ancient moral principle of all—the precept developed in one form or another by nearly every culture because it acts, in its legitimate appeal to self-interest, as a doctrine of stability based upon mutual respect. No one has ever improved upon the golden rule. If we execute such a compact with our planet, pledging to cherish the earth as we would wish to be treated ourselves, she may relent and allow us to muddle through. Such a limited goal may strike some readers as cynical or blinkered. But remember that, to an evolutionary biologist, persistence is the ultimate reward. And human brainpower, for reasons quite unrelated to its evolutionary origin, has the damnedest capacity to discover the most fascinating things, and think the most peculiar thoughts. So why not keep this interesting experiment around, at least for another planetary second or two?

# 1 | History in Evolution

# 1 | George Canning's Left Buttock and the Origin of Species

I KNOW the connection between Charles Darwin and Abraham Lincoln. They conveniently contrived to enter the world on the same day, February 12, 1809, thus providing forgetful humanity with a mnemonic for ordering history. (Thanks also to John Adams and Thomas Jefferson for dying on the same momentous day, July 4, 1826, exactly fifty years after our nation's official birthdate.)

But what is the connection between Charles Darwin and Andrew Jackson? What can an English gentleman who mastered the abstractions of science hold in common with Old Hickory, who inaugurated the legend (later exploited by Lincoln) of the backwoodsman with little formal education fighting his way to the White House? (Jackson was born on the western frontier of the Carolinas in 1767, but later set up shop in the pioneer territory of Nashville.) This more difficult question requires a long string of connections more worthy of Rube Goldberg than of logical necessity. But let's have a try, in nine easy steps.

1. Andy Jackson, as a result of his military exploits in and around the ill-fated War of 1812, became a national figure, and ultimately, on this basis, a presidential contender. In a conflict conspicuously lacking in good news, Jackson provided much solace by winning the Battle of New Orleans, our only major victory on land after so many defeats and stalemates. With help from the privateer Jean Lafitte (who was then pardoned by President Madison but soon resumed his old ways), Jackson decisively defeated the British forces on January 8, 1815, and compelled their withdrawal from Louisiana. Cynics often point out, perhaps ungener-

21

ously, that Jackson's victory occurred more than two weeks after the war had officially ended, but no one had heard the news down in the bayous because the treaty had been signed in Ghent and word then traveled no faster than ship.

2. When we were about to withdraw from Vietnam and acknowledge (at least privately) that the United States had lost the war, some supporters of that venture (I was not among them) drew comfort from recalling that, patriotic cant aside, this was not our first military defeat. Polite traditions depict the War of 1812 as a draw, but let's face it, basically we lost—at least in terms of the larger goal espoused by hawks of that era: the annexation of Canada, at least in part. But we did manage to conserve both territory and face, an important boon to America's future and a crucial ingredient in Jackson's growing reputation. Washington, so humiliated just a few months before when British troops burned the White House and the Capitol, rejoiced in two items of news, received in early 1815 in reverse order of their actual occurrence: Jackson's victory at New Orleans, and the favorable terms of the Treaty of Ghent, signed on December 24, 1814.

3. The Treaty of Ghent restored all national boundaries to their positions before the war; thus, we could claim that we had lost not an inch of territory, even though expansion into Canada had been the not-so-hidden aim of the war's promoters. The treaty provided for commissions of arbitration to settle other points of dispute between the United States and Canada; all remaining controversies were negotiated peacefully under these provisions, including the establishment of our unfortified boundary, the elimination of naval forces from the Great Lakes, and the settlement of the Saint Lawrence boundary. Thomas Boylston Adams, descendant of John Quincy Adams (who negotiated and signed the treaty), recently wrote of that exemplary document (in his wonderful column "History Looks Ahead," appearing twice a month in the *Boston Globe*): "The treaty . . . ended a war that never should have been begun. Yet its consummation was unbounded good. The peace then confirmed . . . has never been broken. Its bounty has been the cheerful coexistence of two friendly nations divided by nothing more tangible than an invisible line that runs for 3,000 miles undefended by armed men or armaments."

4. If the war had not ended, fortunately for us, on such an upbeat, Andy Jackson's belated victory at New Orleans might

have emerged as a bitter joke rather than a symbol of (at least muted) success—and Jackson, deprived of status as a military hero, might never have become president. But why did Britain, in a fit of statesmanship, agree to such a conciliatory treaty, when they held the upper hand militarily? The reasons are complex and based, in part, on expediency (the coalition that had exiled Napoleon to Elba was coming apart, and more troops might soon be needed in Europe). But much credit must also go to the policies of Britain's remarkable foreign secretary, Robert Stewart, Viscount Castlereagh. In a secret dispatch sent to the British minister in Washington in 1817, Castlereagh set out his basic policy for negotiation, a stance that had guided the restructuring of Europe at the Congress of Vienna, following the final defeat of Napoleon: "The avowed and true policy of Great Britain in the existing State of the World is to secure if possible, for all states a long interval of repose."

Three years earlier, Castlereagh had put flesh on these brave words by helping to break the deadlock at Ghent and facilitate a peace treaty that did not take all that Britain could have demanded, thereby leaving the United States with both pride and flexibility for a future and deeper peace with Britain. Negotiations had gone badly at Ghent; anger and stalemate ruled. Then, on his way to Vienna, Castlereagh stopped for two days in Ghent, where, in secret meetings with his negotiators, he advocated conciliation and helped to break the deadlock.

5. We must thank the fortunate tides of history that Castlereagh, rather than his counterpart and rival, the hawkish and uncompromising George Canning, was presiding over Britain's foreign affairs in 1814. (And so you see, dear reader, we are finally getting to Mr. Canning's rear end, as promised in the title.) The vagaries of a key incident in 1809 led to this favorable outcome. Canning, then foreign secretary, had been pushing for Castlereagh's ouster as secretary of war. Castlereagh had sent a British expedition against Napoleon's naval base at Antwerp, but nature had intervened (through no fault of Castlereagh's), and the troops were boxed in on the island of Walcheren, dying in droves of typhoid fever. Canning used this disaster to press his advantage.

Meanwhile (this does get complicated), the prime minister, the duke of Portland, suffered a paralytic stroke and eventually had to resign. In the various reshufflings and explanations that follow such an event, Perceval, the new prime minister, showed Castlereagh some of Canning's incriminating letters. Castlereagh did not challenge Canning's right to lobby for his removal, but he exploded in fury at Canning's apparent secrecy in machination. Canning, for his part (and not without justice), replied that he had urged open confrontation of the issue, but that higher-ups (including the king) had imposed secrecy, hoping to paper over the affair and somehow preserve the obvious talents of both men in government.

Castlereagh, to say the least, was not satisfied and, in the happily abandoned custom of his age, insisted upon a duel. The two men and their seconds met on Putney Heath at 6 A.M. on September 21. They fired a first round to no effect, but Castlereagh insisted on a second, of much greater import. Castlereagh was spared the fate of Alexander Hamilton by inches, as Canning's bullet removed a button from his coat but missed his person. Canning was not so fortunate; though more embarrassed than seriously injured, he took Castlereagh's second bullet in his left buttock. (Historians have tended to euphemism at this point. The latest biography of Castlereagh holds that Canning got it "through the fleshy part of the thigh," but I have it on good authority that Canning was shot in the ass.) In any case, both men subsequently resigned.

As the world turns and passions cool, both Canning and Castlereagh eventually returned to power. Canning achieved his burning ambition (cause of his machinations against Castlereagh) to become prime minister, if only briefly, in 1827. Castlereagh came back in Canning's old job of foreign secretary, where he assured the Treaty of Ghent and presided for Britain at the Congress of Vienna.

6. Suppose Canning had fired more accurately and killed Castlereagh on the spot? Canning, or another of his hawkish persuasion, might have imposed stiffer terms upon the United States and deprived Andy Jackson of his hero's role. More important for our tale, Castlereagh would have been denied the opportunity to die as he actually did, by his own hand, in 1822. Castlereagh had suffered all his life from periods of acute and debilitating "melancholy" and would, today, almost surely be diagnosed as a severe manic depressive. Attacked by the likes of Lord Byron, Shelley,

and Thomas Moore for his foreign policies, and suffering from both overwork and parliamentary reverses, Castlereagh became unreasonably suspicious and downright paranoid. He thought that he was being blackmailed for supposed acts of homosexuality (neither the blackmail nor the sexual orientation has ever been proved). His two closest friends, King George IV and the duke of Wellington, failed to grasp the seriousness of his illness and did not secure adequate protection or treatment. On August 12, 1822, though his wife (fearing the worst) had removed all knives and razors from his vicinity, Castlereagh rushed into his dressing room, seized a small knife that had been overlooked, and slit his throat.

7. Yes, we are getting to Darwin, but it takes a while. Point seven is a simple statement of genealogy: Lord Castlereagh's sister was the mother of Robert FitzRoy, captain of HMS *Beagle* and host to Charles Darwin on a five-year voyage that bred the greatest revolution in the history of biology.

8. Robert FitzRoy took command of the *Beagle* at age twenty-three, after the previous captain had suffered a mental breakdown and shot himself. FitzRoy was a brilliant and ambitious man. He had been instructed to take the *Beagle* on a surveying voyage of the South American coast. But FitzRoy's own plans extended far beyond a simple mapping trip, for he hoped to set a new standard of scientific observation on a much broader scale. To accomplish his aim, he needed more manpower than the Admiralty was willing to supply. As a person of wealth, he decided to take some extra passengers at his own expense, to beef up the *Beagle's* scientific mettle.

A popular scientific myth holds that Darwin sailed on the *Beagle* as official ship's naturalist. This is not true. The official naturalist was the ship's surgeon, Robert McKormick. Darwin, who disliked McKormick and did eventually succeed him as naturalist (after the disgruntled McKormick "invalided out," to use the euphemism of his time), originally sailed as a supernumerary passenger at FitzRoy's discretion.

Why, then, did FitzRoy tap Darwin? The obvious answer—that Darwin was a promising young scientist who could aid FitzRoy's plans for improved observation—may be partly true, but does not get to the heart of FitzRoy's reasons. First of all, Darwin may have possessed abundant intellectual promise, but he had no scientific credentials when he sailed on the *Beagle*—a long-standing interest in natural history and bug collecting to be sure, but neither a degree in science nor an intention to enter the profession (he was preparing for the ministry at the time).

FitzRoy took Darwin along primarily for a much different, and personal, reason. As an aristocratic captain, and following the naval customs of his time, FitzRoy could have no social contact with officers or crew during long months at sea. He dined alone and conversed with his men only in an official manner. FitzRoy understood the psychological toll that such enforced solitude could impose, and he remembered the fate of the *Beagle*'s previous skipper. He decided on a course of action that others had followed in similar circumstances: He decided to take along, at his own expense, a supernumerary passenger to serve, in large part, as a mealtime companion for conversation. He therefore advertised discreetly among his friends for a young man of appropriate social status who could act as both social companion and scientific aid. Charles Darwin, son of a wealthy physician and grandson of the great scholar Erasmus Darwin, fitted the job description admirably.

But most captains did not show such solicitude for their own mental health. Why did FitzRoy so dread the rigors of solitude? We cannot know for sure, but the answer seems to lie, in good part, with the suicide of his uncle, Lord Castlereagh. FitzRoy, by Darwin's own account, was fearful of a presumed hereditary predisposition to madness, an anxiety that he embodied in the suicide of his famous uncle, whom he so much resembled in looks as well as temperament. Moreover, FitzRoy's fears proved well founded, for he did break down and temporarily relinquish his command in Valparaiso during a period of overwork and tension. On November 8, 1834, Darwin wrote to his sister Catherine: "We have had some strange proceedings on board the *Beagle* ... Capt. FitzRoy has for the last two months, been working *extremely* hard and at the same time constantly annoyed. ... This was accompanied by a morbid depression of spirits, and a loss of all decision and resolution. The Captain was afraid that his mind was becoming deranged (being aware of his hereditary predisposition). ... He invalided and Wickham was appointed to the command."

Late in life, and with some hindsight, Darwin mused on the character of Captain FitzRoy in his autobiography:

FitzRoy's character was a singular one, with many very noble features: he was devoted to his duty, generous to a fault, bold, determined, indomitably energetic, and an ardent friend to all under his sway. . . . He was a handsome man, strikingly like a gentleman, with highly courteous manners, which resembled those of his maternal uncle, the famous Lord Castlereagh. . . . FitzRoy's temper was a most unfortunate one. This was shown not only by passion but by fits of long-continued moroseness. . . . He was also somewhat suspicious and occasionally in very low spirits, on one occasion bordering on insanity. He was extremely kind to me, but was a man very difficult to live with on the intimate terms which necessarily followed from our messing by ourselves in the same cabin. [Darwin does mean "eating," and we find no sexual innuendo either here or anywhere else in their relationship.]

I am struck by the similarity, according to Darwin's description, between FitzRoy and his uncle, Lord Castlereagh, not only in physical characteristics and social training, but especially in the chronicle of a mental history so strongly implying a lifelong pattern of severe manic depression. In other words, I think that FitzRoy was correct in his self-diagnosis of a tendency to hereditary mental illness. Castlereagh's dramatic example had served him well as a warning, and his decision, so prompted, to take Darwin on the *Beagle* was history's reward.

But suppose Canning had killed Castlereagh, rather than just removing a button from his coat? Would FitzRoy have developed so clear a premonition about his own potential troubles without the terrible example of his beloved uncle's suicide during his most impressionable years (FitzRoy was seventeen when Castlereagh died)? Would Darwin have secured his crucial opportunity if Canning's bullet had been on the mark?

Tragically, FitzRoy's premonition eventually came to pass in almost eerie consonance with his own nightmare and memory of Castlereagh. FitzRoy's later career had its ups and downs. He suffered from several bouts of prolonged depression, accompanied by increasing suspicion and paranoia. In his last post, FitzRoy served as chief of the newly formed Meteorological Office and became a pioneer in weather forecasting. FitzRoy is much admired today for his cautious and excellent work in a most difficult field. But he encountered severe criticism during his own tenure, and for the obvious reason. Weathermen take enough flak today for incorrect predictions. Imagine the greater uncertainties more than a century ago. FitzRoy was stung by criticism of his imprecision. With a healthy mind, he would have parried the blows and come out fighting. But he sank into even deeper despair and eventually committed suicide by slitting his throat on April 20, 1865. Darwin mourned for his former friend (and more recent enemy of evolution), noting the fulfillment of the prophecy that had fostered his own career: "His end," Darwin wrote, "was a melancholy one, namely suicide, exactly like that of his uncle Ld. Castlereagh, whom he resembled closely in manner and appearance."

9. Finally, the other short and obvious statement: We must reject the self-serving historical myth that Darwin simply "saw" evolution in the raw when he broke free from the constraints of his culture and came face to face with nature all around the world. Darwin, in fact, did not become an evolutionist until he returned to England and struggled to make sense of what he had observed in the light of his own heritage: of Adam Smith, William Wordsworth, and Thomas Malthus, among others. Nonetheless, without the stimulus of the *Beagle,* I doubt that Darwin would have concerned himself with the origin of species or even entered the profession of science at all. Five years aboard the *Beagle* did serve as the sine qua non of Darwin's revolution in thought.

My chain of argument runs in two directions from George Canning's left buttock: on one branch, to Castlereagh's survival, his magnanimous approach to the face-saving Treaty of Ghent, the consequent good feeling that made the Battle of New Orleans a heroic conquest rather than a bitter joke, to Andrew Jackson's emergence as a military hero and national figure ripe for the presidency; on the other branch, to Castlereagh's survival and eventual death by his own hand, to the example thus provided to his similarly afflicted nephew Robert FitzRoy, to FitzRoy's consequent decision to take a social companion aboard the *Beagle,* to the choice of Darwin, to the greatest revolution in the history of biological thought. The duel on Putney Heath branches out in innumerable directions, but one leads to Jackson's presidency and the other to Darwin's discovery.

I don't want to push this style of argument too far, and this essay is meant primarily as comedy (however feeble the attempt). Anyone can set out a list of contrary proposals. Jackson was a tough customer and might have made his way to the top without a boost from New Orleans. Perhaps FitzRoy didn't need the drama of Castlereagh's death to focus a legitimate fear for his own sanity. Perhaps Darwin was so brilliant, so purposeful, and so destined that he needed no larger boost from nature than a beetle collection in an English parsonage.

No connections are certain (for we cannot perform the experiment of replication), but history presents, as its primary fascination, this feature of large and portentous movements arising from tiny quirks and circumstances that appear insignificant at the time but cascade into later, and unpredictable, prominence. The chain of events makes sense after the fact, but would never occur in the same way again if we could rerun the tape of time.

I do not, of course, claim that history contains nothing predictable. Many broad directions have an air of inevitability. A theory of evolution would have been formulated and accepted, almost surely in the mid-nineteenth century, if Charles Darwin had never been born, if only for the simple reason that evolution is true, and not so veiled from our sight (and insight) that discovery could long have tarried behind the historical passage of cultural barriers to perception.

But we are creatures of endless and detailed curiosity. We are not sufficiently enlightened by abstractions devoid of flesh and bones, idiosyncrasies and curiosities. We cannot be satisfied by concluding that a thrust of Western history, and a dollop of geographic separation, virtually guaranteed the eventual independence of the United States. We want to know about the tribulations at Valley Forge, the shape of the rude bridge that arched the flood at Concord, the reasons for crossing out "property" and substituting "pursuit of happiness" in Jefferson's great document. We care deeply about Darwin's encounter with Galápagos tortoises and his studies of earthworms, orchids, and coral reefs, even if a dozen other naturalists would have carried the day for evolution had Canning killed Castlereagh, FitzRoy sailed alone, and Darwin become a country parson. The details do not merely embellish an abstract tale moving in an inexorable way. The details are the story itself; the underlying predictability, if discernible at all, is too nebulous, too far in the background, and too devoid of hooks upon actual events to count as an explanation in any satisfying sense.

Darwin, that great beneficiary of a thousand chains of improbable circumstance, came to understand this principle and to grasp thereby the essence of history in its largest domain of geology and life. When America's great Christian naturalist Asa Gray told Darwin that he was prepared to accept the logic of natural selection but recoiled at the moral implications of a world without divine guidance, Darwin cited history as a resolution. Gray, in obvious distress, had posed the following argument: Science implies lawfulness; laws (like the principle of natural selection) are instituted by God to ensure his benevolent aims in the results of nature; the path of history, however full of apparent sorrow and death, must therefore include purpose. Darwin replied that laws surely exist and that, for all he knew, they might well embody a purpose legitimately labeled divine. But, Darwin continued, laws only regulate the broad outlines of history, "with the details, whether good or bad, left to the working out of what we may call chance." (Note Darwin's careful choice of words. He does not mean "random" in the sense of uncaused; he speaks of events so complex and contingent that they fall, by their unpredictability and unrepeatability, into the domain of "what we may *call* chance.")

But where shall we place the boundary between lawlike events and contingent details? Darwin presses Gray further. If God be just, Darwin holds, you could not claim that the improbable death of a man by lightning or the birth of a child with serious mental handicaps represents the general and inevitable way of our world (even though both events have demonstrable physical causes). And if you accept "what we may call chance" (the presence of this man under that tree at that moment) as an explanation for a death, then why not for a birth? And if for the birth of an individual, why not for the origin of a species? And if for the origin of a species, then why not for the evolution of *Homo sapiens* as well?

You can see where Darwin's chain of argument is leading: Human intelligence itself—the transcendent item that, above all else, supposedly reflected God's benevolence, the rule of law, and the necessary progress of history—might be a detail, and not the predictable outcome of first principles. I wouldn't push this argument to an absurd extreme. Consciousness in some form might lie in the realm of predictability, or at least reasonable probability. But we care about details. Consciousness in *human* form—by means of a brain plagued with inherent paths of illogic, and weighted down by odd and dysfunctional inheritances, in a body with two eyes, two legs, and a fleshy upper thigh—is a detail of history, an outcome of a million improbable events, never destined to repeat. We care about George Canning's sore behind because we sense, in the cascade of consequences, an analogy to our own tenuous existence. We revel in the details of history because they are the source of our being.

# 2 | Grimm's Greatest Tale

WITH THE POSSIBLE EXCEPTION of Eng and Chang, who had no choice, no famous brothers have ever been closer than Wilhelm and Jacob Grimm, who lived and worked together throughout their long and productive lives. Wilhelm (1786–1859) was the prime mover in collecting the *Kinder- und Hausmärchen* (fables for the home and for children) that have become a pillar and icon of our culture. (Can you even imagine a world without Rapunzel or Snow White?) Jacob, senior member of the partnership (1785–1863), maintained a primary interest in linguistics and the history of human speech. His *Deutsche Grammatik,* first published in 1819, became a cornerstone for documenting relationships among Indo-European languages. Late in their lives, after a principled resignation from the University of Göttingen (prompted by the king of Hanover's repeal of the 1833 constitution as too liberal), the brothers Grimm settled in Berlin where they began their last and greatest project, the *Deutsches Wörterbuch*—a gigantic German dictionary documenting the history, etymology, and use of every word contained in three centuries of literature from Luther to Goethe. Certain scholarly projects are, like medieval cathedrals, too vast for completion in the lifetimes of their architects. Wilhelm never got past *D; * Jacob lived to see the letter *F.*

Speaking in Calcutta, during the infancy of the British raj in 1786, the philologist William Jones first noted impressive similarities between Sanskrit and the classical languages of Greece and Rome (an Indian king, or raja, matches *rex,* his Latin counterpart). Jones's observation led to the recognition of a great Indo-European family of languages, now spread from the British Isles and Scandinavia to India, but clearly rooted in a single, ancient origin. Jones may have marked the basic similarity, but the brothers Grimm were among the first to codify regularities of change that underpin the diversification of the rootstock into its major subgroups (Romance languages, Germanic tongues, and so on). Grimm's law, you see, does not state that all frogs shall turn into princes by the story's end, but specifies the characteristic changes in consonants between Proto–Indo-European (as retained in Latin) and the Germanic languages. Thus, for example, Latin *p*'s become *f*'s in Germanic cognates (voiceless stops become voiceless fricatives in the jargon). The Latin *plēnum* becomes "full" (*voll,* pronounced "foll" in German); *piscis* becomes "fish" (*Fisch* in German); and *pēs* becomes "foot" (*Fuss* in German). (Since English is an amalgam of a Germanic stock with Latin-based imports from the Norman conquest, our language has added Latin cognates to Anglo-Saxon roots altered according to Grimm's law—*plenty, piscine,* and *podiatry.* We can even get both for the price of one in *plentiful.*)

I first learned about Grimm's law in a college course more than twenty-five years ago. Somehow, the idea that the compilers of Rapunzel and Rumpelstiltskin also gave the world a great scholarly principle in linguistics struck me as one of the sweetest little facts I ever learned—a statement, symbolic at least, about interdisciplinary study and the proper contact of high and vernacular culture. I have wanted to disgorge this tidbit for years and am delighted that this essay finally provided an opportunity.

A great dream of unification underlay the observations of Jones and the codification of systematic changes by Jacob Grimm. Nearly all the languages of Europe (with such fascinating exceptions as Basque, Hungarian, and Finnish) could be joined to a pathway that spread through Persia all the way to India via Sanskrit and its derivatives. An origin in the middle, somewhere in the Near East, seemed indicated, and such "fossil" Indo-European tongues as Hittite support this interpretation. Whether the languages were spread, as convention dictates, by conquering nomadic tribes on horseback or, as Colin Renfrew argues in his recent book (*Archaeology and Language,* 1987), more gently and passively by the advantages of agriculture, evidence points to a single source with a complex history of proliferation in many directions.

Might we extend the vision of unity even further? Could we link Indo-European with the Semitic (Hebrew, Arabic) languages of the so-called Afro-Asiatic stock; the Altaic languages of Tibet, Mongolia, Korea, and Japan; the Dravidian tongues of southern India; even to the native Amerindian languages of the New World? Could the linkages extend even further to the languages of southeastern Asia (Chinese, Thai, Malay, Tagalog), the Pacific Islands, Australia, and New Guinea, even (dare one dream) to the most different tongues of southern Africa, including the Khoisan family with its complex clicks and implosions?

Most scholars balk at the very thought of direct evidence for connections among these basic "linguistic phyla." The peoples were once united, of course, but the division and spread occurred so long ago (or so the usual argument goes) that no traces of linguistic similarity should be left according to standard views about rates of change in such volatile aspects of human culture. Yet a small group of scholars, including some prominent émigrés from the Soviet Union (where theories of linguistic unification are not so scorned), persists in arguing for such linkages, despite acrimonious rebuttal and dismissal from most Western colleagues. One heterodox view tries to link Indo-European with linguistic phyla of the Near East and northern Asia (from Semitic at the southwest, to Dravidian at the southeast, all the way to Japanese at the northeast) by reconstructing a hypothetical ancestral tongue called Nostratic (from the Latin *noster,* meaning "our"). An even more radical view holds that modern tongues still preserve enough traces of common ancestry to link Nostratic with the native languages of the Americas (all the way to South America via the Eskimo tongues, but excluding the puzzling Na-Dene languages of northwestern America).

The vision is beguiling, but I haven't the slightest idea whether any of these unorthodox notions has a prayer of success. I have no technical knowledge of linguistics, only a hobbyist's interest in language. But I can report, from my own evolutionary domain, that the usual biological argument, invoked a priori against the possibility of direct linkage among linguistic phyla, no longer applies. This conventional argument held that *Homo sapiens* arose and split (by geographical migration) into its racial lines far too long ago for any hope that ancestral linguistic similarities might be retained by modern speakers. (A stronger version held that various races of *Homo sapiens* arose separately and in parallel from different stocks of *Homo erectus,* thus putting the point of common linguistic ancestry even further back into a truly inaccessible past. Indeed, according to this view, the distant common ancestor of all modern people might not even have possessed language. Some linguistic phyla might have arisen as separate evolutionary inventions, scotching any hope for theories of unification.)

The latest biological evidence, mostly genetic but with some contribution from paleontology, strongly indicates a single and discrete African origin for *Homo sapiens* at a date much closer to the present than standard views would have dared to imagine— perhaps only 200,000 years ago or so, with all non-African diversity perhaps no more than 100,000 years old. Within this highly compressed framework of common ancestry, the notion that conservative linguistic elements might still link existing phyla no longer seems so absurd a priori. The idea is worth some serious testing, even if absolutely nothing positive eventually emerges.

This compression of the time scale also suggests possible success for a potentially powerful research program into the great question of historical linkages among modern peoples. Three major and entirely independent sources of evidence might be used to reconstruct the human family tree: (1) direct but limited evidence of fossil bones and artifacts by paleontology and archaeology; (2) indirect but copious data on degrees of genetic relationship among living peoples; (3) relative similarities and differences among languages, as discussed above. We might attempt to correlate these separate sources, searching for similarities in pattern. I am delighted to report some marked successes in this direction ("Reconstruction of Human Evolution: Bringing Together Genetic, Archaeological, and Linguistic Data," by L. L. Cavalli-Sforza, A. Piazza, P. Menozzi, and J. Mountain, *Proceedings of the National Academy of Sciencs,* 1988). The reconstruction of the human family tree—its branching order, its timing, and its geography—may be within our grasp. Since this tree is the basic datum of history, hardly anything in intellectual life could be more important.

Our recently developed ability to measure genetic distances for large numbers of protein or DNA sequences provides the keystone for resolving the human family tree. As I have argued many times, such genetic data take pride of place not because genes are "better" or "more fundamental" than data of morphology, geography, and language, but only because genetic data are so copious and so comparable. We all shared a common origin, and therefore a common genetics and morphology, as a single ancestral population some quarter of a million years ago. Since then, differences have accumulated as populations separated and diversified. As a rough guide, the more extensive the measured differences, the greater the time of separation. This correlation between extent of difference and time of separation becomes our chief tool for reconstructing the human family tree.

But this relationship is only rough and very imperfect. So many factors can distort and disrupt a strict correlation of time and difference. Similar features can evolve independently—black skin in Africans and Australians, for example, since these groups stand as far apart genealogically as any two peoples on earth. Rates of change need not be constant. Tiny populations, in particular, can undergo marked increases in rate, primarily by random forces of genetic drift. The best way to work past these difficulties lies in a "brute force" approach: The greater the quantity of measured differences, the greater the likelihood of a primary correlation between time and overall distance. Any single measure of distance may be impacted by a large suite of forces that can disrupt the correlation of time and difference—natural selection, convergence, rapid genetic drift in small populations. But time is the only common factor underlying all measures of difference; when two populations split, all potential measures of distance become free to diverge. Thus, the more independent measures of distance we compile, the more likely we are to recover the only common signal of diversification: time itself. Only genetic data (at least for now) can supply this required richness in number of comparisons.

Genetic data on human differences are flowing in from laboratories throughout the world, and this essay shall be obsolete before it hits the presses. Blood groups provided our first crude insights during the 1960s, and Cavalli-Sforza was a pioneer in these studies. When techniques of electrophoresis permitted us to survey routinely for variation in the enzymes and proteins coded directly by genes, then data on human differences began to accumulate in useful cascades. More recently, our ability to sequence DNA itself has given us even more immediate access to the sources of variation.

The methodologically proper and powerful brute force comparisons are, for the moment, best made by studying differing states and frequencies of genes as revealed in the amino acid sequences of enzymes and proteins. Cavalli-Sforza and colleagues used information from alleles (varying states of genes, as in tall versus short for Mendel's peas) to construct a tree for human populations least affected by extensive interbreeding. (Few human groups are entirely aboriginal, and most populations are interbred to various degrees, given the two most characteristic attributes of *Homo sapiens:* wanderlust and vigorous sexuality. Obviously, if we wish to reconstruct the order of diversified branching from a common point of origin, historically mixed populations will confuse our quest. The Cape Colored, living disproof from their own ancestors for the Afrikaner "ideal" of apartheid, would join Khoisan with Caucasian. One town in Brazil might well join everyone.)

Cavalli-Sforza's consensus tree, based on overall genetic distances among 120 alleles for 42 populations—probably the best we can do for now, based on the maximal amount of secure and consistent information—divides modern humans into seven major groups, as shown in the accompanying chart. Only branching order counts in assessing relative similarity, not the happenstance of alignment along the bottom of the chart. Africans are not closer to Caucasians than to Australians just because the two groups are adjacent; rather, Africans are equally far from all other peoples by virtue of their common branching point with the ancestor of all six additional groups. (Consider the diagram as a mobile, free to rotate about each vertical "string." We could turn around the entire array of Groups II to VII, placing Australians next to Africans and Caucasians at the far right, without altering the branching order.)

These seven basic groups, established solely on genetic distances, make excellent sense when we consider the geographic distribution of *Homo sapiens.* Humans presumably evolved in Africa, and the first great split separates Africans from all other groups—representing the initial migration of some *Homo sapiens*

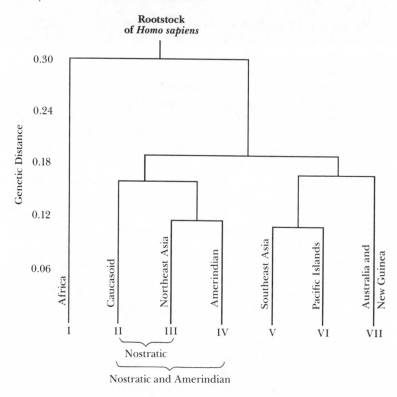

Cavalli-Sforza's consensus tree for the evolutionary relationships of human groups based on overall genetic distances. Postulated relationships among language families match this pattern remarkably well. See text for details. IROMIE WEERAMANTRY. COURTESY OF *NATURAL HISTORY*.

out of the mother continent. The next split separates the coherent region of the Pacific and Southeast Asia from the rest of the world. One group reached Australia and New Guinea, perhaps 40,000 years ago, forming the aboriginal populations of this region. A later division separated the Pacific island peoples (Group VI, including Polynesians, Micronesians, and Melanesians) from the southeastern Asiatics (Group V, including southern Chinese, Thai, Malayan, and Filipino).

Meanwhile, the second great branch divided to split the northern Oriental stocks from the Caucasians (Group II, including

Europeans, Semitic peoples of southwest Asia, Iranians, and Indians). A second division separated the Native American peoples (Group IV) from the northeast Asian family (Group III, including the Uralic peoples who left Hungarian, Finnish, and Estonian as their non–Indo-European calling cards from invasions into Caucasian territories, and the Altaic peoples of Mongolia, Korea, and Japan).

This good and sensible order indicates that genetic data are not betraying our efforts to reconstruct the human family tree. But Cavalli-Sforza and colleagues go further toward the great promise of extending this correlation between genes and geography to the other great sources of independent information—the geological and linguistic records.

I find the linguistic correlations more exciting than anything else in the work of Cavalli-Sforza and colleagues. Language is so volatile. Conquerors can impose their language as well as their will. Tongues interpenetrate and merge with an explosive ease not granted to genes or morphology. Look at English; look at any of us. I, for example, live in America, the indigenous home of very different people. I speak English, and consider the cathedral of Chârtres the world's most beautiful building. But my grandparents spoke Hungarian, a non–Indo-European language. And, along with Disraeli, my more distant ancestors were priests in the Temple of Solomon when the physical forebears of the original English people still lived as "brutal savages in an unknown island." One might have anticipated very little correlation between language and the tree of human ancestry.

Yet the mapping of linguistic upon genetic tree is remarkable in its degree of overlap. Exceptions exist, of course, and for the reasons mentioned above. Ethiopians speak an Afro-Asiatic language (in the phylum of Hebrew and Arabic), but belong to the maximally distant African group by genes. The Tibetan language links with Chinese in Group V, although the Tibetan people belong with northeast Asians in Group III. But Tibetans migrated from the steppes north of China, and Ethiopians have maintained primary contact and admixture with Semitic speakers for millennia. The correlations, however, are striking. Each genetic group also defines either a single linguistic phylum or a few closely related phyla. The Pacific island languages, with their mellifluous vowels and nearly nonexistent consonants, define

Group VI almost as well as the genetic distances. The Indo-European languages set the borders of Caucasian affinity, while the other major tongues of Caucasian peoples (Afro-Asiatic of the Semitic group) belong to a related linguistic phylum.

I am especially intrigued that the heterodox hypotheses for linkages among linguistic phyla, and for potential reconstructions of human languages even closer to the original tongue, follow the genetic connections so faithfully. Nostratic would link Groups II and III. The even more heterodox connection of Nostratic with Amerindian tongues would include Group IV as well. Note that Groups II to IV form a coherent limb of the human family tree. The Tower of Babel may emerge as a strikingly accurate metaphor. We probably did once speak the same language, and we did diversify into incomprehension as we spread over the face of the earth. But this original tongue was not an optimal construction given by a miracle to all people. Our original linguistic unity is only historical happenstance, not crafted perfection. We were once a small group of Africans, and the mother tongue is whatever these folks said to each other, not the Holy Grail.

This research has great importance for the obvious and most joyously legitimate parochial reason—our intense fascination with ourselves and the details of our history. We really do care that our species arose closer to 250,000 than to 2 million years ago, that Basque is the odd man out of European languages, and that the peopling of the Americas is not mysterious for its supposed "delay," but part of a regular process of expansion from an African center, and basically "on time" after all.

But I also sense a deeper importance in this remarkable correlation among all major criteria for reconstructing our family tree. This high correspondence can only mean that a great deal of human diversity, far more than we ever dared hope, achieves a remarkably simple explanation in history itself. If you know when a group split off and where it spread, you have the basic outline (in most cases) of its relationships with others. The primary signature of time and history is not effaced, or even strongly overlain in most cases, by immediate adaptation to prevailing circumstances or by recent episodes of conquest and amalgamation. We remain the children of our past—and we might even be able to pool our differences and to extract from inferred pathways of change a blurred portrait of our ultimate parents.

The path is tortuous and hard to trace, as the sister of the seven ravens learned when she went from the sun to the moon to the glass mountain in search of her brothers. History is also a hard taskmaster, for she covers her paths by erasing so much evidence from her records—as Hansel and Gretel discovered when birds ate their Ariadne's thread of bread crumbs. Yet the potential rewards are great, for we may recover the original state so hidden by our later changes—the prince behind the frog or the king that became the bear companion of Snow White and Rose Red. And the criteria that may lead to success are many and varied—not only the obvious data of genes and fossils but also the clues of language. For we must never doubt the power of names, as Rumpelstiltskin learned to his sorrow.

# 3 | The Creation Myths of Cooperstown

YOU MAY EITHER LOOK upon the bright side and say that hope springs eternal or, taking the cynic's part, you may mark P.T. Barnum as an astute psychologist for his proclamation that suckers are born every minute. The end result is the same: You can, Honest Abe notwithstanding, fool most of the people all of the time. How else to explain the long and continuing compendium of hoaxes—from the medieval shroud of Turin to Edwardian Piltdown Man to an ultramodern array of flying saucers and astral powers—eagerly embraced for their consonance with our hopes or their resonance with our fears.

Some hoaxes make a sufficient mark upon history that their products acquire the very status initially claimed by fakery—legitimacy (although as an object of human or folkloric, rather than natural, history; I once held the bones of Piltdown Man and felt that I was handling an important item of Western culture).

The Cardiff Giant, the best American entry for the title of paleontological hoax turned into cultural history, now lies on display in a shed behind a barn at the Farmer's Museum in Cooperstown, New York. This gypsum man, more than ten feet tall, was "discovered" by workmen digging a well on a farm near Cardiff, New York, in October 1869. Eagerly embraced by a gullible public, and ardently displayed by its creators at fifty cents a pop, the Cardiff Giant caused quite a brouhaha around Syracuse, and then nationally, for the few months of its active life between exhumation and exposure.

The Cardiff Giant was the brainchild of George Hull, a cigar manufacturer (and general rogue) from Binghamton, New York.

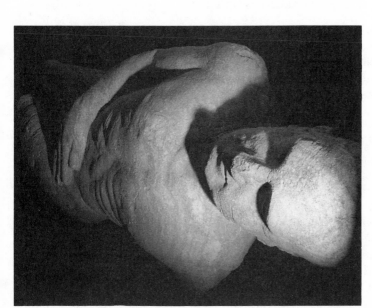

The Cardiff Giant as now on display in the Farmer's Museum in Cooperstown, New York. NEW YORK STATE HISTORICAL ASSOCIATION, COOPERSTOWN, NY.

## THE GREAT
# CARDIFF GIANT!

Discovered at Cardiff, Onondaga Co. N. Y. is now on Exhibition in the

## Geological Hall, Albany,

### For a few days only.

#### HIS DIMENSIONS.

| | | |
|---|---|---|
| Length of Body, | 10 feet, | 4 1-2 inches. |
| Length of Head from Chin to Top of Head. | 21 | : |
| Length of Nose, | 6 | : |
| Across the Nostrils, | 3 1-2 | : |
| Width of Mouth, | 5 | : |
| Circumference of Neck, | 37 | : |
| Shoulders, from point to point, | 3 feet, 1 1-2 | : |
| Length of Right Arm. | 4 feet, 9 1-2 | : |
| Across the Wrist, | 5 | : |
| Across the Palm of Hand. | 7 | : |
| Length of Second Finger. | 8 | : |
| Around the Thighs, | 6 feet. 3 1-2 | : |
| Diameter of the Thigh. | 13 | : |
| Through the Calf of Leg. | 9 1-2 | : |
| Length of Foot. | 21 | : |
| Across the Ball of Foot. | 8 | : |
| Weight. | | 2990 pounds. |

ALBANY November 19th. 1869.

A broadsheet from 1869 giving vital statistics of the Cardiff Giant. NEW YORK STATE HISTORICAL ASSOCIA-TION, COOPERSTOWN, NY.

He quarried a large block of gypsum from Fort Dodge, Iowa, and shipped it to Chicago, where two marble cutters fashioned the rough likeness of a naked man. Hull made some crude and minimal attempts to give his statue an aged appearance. He chipped off the carved hair and beard because experts told him that such items would not petrify. He drove darning needles into a wooden block and hammered the statue, hoping to simulate skin pores. Finally, he dumped a gallon of sulfuric acid all over his creation to simulate extended erosion. Hull then shipped his giant in a large box back to Cardiff.

Hull, as an accomplished rogue, sensed that his story could not hold for long and, in that venerable and alliterative motto, got out while the getting was good. He sold a three-quarter interest in the Cardiff Giant to a consortium of highly respectable businessmen, including two former mayors of Syracuse. These men raised the statue from its original pit on November 5 and carted it off to Syracuse for display.

The hoax held on for a few more weeks, and Cardiff Giant fever swept the land. Debate raged in newspapers and broadsheets between those who viewed the giant as a petrified fossil and those who regarded it as a statue wrought by an unknown and wondrous prehistoric race. But Hull had left too many tracks—at the gypsum quarries in Fort Dodge, at the carver's studio in Chicago, along the roadways to Cardiff (several people remembered seeing an awfully large box passing by on a cart). By December, Hull was ready to recant, but held his tongue a while longer. Three months later, the two Chicago sculptors came forward, and the Cardiff Giant's brief rendezvous with fame and fortune ended.

The common analogy of the Cardiff Giant with Piltdown Man works only to a point (both were frauds passed off as human fossils) and fails in one crucial respect. Piltdown was cleverly wrought and fooled professionals for forty years, while the Cardiff Giant was preposterous from the start. How could a man turn to solid gypsum, while preserving all his soft anatomy, from cheeks to toes to penis? Geologists and paleontologists never accepted Hull's statue. O. C. Marsh, later to achieve great fame as a discoverer of dinosaurs, echoed a professional consensus in his unambiguous pronouncement: "It is of very recent origin and a decided humbug."

Why, then, was the Cardiff Giant so popular, inspiring a wave of interest and discussion as high as any tide in the affairs of men during its short time in the sun? If the fraud had been well executed, we might attribute this great concern to the dexterity of the hoaxers (just as we grant grudging attention to a few of the most accomplished art fakers for their skills as copyists). But since the Cardiff Giant was so crudely done, we can only attribute its fame to the deep issue, the raw nerve, touched by the subject of its fakery—human origins. Link an absurd concoction to a noble and mysterious subject and you may prevail, at least for a while. My opening reference to P.T. Barnum was not meant sarcastically; he was one of the great practical psychologists of the nineteenth century—and his motto applies with special force to the Cardiff Giant: "No humbug is great without truth at bottom." (Barnum made a copy of the Cardiff Giant and exhibited it in New York City. His mastery of hype and publicity assured that his model far outdrew the "real" fake when the original went on display at a rival establishment in the same city.)

For some reason (to be explored, but not resolved, in this essay), we are powerfully drawn to the subject of beginnings. We yearn to know about origins, and we readily construct myths when we do not have data (or we suppress data in favor of legend when a truth strikes us as too commonplace). The hankering after an origin myth has always been especially strong for the closest subject of all—the human race. But we extend the same psychic need to our accomplishments and institutions—and we have origin myths and stories for the beginning of hunting, of language, of art, of kindness, of war, of boxing, bow ties, and brassieres. Most of us know that the Great Seal of the United States pictures an eagle holding a ribbon reading *e pluribus unum*. Fewer would recognize the motto on the other side (check it out on the back of a dollar bill): *annuit coeptis*—"he smiles on our beginnings."

Cooperstown may house the Cardiff Giant, but the fame of this small village in central New York does not rest upon its celebrated namesake, author James Fenimore, or its lovely Lake Otsego or the Farmer's Museum. Cooperstown is "on the map" by virtue of a different origin myth—one more parochial but no less powerful for many Americans than the tales of human beginnings that gave life to the Cardiff Giant. Cooperstown is the sacred founding place in the official myth about the origin of baseball.

Origin myths, since they are so powerful, can engender enormous practical problems. Abner Doubleday, as we shall soon see, most emphatically did not invent baseball at Cooperstown in 1839 as the official tale proclaims; in fact, no one invented baseball at any moment or in any spot. Nonetheless, this creation myth made Cooperstown the official home of baseball, and the Hall of Fame, with its associated museum and library, set its roots in this small village, inconveniently located near nothing in the way of airports or accommodations. We all revel in bucolic imagery on the field of dreams, but what a hassle when tens of thousands line the roads, restaurants, and Port-a-potties during the annual Hall of Fame weekend, when new members are enshrined and two major league teams arrive to play an exhibition game at Abner Doubleday Field, a sweet little 10,000-seater in the middle of town. Put your compass point at Cooperstown, make your radius at Albany—and you'd better reserve a year in advance if you want any accommodation within the enormous resulting circle.

After a lifetime of curiosity, I finally got the opportunity to witness this annual version of forty students in a telephone booth or twenty circus clowns in a Volkswagen. Since Yaz (former Boston star Carl Yastrzemski to the uninitiated) was slated to receive baseball's Nobel in 1989, and his old team was playing in the Hall of Fame game, and since I'm a transplanted Bostonian (although still a New Yorker and not-so-secret Yankee fan at heart), Tom Heitz, chief of the wonderful baseball library at the Hall of Fame, kindly invited me to join the sardines in this most lovely of all cans.

The silliest and most tendentious of baseball writing tries to wrest profundity from the spectacle of grown men hitting a ball with a stick by suggesting linkages between the sport and deep issues of morality, parenthood, history, lost innocence, gentleness, and so on, seemingly *ad infinitum.* (The effort reeks of silliness because baseball is profound all by itself and needs no excuses; people who don't know this are not fans and are therefore unreachable anyway.) When people ask me how baseball imitates life, I can only respond with what the more genteel newspapers used to call a "barnyard epithet," but now, with growing bravery, usually render as "bullbleep." Nonetheless, baseball is a major item of our culture, and the sport does have a long and interesting history. Any item or institution with these two properties must generate a set of myths and stories (perhaps even some truths) about beginnings. And the subject of beginnings is the bread and butter of these essays on evolution in the broadest sense. I shall make no woolly analogies between baseball and life; this is an essay on the origins of baseball, with some musings on why beginnings of all sorts hold such fascination for us. (I thank

A.G. Spalding, promoter of the Doubleday creation myth. NATIONAL BASEBALL LIBRARY, COOPERSTOWN, NY.

Tom Heitz not only for the invitation to Cooperstown at its yearly acme but also for drawing the contrast between creation and evolution stories of baseball, and for supplying much useful information from his unparalleled storehouse.)

Stories about beginnings come in only two basic modes. An entity either has an explicit point of origin, a specific time and place of creation, or else it evolves and has no definable moment of entry into the world. Baseball provides an interesting example of this contrast because we know the answer and can judge received wisdom by the two chief criteria, often opposed, of external fact and internal hope. Baseball evolved from a plethora of previous stick-and-ball games. It has no true Cooperstown and no Doubleday. Yet we seem to prefer the alternative model of origin by a moment of creation—for then we can have heroes and sacred places. By contrasting the myth of Cooperstown with the fact of evolution, we can learn something about our cultural practices and their frequent disrespect for truth.

The official story about the beginning of baseball is a creation myth, and a review of the reasons and circumstances of its fabrication may give us insight into the cultural appeal of stories in this mode. A. G. Spalding, baseball's first great pitcher during his early career, later founded the sporting goods company that still bears his name and became one of the great commercial moguls of America's gilded age. As publisher of the annual *Spalding's Official Base Ball Guide*, he held maximal power in shaping both public and institutional opinion on all facets of baseball and its history. As the sport grew in popularity, and the pattern of two stable major leagues coalesced early in our century, Spalding and others felt the need for clarification (or merely for codification) of opinion on the hitherto unrecorded origin of an activity that truly merited its common designation as America's "national pastime."

In 1907, Spalding set up a blue ribbon committee to investigate and resolve the origin of baseball. The committee, chaired by A. G. Mills and including several prominent businessmen and two senators who had also served as presidents of the National League, took much testimony but found no smoking gun. Then, in July 1907, Spalding himself transmitted to the committee a letter from an Abner Graves, then a mining engineer in Denver, who reported that Abner Doubleday had, in 1839, interrupted a

Abner Doubleday, who fired the first Union volley at Fort Sumter, but who, in the words of one historian, didn't know a baseball from a kumquat. NATIONAL BASEBALL LIBRARY, COOPERSTOWN, NY.

marbles game behind the tailor's shop in Cooperstown, New York, to draw a diagram of a baseball field, explain the rules of the game, and designate the activity by its modern name of "base ball" (then spelled as two words).

Such "evidence" scarcely inspired universal confidence, but the commission came up with nothing better—and the Double-

day myth, as we shall soon see, was eminently functional. Therefore, in 1908, the Mills Commission reported its two chief findings: first, "that base ball had its origins in the United States";
and second, "that the first scheme for playing it, according to the
best evidence available to date, was devised by Abner Doubleday,
at Cooperstown, New York, in 1839." This "best evidence" consisted only of "a circumstantial statement by a reputable gentleman"—namely Grave's testimony as reported by Spalding
himself.

Henry Chadwick, who knew that baseball had evolved from English
stick-and-ball games. NATIONAL BASEBALL LIBRARY, COOPERSTOWN, NY.

When cited evidence is so laughably insufficient, one must seek motivations other than concern for truth. The key to underlying reasons stands in the first conclusion of Mills's committee: Hoopla and patriotism (cardboard version) decreed that a national pastime must have an indigenous origin. The idea that baseball had evolved from a wide variety of English stick-and-ball games—although true—did not suit the mythology of a phenomenon that had become so quintessentially American. In fact, Spalding had long been arguing, in an amiable fashion, with Henry Chadwick, another pioneer and entrepreneur of baseball's early years. Chadwick, born in England, had insisted for years that baseball had developed from the British stick-and-ball game called rounders; Spalding had vociferously advocated a purely American origin, citing the colonial game of "one old cat" as a distant precursor, but holding that baseball itself represented something so new and advanced that a pinpoint of origin—a creation myth—must be sought.

Chadwick considered the matter of no particular importance, arguing (with eminent justice) that an English origin did not "detract one iota from the merit of its now being unquestionably a thoroughly American field sport, and a game too, which is fully adapted to the American character." (I must say that I have grown quite fond of Mr. Chadwick, who certainly understood evolutionary change and its chief principle that historical origin need not match contemporary function.) Chadwick also viewed the committee's whitewash as a victory for his side. He labeled the Mills report as "a masterful piece of special pleading which lets my dear old friend Albert [Spalding] escape a bad defeat. The whole matter was a joke between Albert and myself."

We may accept the psychic need for an indigenous creation myth, but why Abner Doubleday, a man with no recorded tie to the game and who, in the words of Donald Honig, probably "didn't know a baseball from a kumquat"? I had wondered about this for years, but only ran into the answer serendipitously during a visit to Fort Sumter in the harbor of Charleston, South Carolina. There, an exhibit on the first skirmish of the Civil War points out that Abner Doubleday, as captain of the Union artillery, had personally sighted and given orders for firing the first responsive volley following the initial Confederate attack on the fort. Doubleday later commanded divisions at Antietam and Fredericksburg, became at least a minor hero at Gettysburg, and retired as a brevet major general. In fact, A. G. Mills, head of the commission, had served as part of an honor guard when Doubleday's body lay in state in New York City, following his death in 1893.

If you have to have an American hero, could anyone be better than the man who fired the first shot (in defense) of the Civil War? Needless to say, this point was not lost on the members of Mills's committee. Spalding, never one to mince words, wrote to the committee when submitting Graves's dubious testimony: "It certainly appeals to an American pride to have had the great national game of base ball created and named by a Major General in the United States Army." Mills then concluded in his report: "Perhaps in the years to come, in view of the hundreds of thousands of people who are devoted to baseball, and the millions who will be, Abner Doubleday's fame will rest evenly, if not quite as much, upon the fact that he was its inventor . . . as upon his brilliant and distinguished career as an officer in the Federal Army."

And so, spurred by a patently false creation myth, the Hall of Fame stands in the most incongruous and inappropriate locale of a charming little town in central New York. Incongruous and inappropriate, but somehow wonderful. Who needs another museum in the cultural maelstroms (and summer doldrums) of New York, Boston, or Washington? Why not a major museum in a beautiful and bucolic setting? And what could be more fitting than the spatial conjunction of two great American origin myths—the Cardiff Giant and the Doubleday Fable? Thus, I too am quite content to treat the myth gently, while honesty requires 'fessing up. The exhibit on Doubleday in the Hall of Fame Museum sets just the right tone in its caption: "In the hearts of those who love baseball, he is remembered as the lad in the pasture where the game was invented. Only cynics would need to know more." Only in the hearts; not in the minds.

Baseball evolved. Since the evidence is so clear (as epitomized below), we must ask why these facts have been so little appreciated for so long, and why a creation myth like the Doubleday story ever gained a foothold. Two major reasons have conspired: first, the positive block of our attraction to creation stories; second, the negative impediment of unfamiliar sources outside the usual purview of historians. English stick-and-ball games of the nineteenth century can be roughly classified into two categories along social lines. The upper and educated classes played cricket, and the history of this sport is copiously documented because literati write about their own interests and because the activities of men in power are well recorded (and constitute virtually all of history, in the schoolboy version). But the ordinary pastimes of rural and urban working people can be well nigh invisible in conventional sources of explicit commentary. Working people played a different kind of stick-and-ball game, existing in various forms and designated by many names, including "rounders" in western England, "feeder" in London, and "base ball" in southern England. For a large number of reasons, forming the essential difference between cricket and baseball, cricket matches can last up to several days (a batsman, for example, need not run after he hits the ball and need not expose himself to the possibility of being put out every time he makes contact). The leisure time of working people does not come in such generous gobs, and the lower-class stick-and-ball games could not run more than a few hours.

Several years ago, at the Victoria and Albert Museum in London, I learned an important lesson from an excellent exhibit on late nineteenth century history of the British music hall. This is my favorite period (Darwin's century, after all), and I consider myself tolerably well informed on cultural trends of the time. I can sing any line from any of the Gilbert and Sullivan operas (a largely middle-class entertainment), and I know the general drift of high cultural interests in literature and music. But the music hall provided a whole world of entertainment for millions, a realm with its heroes, its stars, its top-forty songs, its gaudy theaters—and I knew nothing, absolutely nothing, about this world. I felt chagrined, but my ignorance had an explanation beyond personal insensitivity (and the exhibit had been mounted explicitly to counteract the selective invisibility of certain important trends in history). The music hall was a chief entertainment of Victorian working classes, and the history of working people is often invisible in conventional written sources. This history must be rescued and reconstituted from different sorts of data; in this case, from posters, playbills, theater accounts, persistence of some songs in the oral tradition (most were never published as sheet music), recollections of old-timers who knew the person who knew the person. . . .

The early history of baseball—the stick-and-ball game of working people—presents the same problem of conventional invisibility, and the same promise of rescue by exploration of unusual sources. Work continues and intensifies as the history of sport becomes more and more academically respectable, but the broad outlines (and much fascinating detail) are now well established. As the upper classes played a codified and well-documented cricket, working people played a largely unrecorded and much more diversified set of stick-and-ball games ancestral to baseball. Many sources, including primers and boys' manuals, depict games recognizable as precursors to baseball well into the eighteenth century. Occasional references even spill over into high culture. In *Northanger Abbey,* written in 1798 or 1799, Jane Austen remarks: "It was not very wonderful that Catherine . . . should prefer cricket, base ball, riding on horseback, and running about the country, at the age of fourteen, to books." As this quotation illustrates, the name of the game is no more Doubleday's than the form of play.

These ancestral styles of baseball came to America with early settlers and were clearly well established by colonial times. But they were driven ever further underground by Puritan proscriptions of sport for adults. They survived largely as children's games and suffered the double invisibility of location among the poor and the young. But two major reasons brought these games into wider repute and led to a codification of standard forms quite close to modern baseball between the 1820s and the 1850s. First, a set of social reasons, from the decline of Puritanism to increased concern about health and hygiene in crowded cities, made sport an acceptable activity for adults. Second, middle-class and professional people began to take up these early forms of baseball, and this upward social drift inspired teams, leagues, written rules, uniforms, stadiums, guidebooks: in short, all the paraphernalia of conventional history.

I am not arguing that these early games could be called baseball with a few trivial differences (evolution means substantial change, after all), but only that they stand in a complex lineage, better designated a nexus, from which modern baseball emerged, eventually in a codified and canonical form. In those days before instant communication, every region had its own version, just as every set of outdoor steps in New York City generated a different form of stoopball in my youth, without threatening the basic identity of the game. These games, most commonly called town ball, differed from modern baseball in substantial ways. In the Massachusetts Game, a codification of the late 1850s drawn up by ball players in New England towns, four bases and three strikes identify the genus, but many specifics are strange by modern standards. The bases were made of wooden stakes projecting four feet from the ground. The batter (called the striker) stood between first and fourth base. Sides changed after a single out. One hundred runs (called tallies), not higher score after a specified number of innings, spelled victory. The field contained no

A.J. Cartwright, a most interesting point in the continuum of baseball's evolution. NATIONAL BASEBALL LIBRARY, COOPERSTOWN, NY.

foul lines, and balls hit in any direction were in play. Most important, runners were not tagged out, but rather dismissed by "plugging," that is, being hit with a thrown ball while running between bases. Consequently, since baseball has never been a game for masochists, balls were soft—little more than rags stuffed into leather covers—and could not be hit far. (Tom Heitz has put together a team of Cooperstown worthies to re-create town ball for interested parties and prospective opponents. Since few other groups are well schooled in this lost art, Tom's team hasn't been defeated in ages, if ever. "We are the New York Yankees of town ball," he told me. His team is called, quite appropriately in general but especially for this essay, the Cardiff Giants.)

Evolution is continual change, but not insensibly gradual transition; in any continuum, some points are always more interesting than others. The conventional nomination for most salient point in this particular continuum goes to Alexander Joy Cartwright, leader of a New York team that started to play in Lower Manhattan, eventually rented some changing rooms and a field in Hoboken (just a quick ferry ride across the Hudson), and finally drew up a set of rules in 1845, later known as the New York Game. Cartwright's version of town ball is much closer to modern baseball, and many clubs followed his rules—for standardization became ever more vital as the popularity of early baseball grew and opportunity for play between regions increased. In particular, Cartwright introduced two key innovations that shaped the disparate forms of town ball into a semblance of modern baseball. First, he eliminated plugging and introduced tagging in the modern sense; the ball could now be made harder, and hitting for distance became an option. Second, he introduced foul lines, again in the modern sense, as his batter stood at a home plate and had to hit the ball within lines defined from home through first and third bases. The game could now become a spectator sport because areas close to the field but out of action could, for the first time, be set aside for onlookers.

The New York Game may be the highlight of a continuum, but it provides no origin myth for baseball. Cartwright's rules were followed in various forms of town ball. His New York Game still included many curiosities by modern standards (twenty-one runs, called aces, won the game, and balls caught on one bounce were outs). Moreover, our modern version is an amalgam of the

New York Game plus other town-ball traditions, not Cartwright's baby grown up by itself. Several features of the Massachusetts Game entered the modern version in preference to Cartwright's rules. Balls had to be caught on the fly in Boston, and pitchers threw overhand, not underhand as in the New York Game (and in professional baseball until the 1880s).

Scientists often lament that so few people understand Darwin and the principles of biological evolution. But the problem goes deeper. Too few people are comfortable with evolutionary modes of explanation in any form. I do not know why we tend to think so fuzzily in this area, but one reason must reside in our social and psychic attraction to creation myths in preference to evolutionary stories—for creation myths, as noted before, identify heroes and sacred places, while evolutionary stories provide no palpable, particular object as a symbol for reverence, worship, or patriotism. Still, we must remember—and an intellectual's most persistent and nagging responsibility lies in making this simple point over and over again, however noxious and bothersome we render ourselves thereby—that truth and desire, fact and comfort, have no necessary, or even preferred, correlation (so rejoice when they do coincide).

To state the most obvious example in our current political turmoil: Human growth is a continuum, and no creation myth can define an instant for the origin of an individual life. Attempts by anti-abortionists to designate the moment of fertilization as the beginning of personhood make no sense in scientific terms (and also violate a long history of social definitions that traditionally focused on the quickening, or detected movement, of the fetus in the womb). I will admit—indeed, I emphasized as a key argument of this essay—that not all points on a continuum are equal. Fertilization is a more interesting moment than most, but it no more provides a clean definition of origin than the most intriguing moment of baseball's continuum—Cartwright's codification of the New York Game—defines the beginning of our national pastime. Baseball evolved and people grow; both are continua without definable points of origin. Probe too far back and you reach absurdity, for you will see Nolan Ryan on the hill when the first ape hit a bird with a stone, or you will define both masturbation and menstruation as murder—and who will then cast the first stone? Look for something in the middle, and you find nothing but continuity—always a meaningful "before," and always a more modern "after." (Please note that I am not stating an opinion on the vexatious question of abortion—an ethical issue that can only be decided in ethical terms. I only point out that one side has rooted its case in an argument from science that is not only entirely irrelevant to the proper realm of resolution but also happens to be flat-out false in trying to devise a creation myth within a continuum.)

And besides, why do we prefer creation myths to evolutionary stories? I find all the usual reasons hollow. Yes, heroes and shrines are all very well, but is there not grandeur in the sweep of continuity? Shall we revel in a story for all humanity that may include the sacred ball courts of the Aztecs, and perhaps, for all we know, a group of *Homo erectus* hitting rocks or skulls with a stick or a femur? Or shall we halt beside the mythical Abner Doubleday, standing behind the tailor's shop in Cooperstown, and say "behold the man"—thereby violating truth and, perhaps even worse, extinguishing both thought and wonder?

# 4 | The Panda's Thumb of Technology

THE BRIEF STORY of Jephthah and his daughter (Judg. 11:30–40) is, to my mind and heart, the saddest of all biblical tragedies. Jephthah makes an intemperate vow, yet all must abide by its consequences. He promises that if God grant him victory in a forthcoming battle, he will sacrifice by fire the first living thing that passes through his gate to greet him upon his return. Expecting (I suppose) a dog or a goat, he returns victorious to find his daughter, and only child, waiting to meet him "with timbrels and with dances."

Handel's last oratorio, *Jephtha,* treats this tale with great power (although his librettist couldn't bear the weight of the original and gave the story a happy ending, with angelic intervention to spare Jephthah's daughter at the price of her lifelong chastity). At the end of Part 2, while all still think that the terrible vow must be fulfilled, the chorus sings one of Handel's wonderful "philosophical" choruses. It begins with a frank account of the tragic circumstance:

How dark, O Lord, are thy decrees! . . .
No certain bliss, no solid peace,
We mortals know on earth below.

Yet the last two lines, in a curious about-face, proclaim (with magnificent musical solidity as well):

Yet on this maxim still obey:
WHATEVER IS, IS RIGHT

59

This odd reversal, from frank acknowledgment to unreasonable acceptance, reflects one of the greatest biases ("hopes" I like to call them) that human thought imposes upon a world indifferent to our suffering. Humans are pattern-seeking animals. We must find cause and meaning in all events (quite apart from the probable reality that the universe both doesn't care much about us and often operates in a random manner). I call this bias "adaptationism"—the notion that everything must fit, must have a purpose, and in the strongest version, must be for the best.

The final line of Handel's chorus is, of course, a quote from Alexander Pope, the last statement of the first epistle of his *Essay on Man*, published twenty years before Handel's oratorio. Pope's text contains (in heroic couplets to boot) the most striking paean I know to the bias of adaptationism. In my favorite lines, Pope chastises those people who may be unsatisfied with the senses that nature bestowed upon us. We may wish for more acute vision, hearing, or smell, but consider the consequences.

> If nature thunder'd in his op'ning ears
> And stunn'd him with the music of the spheres
> How would he wish that Heav'n had left him still
> The whisp'ring zephyr, and the purling rill!

And my favorite couplet, on olfaction:

> Or, quick effluvia darting thro' the brain,
> Die of a rose in aromatic pain.

What we have is best for us—whatever is, is right.

By 1859, most educated people were prepared to accept evolution as the reason behind similarities and differences among organisms—thus accounting for Darwin's rapid conquest of the intellectual world. But they were decidedly not ready to acknowledge the radical implications of Darwin's proposed mechanism of change, natural selection, thus explaining the brouhaha that the *Origin of Species* provoked—and still elicits (at least before our courts and school boards).

Darwin's world is full of "terrible truths," two in particular. First, when things do fit and make sense (good design of organisms, harmony of ecosystems), they did not arise because the laws of nature entail such order as a primary effect. They are, rather, only epiphenomena, side consequences of the basic causal process at work in natural populations—the purely "selfish" struggle among organisms for personal reproductive success. Second, the complex and curious pathways of history guarantee that most organisms and ecosystems cannot be designed optimally. Indeed, to make an even stronger statement, imperfections are the primary proofs that evolution has occurred, since optimal designs erase all signposts of history.

This principle of imperfection has been a major theme of my essays for several years. I call it the panda principle to honor my favorite example, the panda's false thumb. Pandas are the herbivorous descendants of carnivorous bears. Their true anatomical thumbs were, long ago during ancestral days of meat eating, irrevocably committed to the limited motion appropriate for this mode of life and universally evolved by mammalian Carnivora. When adaptation to a diet of bamboo required more flexibility in manipulation, pandas could not redesign their thumbs but had to make do with a makeshift substitute—an enlarged radial sesamoid bone of the wrist, the panda's false thumb. The sesamoid thumb is a clumsy, suboptimal structure, but it works. Pathways of history (commitment of the true thumb to other roles during an irreversible past) impose such jury-rigged solutions upon all creatures. History inheres in the imperfections of living organisms—and thus we know that modern creatures had a different past, converted by evolution to their current state.

We can accept this argument for organisms (we know, after all, about our own appendixes and aching backs). But is the panda principle more pervasive? Is it a general statement about all historical systems? Will it apply, for example, to the products of technology? We might deem this principle irrelevant to the manufactured objects of human ingenuity—and for good reason. After all, constraints of genealogy do not apply to steel, glass, and plastic. The panda cannot shuck its digits (and can only build its future upon an inherited ground plan), but we can abandon gas lamps for electricity and horse carriages for motor cars. Consider, for example, the difference between organic architecture and human buildings. Complex organic structures cannot be re-evolved following their loss; no snake will redevelop front legs. But the apostles of post-modern architecture, in reaction to the sterility of so many glass-box buildings of the international style, have juggled together all the classical forms of history in a cascading effort to rediscover the virtues of ornamentation. Thus, Philip Johnson could place a broken pediment atop a New York skyscraper and raise a medieval castle of plate glass in downtown Pittsburgh. Organisms cannot recruit the virtues of their lost pasts.

Yet I am not so sure that technology is exempt from the panda principle of history, for I am now sitting face to face with the best example of its application. Indeed, I am in most intimate (and striking) contact with this object—the typewriter keyboard.

I could type before I could write. My father was a court stenographer, and my mother is a typist. I learned proper eight-finger touch-typing when I was about nine years old and still endowed with small hands and weak, tiny pinky fingers. I was thus, from the first, in a particularly good position to appreciate the irrationality of placement for letters on the standard keyboard—called QWERTY by all aficionados in honor of the first six letters on the top letter row.

Clearly, QWERTY makes no sense (beyond the whiz and joy of typing QWERTY itself). More than 70 percent of English words can be typed with the letters DHIATENSOR, and these should be on the most accessible second, or home, row—as they were in a failed competitor to QWERTY introduced as early as 1893. But in QWERTY, the most common English letter, E, requires a reach to the top row, as do the vowels U, I, and O (with O struck by the weak fourth finger), while A remains in the home row but must be typed with the weakest finger of all (at least for the dexterous majority of right-handers)—the left pinky. (How I struggled with this as a boy. I just couldn't depress that key. I once tried to type the Declaration of Independence and ended up with: th t  ll men  re cre ted equ l.)

As a dramatic illustration of this irrationality, consider the accompanying photograph, the keyboard of an ancient Smith-Corona upright, identical with the one (my dad's original) that I use to type these essays (a magnificent machine—no breakdown in twenty years and a fluidity of motion unmatched by any manual typewriter since). After more than half a century of use, some of the most commonly struck keys have been worn right through the surface into the soft pad below (they weren't solid plastic in those days). Note that E, A, and S are worn in this way—but note also that all three are either not in the home row or are struck with the weak fourth and pinky fingers in QWERTY.

This claim is not just a conjecture based on idiosyncratic personal experience. Evidence clearly shows that QWERTY is drastically suboptimal. Competitors have abounded since the early days of typewriting, but none has supplanted or even dented the universal dominance of QWERTY for English type-writers. The best-known alternative, DSK, for Dvorak Simplified Keyboard, was introduced in 1932. Since then, virtually all records for speed typing have been held by DSK, not QWERTY, typists. During the 1940s, the U.S. Navy, ever mindful of efficiency, found that the increased speed of DSK would amortize the cost of retraining typists within ten days of full employment. (Mr. Dvorak was not Anton of the *New World Symphony*, but August, a professor of education at the University of Washington, who died disappointed in 1975. Dvorak was a disciple of Frank B. Gilbreth, pioneer of time and motion studies in industrial management.)

Since I have a special interest in typewriters (my affection for them dates to childhood days of splendor in the grass and glory in the flower), I have wanted to write such an essay for years. But I never had the data I needed until Paul A. David, Coe Professor of American Economic History at Stanford University, kindly sent me his fascinating article, "Understanding the Economics of QWERTY: The Necessity of History" (in *Economic History and the Modern Economist,* edited by W. N. Parker, New York, Basil Blackwell Inc., 1986, pp. 30–49). Virtually all the nonidiosyncratic data in this essay come from David's work, and I thank him for this opportunity to satiate an old desire.

The puzzle of QWERTY's dominance resides in two separate questions: Why did QWERTY ever arise in the first place? And why has QWERTY survived in the face of superior competitors?

My answers to these questions will invoke analogies to principles of evolutionary theory. Let me, then, state some ground rules for such a questionable enterprise. I am convinced that comparisons between biological evolution and human cultural or technological change have done vastly more harm than good—and examples abound of this most common of all intellectual traps. Biological evolution is a bad analogue for cultural change

A classic upright typewriter of World War I vintage. Brother to the machine that I use to write these essays.

Notice the patterns of wear for most frequently used keys, as illustrated by breakage through the surface after so many years of striking. In QWERTY, all the most common keys are either not in the home row, or are hit by weak fingers in the home row—thus illustrating the suboptimality of this standard arrangement.

A keyboard for a typewriter made in the 1880's, illustrating one of the many competing non-QWERTY arrangements so common at the time.

because the two systems are so different for three major reasons that could hardly be more fundamental.

First, cultural evolution can be faster by orders of magnitude than biological change at its maximal Darwinian rate—and questions of timing are of the essence in evolutionary arguments. Second, cultural evolution is direct and Lamarckian in form: The achievements of one generation are passed by education and publication directly to descendants, thus producing the great potential speed of cultural change. Biological evolution is indirect and Darwinian, as favorable traits do not descend to the next generation unless, by good fortune, they arise as products of genetic change. Third, the basic topologies of biological and cultural change are completely different. Biological evolution is a system of constant divergence without subsequent joining of branches. Lineages, once distinct, are separate forever. In human history, transmission across lineages is, perhaps, the major source of cultural change. Europeans learned about corn and potatoes from Native Americans and gave them smallpox in return.

So, when I compare the panda's thumb with a typewriter keyboard, I am not attempting to derive or explain technological change by biological principles. Rather, I ask if both systems might not record common, deeper principles of organization. Biological evolution is powered by natural selection, cultural evolution by a different set of principles that I understand but dimly. But both are systems of historical change. More general principles of structure must underlie all systems that proceed through history (perhaps I now only show my own bias for intelligibility in our complex world)—and I rather suspect that the panda principle of imperfection might reside among them.

My main point, in other words, is not that typewriters are like biological evolution (for such an argument would fall right into the nonsense of false analogy), but that both keyboards and the panda's thumb, as products of history, must be subject to some regularities governing the nature of temporal connections. As scientists, we must believe that general principles underlie structurally related systems that proceed by different overt rules. The proper unity lies not in false applications of these overt rules (like natural selection) to alien domains (like technological change), but in seeking the more general rules of structure and change themselves.

*The Origin of QWERTY:* True randomness has limited power to intrude itself into the forms of organisms. Small and unimportant changes, unrelated to the working integrity of a complex creature, may drift in and out of populations by a process akin to throwing dice. But intricate structures, involving the coordination of many separate parts, must arise for an active reason—since the bounds of mathematical probability for fortuitous association are soon exceeded as the number of working parts grows.

But if complex structures must arise for a reason, history may soon overtake the original purpose—and what was once a sensible solution becomes an oddity or imperfection in the altered context of a new future. Thus, the panda's true thumb permanently lost its ability to manipulate objects when carnivorous ancestors found a better use for this digit in the limited motions appropriate for creatures that run and claw. This altered thumb then becomes a constraint imposed by past history upon the panda's ability to adapt in an optimal way to its new context of herbivory. The panda's thumb, in short, becomes an emblem of its different past, a sign of history.

Similarly, QWERTY had an eminently sensible rationale in the early technology of typewriting but soon became a constraint upon faster typing as advances in construction erased the reason for QWERTY's origin. The key (pardon the pun) to QWERTY's origin lies in another historical vestige easily visible on the second row of letters. Note the sequence: DFGHJKL—a good stretch of the alphabet in order, with the vowels E and I removed. The original concept must have simply arrayed the letters in alphabetical order. Why were the two most common letters of this sequence removed from the most accessible home row? And why were other letters dispersed to odd positions?

Those who remember the foibles of manual typewriters (or, if as hidebound as yours truly, still use them) know that excessive speed or unevenness of stroke may cause two or more keys to jam near the striking point. You also know that if you don't reach in and pull the keys apart, any subsequent stroke will type a repetition of the key leading the jam—as any key subsequently struck will hit the back of the jammed keys and drive them closer to the striking point.

These problems were magnified in the crude technology of early machines—and too much speed became a hazard rather than a blessing, as key jams canceled the benefits of celerity. Thus, in the great human traditions of tinkering and pragmatism, keys were moved around to find a proper balance between speed and jamming. In other words—and here comes the epitome of the tale in a phrase—QWERTY arose in order to slow down the maximal speed of typing and prevent jamming of keys. Common letters were either allotted to weak fingers or dispersed to positions requiring a long stretch from the home row.

This basic story has gotten around, thanks to short takes in *Time* and other popular magazines, but the details are enlightening, and few people have the story straight. I have asked nine typists who knew this outline of QWERTY's origin and all (plus me for an even ten) had the same misconception. The old machines that imposed QWERTY were, we thought, of modern design—with keys in front typing a visible line on paper rolled around a platen. This leads to a minor puzzle: Key jams may be a pain in the butt, but you see them right away and can easily reach in and pull them apart. So why QWERTY?

As David points out, the prototype of QWERTY, a machine invented by C. L. Sholes in the 1860s, was quite different in form from modern typewriters. It had a flat paper carriage and did not roll paper right around the platen. Keys struck the paper invisibly from beneath, not patently from the front as in all modern typewriters. You could not view what you were typing unless you stopped to raise the carriage and inspect your product. Keys jammed frequently, but you could not see (and often did not feel) the aggregation. Thus, you might type a whole page of deathless prose and emerge only with a long string of E's.

Sholes filed for a patent in 1867 and spent the next six years in trial-and-error efforts to improve his machine. QWERTY emerged from this period of tinkering and compromise. As another added wrinkle (and fine illustration of history's odd quirks), R joined the top row as a last-minute entry, and for a somewhat capricious motive according to one common tale (perhaps apocryphal)—for salesmen could then impress potential buyers by smooth and rapid production of the brand name TYPE WRITER, all on one row. (Although I wonder how many sales were lost when TYPE EEEEEE appeared after a jam!)

*The Survival of QWERTY:* We can all accept this story of QWERTY's origin, but why did it persist after the introduction of the modern platen roller and front-stroke key? (The first typewriter with a fully visible printing point was introduced in 1890.) In fact, the situation is even more puzzling. I thought that alternatives to keystroke typing only became available with the IBM electric ball, but none other than Thomas Edison filed a patent for an electric print-wheel machine as early as 1872, and L. S. Crandall marketed a writing machine without typebars in 1879. (Crandall arranged his type on a cylindrical sleeve and made the sleeve revolve to the required letter before striking the printing point.)

The 1880s were boom years for the fledgling typewriter industry, a period when a hundred flowers bloomed and a hundred schools of thought contended. Alternatives to QWERTY were touted by several companies, and both the variety of printing designs (several without typebars) and the improvement of key-stroke typewriters completely removed the original rationale for QWERTY. Yet during the 1890s, more and more companies made the switch to QWERTY, which became an industry standard by the early years of our century. And QWERTY has held on stubbornly, through the introduction of the IBM Selectric and the Hollerith punch card machine to that ultimate example of its nonnecessity, the microcomputer terminal.

To understand the survival (and domination to this day) of drastically suboptimal QWERTY, we must recognize two other commonplaces of history, as applicable to life in geological time as to technology over decades—contingency and incumbency. We call a historical event—the rise of mammals or the dominance of QWERTY—contingent when it occurs as the chancy result of a long string of unpredictable antecedents, rather than as a necessary outcome of nature's laws. Such contingent events often depend crucially upon choices from a distant past that seemed tiny and trivial at the time. Minor perturbations early in the game can nudge a process into a new pathway, with cascading consequences that produce an outcome vastly different from any alternative.

Incumbency also reinforces the stability of a pathway once the little quirks of early flexibility push a sequence into a firm channel. Suboptimal politicians often prevail nearly forever once they gain office and grab the reins of privilege, patronage, and visibility. Mammals waited 100 million years to become the dominant animals on land and only got a chance because dinosaurs succumbed during a mass extinction. If every typist in the world stopped using QWERTY tomorrow and began to learn Dvorak, we would all be winners, but who will bell the cat or start the ball rolling? (Choose your cliché, for they all record this evident truth.) Stasis is the norm for complex systems; change, when provoked at all, is usually rapid and episodic.

QWERTY's fortunate and improbable ascent to incumbency occurred by a concatenation of circumstances, each indecisive in itself, but all probably necessary for the eventual outcome. Remington had marketed the Sholes machine with its QWERTY keyboard, but this early tie with a major firm did not secure QWERTY's victory. Competition was tough, and no lead meant much with such small numbers in an expanding market. David estimates that only 5,000 or so QWERTY machines existed at the beginning of the 1880s.

The push to incumbency was complex and multifaceted, dependent more upon the software of teachers and promoters than upon the hardware of improving machines. Most early typists used idiosyncratic hunt-and-peck, few-fingered methods. In 1882, Ms. Longley, founder of the Shorthand and Typewriter Institute in Cincinnati, developed and began to teach the eight-finger typing that professionals use today. She happened to teach with a QWERTY keyboard, although many competing arrangements would have served her purposes as well. She also published a popular do-it-yourself pamphlet. At the same time, Remington began to set up schools for typewriting using (of course) its QWERTY standard. The QWERTY ball was rolling but this head start did not guarantee a place at the summit. Many other schools taught rival methods on different machines and might have gained an edge.

Then a crucial event in 1888 probably added the decisive increment to QWERTY's small advantage. Longley was challenged to prove the superiority of her eight-finger method by Louis Taub, another Cincinnati typing teacher, who worked with four fingers on a rival non-QWERTY keyboard with six rows, no shift action, and (therefore) separate keys for upper- and lowercase letters. As her champion, Longley engaged Frank E. McGurrin, an experienced QWERTY typist who had given himself a decisive advantage that, apparently, no one had utilized before. He had memorized the QWERTY keyboard and could therefore operate his machine as all competent typists do today—by what we now call touch-typing. McGurrin trounced Taub in a well-advertised and well-reported public competition.

In public perception, and (more important) in the eyes of those who ran typing schools and published typing manuals, QWERTY had proved its superiority. But no such victory had really occurred. The tie of McGurrin to QWERTY was fortuitous and a good break for Longley and for Remington. We shall never know why McGurrin won, but reasons quite independent of QWERTY cry out for recognition: touch-typing over hunt-and-peck, eight fingers over four fingers, the three-row letter board with a shift key versus the six-row board with two separate keys for each letter. An array of competitions that would have tested QWERTY were never held—QWERTY versus other arrangements of letters with both contestants using eight-finger touch-typing on a three-row keyboard, or McGurrin's method of eight-finger touch-typing on a non-QWERTY three-row keyboard versus Taub's procedure to see whether the QWERTY arrangement (as I doubt) or McGurrin's method (as I suspect) had secured his success.

In any case, the QWERTY steamroller now gained crucial momentum and prevailed early in our century. As touch-typing by QWERTY became the norm in America's typing schools, rival manufacturers (especially in a rapidly expanding market) could adapt their machines more easily than people could change their habits—and the industry settled upon the wrong standard.

If Sholes had not gained his tie to Remington, if the first typist who decided to memorize a keyboard had used a non-QWERTY design, if McGurrin had a bellyache or drank too much the night before, if Longley had not been so zealous, if a hundred other perfectly possible things had happened, then I might be typing this essay with more speed and much greater economy of finger motion.

But why fret over lost optimality. History always works this way. If Montcalm had won a battle on the Plains of Abraham, perhaps I would be typing *en français.* If a portion of the African jungles had not dried to savannas, I might still be an ape up a tree. If some comets had not struck the earth (if they did) some 60 million years ago, dinosaurs might still rule the land, and all mammals would be rat-sized creatures scurrying about in the dark corners of their world. If *Pikaia,* the only chordate of the Burgess Shale, had not survived the great sorting out of body plans after the Cambrian explosion, mammals might not exist at all. If multicellular creatures had never evolved after five-sixths of life's history had yielded nothing more complicated than an algal mat, the sun might explode a few billion years hence with no multicellular witness to the earth's destruction.

Compared with these weighty possibilities, my indenture to QWERTY seems a small price indeed for the rewards of history.

For if history were not so maddeningly quirky, we would not be here to enjoy it. Streamlined optimality contains no seeds for change. We need our odd little world, where QWERTY rules and the quick brown fox jumps over the lazy dog.*

---

## *Postscript*

Since typing falls into the category of things that many, if not most of us, can do (like walking and chewing gum simultaneously) this essay elicited more commentary than most of my more obscure ramblings.

Some queried the central premises and logic. An interesting letter from Folsom Prison made a valid point in the tough humor of such institutions. (I receive many letters from prisoners and am always delighted by such reminders that, at least for many people, the quest for knowledge never abates, even in most uncongenial temporary domiciles):

> Some of us were left with a nagging question: If the hunt 'n peck method prevailed until around 1882, how could Sholes or his cohorts have "relegated common letters to weak fingers" when there were no weak fingers, just hunt 'n peck type fingers? At least none of the hunt 'n peck typing clerks or cops around here use the weak fingers. If you could find the time to answer this it would really be appreciated and could serve to reduce the likelihood of increased violence at Folsom between opposing QWERTY origin factions.

My correspondent is quite right, and I misspoke (I also trust that recent tension at Folsom had sources other than the great typewriter wars—yes, I did answer the letter promptly). Fortu-

---

*I must close with a pedantic footnote, lest nonaficionados be utterly perplexed by this ending. This quirky juxtaposition of uncongenial carnivores is said to be the shortest English sentence that contains all twenty-six letters. It is, as such, *de rigueur* in all manuals that teach typing.

nately, my hypothesis is secure against my own carelessness—for Sholes needed simply to separate frequently struck keys to avoid jamming. The finger used to strike mattered little (I also rather suspect that many people were experimenting with many-fingered typing before the full four-fingered methods became canonical).

But the vast bulk of correspondence, more than 80 percent, took issue with my throwaway and tangential last line—thanks to our long-standing and happy fascination with words and word games. I gave the conventional typist's sentence as being the shortest phrase using all letters:

The quick brown fox jumps over the lazy dog.

I have since learned that sentences containing all letters of the alphabet are called "pangrams," and that the quest for the shortest represents at least a minor industry, with much effort spent, and opposing factions with strong passions. Many readers suggested, as a well-known alternative with three fewer letters (32 versus 35),

Pack my box with five dozen liquor jugs.

Zoological enthusiasts and prohibitionists then retort that the fox-dog classic can still tie by dropping the first article and becoming only slightly less grammatical:

Quick brown fox jumps over the lazy dog.

But Ted Leather wins this limited derby for shortest sensible pangram with the 31-stroke

Jackdaws love my big sphinx of quartz.

We now enter the world of arcana. Can shorter pangrams be made? Can the ultimate 26-letter sentence be constructed? This quest has so far stymied all wordsmiths. Using common words only, we can get down to 28 (but only by the slightly dishonorable route of using proper names):

Waltz, nymph, for quick jigs vex Bud.

And to 27, with some archaic orthography:

Frowzy things plumb vex'd Jack Q.

But for the ultimate of 26, we either use initials in abundance (which doesn't seem quite fair),

J. Q. Schwartz flung V. D. Pike my box,

or we avoid names and initials, but employ such unfamiliar and marginally admissable words that an equal feeling of dissatisfaction arises,

Zing! Vext cwm fly jabs Kurd qoph.

A *cwm* is a mountain hollow in Wales, while *qoph,* the nineteenth letter of the Hebrew alphabet, has been drawn (and has attracted the ire of an immigrant fly) by a member of an Iranian minority. Sounds awfully improbable.

My favorite proposal for a 26-letter pangram requires an entire story for comprehension (thanks to Dan Lufkin of Hood College):

During World War I, Lawrence's Arab Legion was operating on the southern flank of the Ottoman Empire. Hampered by artillery fire from across a river, Lawrence asked for a volunteer to cross the river at night and locate the enemy guns. An Egyptian soldier stepped forward. The man was assigned to Lawrence's headquarters [G.H.Q. for "general headquarters"—this becomes important later] and had a reputation for bringing bad luck. But Lawrence decided to send him. The mission was successful and the soldier appeared, at dawn the next morning, at a remote sentry post near the river, dripping wet, shivering, and clad in nothing but his underwear and native regimental headgear. The sentry wired to Lawrence for instructions, and he replied:

Warm plucky G.H.Q. jinx, fez to B.V.D.'s.

A free copy of this and all my subsequent books to anyone who can construct a 26-letter pangram with common words only and no proper names.

# 2 | Dinomania

# 5 | Bully for Brontosaurus

QUESTION: What do Catherine the Great, Attila the Hun, and Bozo the Clown have in common? Answer: They all have the same middle name.

Question: What do San Marino, Tannu Tuva, and Monaco have in common? Answer: They all realized that they could print pretty pieces of perforated paper, call them stamps, and sell them at remarkable prices to philatelists throughout the world. (Did these items ever bear any relationship to postage or utility? Does anyone own a canceled stamp from Tannu Tuva?) Some differences, however, must be admitted. Although San Marino (a tiny principality within Italy) and Tannu Tuva (a former state adjacent to Mongolia but now annexed to the Soviet Union) may rely on stamps for a significant fraction of their GNP, Monaco, as we all know, has another considerable source of outside income—the casino of Monte Carlo (nurtured by all the hype and elegance of the Grimaldis—Prince Rainier, Grace Kelly, and all that).

So completely do we identify Monaco with Monte Carlo that we can scarcely imagine any other activity, particularly something productive, taking place in this little land of fantasy and fractured finances.

Nonetheless, people are born, work, and die in Monaco. And this tiny nation boasts, among other amenities, a fine station for oceanographic research. This combination of science and hostelry makes Monaco an excellent place for large professional meetings. In 1913, Monaco hosted the International Zoological Congress, the largest of all meetings within my clan. This 1913 gathering adopted the important Article 79, or "plenary powers

decision," stating that "when stability of nomenclature is threatened in an individual case, the strict application of the Code may
under specified conditions be suspended by the International
Commission on Zoological Nomenclature."

Now I will not blame any reader for puzzlement over the last
paragraph. The topic—rules for giving scientific names to organisms—is easy enough to infer. But why should we be concerned
with such legalistic arcana? Bear with me. We shall detour around
the coils of *Boa constrictor,* meet the International Code of Zoological Nomenclature head-on, and finally arrive at a hot issue now
generating much passion and acrimony at the heart of our greatest contemporary fad. You may deny all concern for rules of taxonomy, our last domain of active Latin (now that Catholicism has
embraced the vernacular), but millions of Americans are now het
up about the proper name of *Brontosaurus,* the canonical dinosaur. And you can't grasp the name of the beast without engaging
the beastly rules of naming.

Nonprofessionals often bridle at the complex Latin titles used
by naturalists as official designations for organisms. Latin is a
historical legacy from the foundation of modern taxonomy in the
mid-eighteenth century—a precomputer age when Romespeak
was the only language shared by scientists throughout the world.
The names may seem cumbersome, now that most of us pass our
youthful years before a television set, rather than declaiming *hic-
haec-hoc* and *amo-amas-amat.* But the principle remains sound. Effective communication demands that organisms have official
names, uniformly recognized in all countries, while a world of
changing concepts and increasing knowledge requires that rules
of naming foster maximal stability and minimal disruption.

New species are discovered every day; old names must often
change as we correct past errors and add new information. If
every change of concept demanded a redesignation of all names
and a reordering of all categories, natural history would devolve
into chaos. Our communications would fail as species, the basic
units of all our discourse, would have no recognized labels. All
past literature would be a tangle of changing designations, and
we could not read without a concordance longer than the twenty
volumes of the *Oxford English Dictionary.*

The rules for naming animals are codified in the *International
Code of Zoological Nomenclature,* as adopted and continually revised

by the International Union of Biological Sciences (plant people have a different code based on similar principles). The latest edition (1985), bound in bright red, runs to 338 pages. I will not attempt to summarize the contents, but only state the primary goal: to promote maximal stability as new knowledge demands revision.

Consider the most prevalent problem demanding a solution in the service of stability: When a single species has been given two or more names, how do we decide which to validate and which to reject? This common situation can arise for several reasons: Two scientists, each unaware of the other's work, may name the same animal; or a single scientist, mistaking a variable species for two or more separate entities, may give more than one name to members of the same species. A simple and commonsensical approach might attempt to resolve all such disputes with a principle of priority—let the oldest name prevail. In practice, such "obvious" solutions rarely work. The history of taxonomy since Linnaeus has featured three sequential approaches to this classic problem.

1. *Appropriateness.* Modern nomenclature dates from the publication, in 1758, of the tenth edition of Linnaeus's *Systema Naturae.* In principle, Linnaeus endorsed the rule of priority. In practice, he and most of his immediate successors commonly changed names for reasons, often idiosyncratic, of supposed "appropriateness." If the literal Latin of an original name ceased to be an accurate descriptor, new names were often devised. (For example, a species originally named *floridensis* to denote a restricted geographic domain might be renamed *americanus* if it later spread throughout the country.)

Some unscrupulous taxonomists used appropriateness as a thinly veiled tactic to place their own stamp upon species by raiding rather than by scientific effort. A profession supposedly dedicated to expanding knowledge about things began to founder into a quagmire of arguments about names. In the light of such human foibles, appropriateness could not work as a primary criterion for taxonomic names.

2. *Priority.* The near anarchy of appropriateness provoked a chorus of demands for reform and codification. The British Association for the Advancement of Science finally appointed a committee to formulate a set of official rules for nomenclature. The Strickland Committee, obedient to the age-old principle that

periods of permissiveness lead to stretches of law 'n' order (before the cycle swings round again), reported in 1842 with a "strict construction" that must have brought joy to all Robert Borks of the day. Priority in publication shall be absolutely and uncompromisingly enforced. No ifs, ands, buts, quibbles, or exceptions.

This decision may have ended the anarchy of capricious change, but it introduced another impediment, perhaps even worse, based on the exaltation of incompetence. When new species are introduced by respected scientists, in widely read publications with clear descriptions and good illustrations, people take notice and the names pass into general use. But when Ignatz Doofus publishes a new name with a crummy drawing and a few lines of telegraphic and muddled description in the *Proceedings of the Philomathematical Society of Pfennighalbpfennig* (circulation 533), it passes into well-deserved oblivion. Unfortunately, under the Strickland Code of strict priority, Herr Doofus's name, if published first, becomes the official moniker of the species—so long as Doofus didn't break any rule in writing his report. The competence and usefulness of his work have no bearing on the decision. The resulting situation is perversely curious. What other field defines its major activity by the work of the least skilled? As Charles Michener, our greatest taxonomist of bees, once wrote: "In other sciences the work of incompetents is merely ignored; in taxonomy, because of priority, it is preserved."

If the Sterling/Doofus ratio were high, priority might pose few problems in practice. Unfortunately, the domain of Doofuses forms a veritable army, issuing cannonade after cannonade of publications filled with new names destined for oblivion but technically constituted in correct form. Since every profession has its petty legalists, its boosters of tidiness and procedure over content, natural history sank into a mire of unproductive pedantry that, in Ernst Mayr's words, "deflected taxonomists from biological research into bibliographic archeology." Legions of technocrats delighted in searching obscure and forgotten publications for an earlier name that could displace some long-accepted and stable usage. Acrimonious arguments proliferated, for Doofus's inadequate descriptions rarely permitted an unambiguous identification of his earlier name with any well-defined species. Thus, a rule introduced to establish stability against capricious change for appropriateness sowed even greater disruption by forcing the abandonment of accepted names for forgotten predecessors.

3. *Plenary Powers.* The abuses of Herr Doofus and his ilk induced a virtual rebellion among natural historians. A poll of Scandinavian zoologists, taken in 1911, yielded 2 in favor and 120 opposed to strict priority. All intelligent administrators know that the key to a humane and successful bureaucracy lies in creative use of the word *ordinarily.* Strict rules of procedure are ordinarily inviolable—unless a damned good reason for disobedience arises, and then flexibility permits humane and rational exceptions. The Plenary Powers Rule, adopted in Monaco in 1913 to stem the revolt against strict priority, is a codification of the estimable principle of *ordinarily.* It provided, as quoted early in this essay, that the first designation shall prevail, unless a later name has been so widely accepted that its suppression in favor of a forgotten predecessor would sow confusion and instability.

Such exceptions to strict priority cannot be asserted by individuals but must be officially granted by the International Commission of Zoological Nomenclature, acting under its plenary powers. The procedure is somewhat cumbersome and demands a certain investment of time and paperwork, but the plenary powers rule has served us well and has finally achieved stability by locating the fulcrum between strict priority and proper exception. To suppress an earlier name under the plenary powers, a taxonomist must submit a formal application and justification to the International Commission (a body of some thirty professional zoologists). The commission then publishes the case, invites commentary from taxonomists throughout the world, considers the initial appeal with all elicited support and rebuttal, and makes a decision by majority vote.

The system has worked well, as two cases may illustrate. The protozoan species *Tetrahymena pyriforme* has long been a staple for biological research, particularly on the physiology of single-celled organisms. John Corliss counted more than 1,500 papers published over a 27-year span—all using this name. However, at least ten technically valid names, entirely forgotten and unused, predate the first publication of *Tetrahymena.* No purpose would be served by resurrecting any of these earlier designations and suppressing the universally accepted *Tetrahymena*. Corliss's petition to the commission was accepted without protest, and *Tetrahymena* has been officially accepted under the plenary powers.

One of my favorite names recently had a much closer brush with official extinction. The generic names of many animals are the same as their common designation: the gorilla is *Gorilla;* the rat, *Rattus.* But I know only one case of a vernacular name identical with both generic *and* specific parts of the technical Latin. The boa constrictor is (but almost wasn't) *Boa constrictor,* and it would be a damned shame if we lost this lovely consonance. Nevertheless, in 1976, *Boa constrictor* barely survived one of the closest contests ever brought before the commission, as thirteen members voted to suppress this grand name in favor of *Boa canina,* while fifteen noble nays stood firm and saved the day. The details are numerous and not relevant to this essay. Briefly, in the founding document of 1758, Linnaeus placed nine species in his genus *Boa,* including *canina* and *constrictor.* As later zoologists divided Linnaeus's overly broad concept of *Boa* into several genera, a key question inevitably arose: Which of Linnaeus's original species should become the "type" (or name bearer) for the restricted version of *Boa,* and which should be assigned to other genera? Many professional herpetologists had accepted *canina* as the best name bearer (and assigned *constrictor* to another genus); but a world of both technical and common usage, from textbooks to zoo labels to horror films, recognized *Boa constrictor.* The commission narrowly opted, in a tight squeeze (sorry, I couldn't resist that one), for the name we all know and love. Ernst Mayr, in casting his decisive vote, cited the virtue of stability in validating common usage—the basis for the plenary powers decision in the first place:

> I think here is clearly a case where stability is best served by following usage in the general zoological literature. I have asked numerous zoologists "what species does the genus *Boa* call to your mind?" and they all said immediately *"constrictor."*... Making *constrictor* the type of *Boa* will remove all ambiguity from the literature.

These debates often strike nonprofessionals as a bit ridiculous—a sign, perhaps, that taxonomy is more wordplay than science. After all, science studies the external world (through the dark glass of our prejudices and perceptions to be sure). Questions of first publication versus common usage raise no issues about the animals "out there," and only concern human conventions for naming. But this is the point, not the problem. These are debates about names, not things—and the arbitrary criteria of human decision-making, not boundaries imposed by the external world, apply to our resolutions. The aim of these debates (although not always, alas, the outcome) is to cut through the verbiage, reach a stable and practical decision, and move on to the world of things.

Which leads—did you think that I had forgotten my opening paragraph?—back to philately. The United States government, jumping on the greatest bandwagon since the hula hoop, recently issued four striking stamps bearing pictures of dinosaurs—and labeled *Tyrannosaurus, Stegosaurus, Pteranodon,* and *Brontosaurus.*

Thrusting itself, with all the zeal of a convert, into the heart of commercial hype, the U.S. Post Office seems committed to shedding its image for stodginess in one fell, crass swoop. Its small brochure, announcing October as "national stamp collecting month," manages to sponsor a contest, establish a tie-in both with T-shirts and a videocassette for *The Land Before Time,* and offer a dinosaur "discovery kit" (a $9.95 value for just $3.95; "Valid while supplies last. Better hurry!"). You will, in this context, probably not be surprised to learn that the stamps were officially launched on October 1, 1989, in Orlando, Florida, at Disney World.

Amidst this maelstrom of marketing, the Post Office also engendered quite a brouhaha about the supposed subject of one stamp—a debate given such prominence in the press that much of the public (at least judging from my voluminous mail) now thinks that an issue of great scientific importance has been raised to the detriment and shame of an institution otherwise making a worthy step to modernity. (We must leave this question for another time, but I confess great uneasiness about such approbation. I appreciate the argument that T-shirts and videos heighten awareness and expose aspects of science to millions of kids otherwise unreached. I understand why many will accept the forceful spigot of hype, accompanied by the watering-down of content— all in the interest of extending contact. But the argument works only if, having made contact, we can then woo these kids to a deeper intellectual interest and commitment. Unfortunately, we are often all too ready to compromise. We hear the blandishments: Dumb it down; hype it up. But go too far and you cannot turn back; you lose your own soul by dripping degrees. The space for wooing disappears down the maw of commercialism. Too many wise people, from Shakespeare to my grandmother, have said that dignity is the only bit of our being that cannot be put up for sale.)

This growing controversy even reached the august editorial pages of the *New York Times* (October 11, 1989), and their description serves as a fine epitome of the supposed mess:

> The Postal Service has taken heavy flak for mislabeling its new 25-cent dinosaur stamp, a drawing of a pair of dinosaurs captioned *"Brontosaurus."* Furious purists point out that the "brontosaurus" is now properly called "apatosaurus." They accuse the stamp's authors of fostering scientific illiteracy, and want the stamps recalled.

*Brontosaurus* versus *Apatosaurus.* Which is right? How important is this issue? How does it rank amidst a host of other controversies surrounding this and other dinosaurs: What head belongs on this dinosaur (whether it be called *Brontosaurus* or *Apatosaurus*); were these large dinosaurs warm-blooded; why did they become extinct? The press often does a good job of reporting basic facts of a dispute, but fails miserably in supplying the context that would allow a judgment about importance. I have tried, in the first part of this essay, to supply the necessary context for grasping *Brontosaurus* versus *Apatosaurus.* I regret to report, and shall now document, that the issue could hardly be more trivial—for the dispute is only about names, not about things. The empirical question was settled to everyone's satisfaction in 1903. To understand the argument about names, we must know the rules of taxonomy and something about the history of debate on the principle of priority. But the exposure of context for *Brontosaurus* versus *Apatosaurus* does provide an interesting story in itself and does raise important issues about the public presentation of science—and thus do I hope to snatch victory (or at least interest) from the jaws of defeat (or triviality).

*Brontosaurus* versus *Apatosaurus* is a direct legacy of the most celebrated feud in the history of vertebrate paleontology—Cope versus Marsh. As E. D. Cope and O. C. Marsh vied for the glory of finding spectacular dinosaurs and mammals in the American West, they fell into a pattern of rush and superficiality born of their intense competition and mutual dislike. Both wanted to bag as many names as possible, so they published too quickly, often with inadequate descriptions, careless study, and poor illustrations. In this unseemly rush, they frequently gave names to fragmentary material that could not be well characterized and sometimes described the same creature twice by failing to make proper distinctions among the fragments. (For a good history of this issue, see D. S. Berman and J. S. McIntosh, 1978. These authors point out that both Cope and Marsh often described and officially named a species when only a few bones had been excavated and most of the skeleton remained in the ground.)

In 1877, in a typically rushed note, O. C. Marsh named and described *Apatosaurus ajax* in two paragraphs without illustrations ("Notice of New Dinosaurian Reptiles from the Jurassic Formation," *American Journal of Science,* 1877). Although he noted that this "gigantic dinosaur . . . is represented in the Yale Museum by a nearly complete skeleton in excellent preservation," Marsh described only the vertebral column. In 1879, he published another page of information and presented the first sketchy illustrations—of pelvis, shoulder blade, and a few vertebrae ("Principal Characters of American Jurassic Dinosaurs, Part II," *American Journal of Science,* 1879). He also took this opportunity to pour some vitriol upon Mr. Cope, claiming that Cope had misnamed and misdescribed several forms in his haste. "Conclusions based on such work," Marsh asserts, "will naturally be received with distrust by anatomists."

In another 1879 article, Marsh introduced the genus *Brontosaurus,* with two paragraphs (even shorter than those initially devoted to *Apatosaurus*), no illustrations, and just a few comments on the pelvis and vertebrae. He did estimate the length of his new beast at seventy to eighty feet, in comparison with some fifty feet for *Apatosaurus* ("Notice of New Jurassic Reptiles," *American Journal of Science,* 1879).

Marsh considered *Apatosaurus* and *Brontosaurus* as distinct but

Marsh's famous illustration of the complete skeleton of *Brontosaurus*.
FROM THE SIXTEENTH ANNUAL REPORT OF THE U.S. GEOLOGICAL SURVEY,
1895. NEG. NO. 328654. COURTESY DEPARTMENT OF LIBRARY SERVICES, AMERI-
CAN MUSEUM OF NATURAL HISTORY.

closely related genera within the larger family of sauropod dinosaurs. *Brontosaurus* soon became everyone's typical sauropod—
indeed *the* canonical herbivorous dinosaur of popular consciousness, from the Sinclair logo to Walt Disney's *Fantasia*— for
a simple and obvious reason. Marsh's *Brontosaurus* skeleton, from
the most famous of all dinosaur localities at Como Bluff Quarry
10, Wyoming, remains to this day "one of the most complete
sauropod skeletons ever found" (quoted from Berman and McIntosh, cited previously). Marsh mounted the skeleton at Yale and
often published his spectacular reconstruction of the entire animal. (*Apatosaurus,* meanwhile, remained a pelvis and some vertebrae.) In his great summary work, *The Dinosaurs of North America,*
Marsh wrote (1896): "The best-known genus of the Atlanto-
sauridae is *Brontosaurus,* described by the writer in 1879, the type
specimen being a nearly entire skeleton, by far the most complete
of any of the Sauropoda yet discovered." *Brontosaurus* also became the source of the old stereotype, now so strongly challenged, of slow, stupid, lumbering dinosaurs. Marsh wrote in

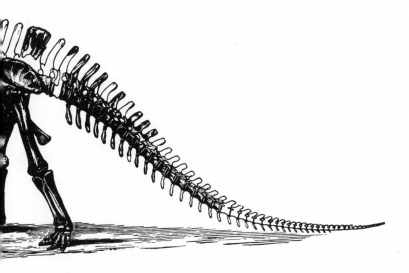

1883, when presenting his full reconstruction of *Brontosaurus* for the first time:

> A careful estimate of the size of *Brontosaurus*, as here restored, shows that when living the animal must have weighed more than twenty tons. The very small head and brain, and slender neural cord, indicate a stupid, slow-moving reptile. The beast was wholly without offensive or defensive weapons, or dermal armature. In habits, *Brontosaurus* was more or less amphibious, and its food was probably aquatic plants or other succulent vegetation.

In 1903, Elmer Riggs of the Field Museum in Chicago restudied Marsh's sauropods. Paleontologists had realized by then that Marsh had been overgenerous in his designation of species (a "splitter" in our jargon), and that many of his names would have to be consolidated. When Riggs restudied *Apatosaurus* and *Brontosaurus*, he recognized them as two versions of the same creature, with *Apatosaurus* as a more juvenile specimen. No big deal; it happens all the time. Riggs rolled the two genera into one in a single paragraph:

The genus *Brontosaurus* was based chiefly upon the structure of the scapula and the presence of five vertebrae in the sacrum. After examining the type specimens of these genera, and making a careful study of the unusually well-preserved specimen described in this paper, the writer is convinced that the Apatosaur specimen is merely a young animal of the form represented in the adult by the Brontosaur specimen. . . . In view of these facts the two genera may be regarded as synonymous. As the term *"Apatosaurus"* has priority, *"Brontosaurus"* will be regarded as a synonym.

In 1903, ten years before the plenary powers decision, strict priority ruled in zoological nomenclature. Thus, Riggs had no choice but to sink the later name, *Brontosaurus,* once he had decided that Marsh's earlier name, *Apatosaurus,* represented the same animal. But then I rather doubt that Riggs would have gone to bat for *Brontosaurus* even if he could have submitted a case on its behalf. After all, *Brontosaurus* was not yet an icon of pop culture in 1903—no Sinclair logo, no Alley-Oop, no *Fantasia,* no *Land Before Time.* Neither name had captured public or scientific fancy, and Riggs probably didn't lament the demise of *Brontosaurus.*

No one has ever seriously challenged Riggs's conclusion, and professionals have always accepted his synonymy. But Publication 82 of the "Geological Series of the Field Columbian Museum" for 1903—the reference for Riggs's article—never gained much popular currency. The name *Brontosaurus,* still affixed to skeletons in museums thoughout the world, still perpetuated in countless popular and semi-technical books about nature, never lost its luster, despite its technical limbo. Anyone could have applied to the commission for suppression of *Apatosaurus* under the plenary powers in recognition of the widespread popularity and stability of *Brontosaurus.* I suspect that such an application would have succeeded. But no one bothered, and a good name remains in limbo. (I also wish that someone had fought for suppression of the unattractive and inappropriate name *Hyracotherium* in favor of the lovely but later *Eohippus,* also coined by Marsh. But again, no one did.)

I'm afraid there's not much more to this story—not nearly the issue hyped by your newspapers as the Great Stamp Flap. No argument of fact arises at all, just a question of names, settled in

1903, but never transferred to a general culture that continues to learn and favor the technically invalid name *Brontosaurus*. But the story does illustrate something troubling about the presentation of science in popular media. The world of *USA Today* is a realm of instant fact and no analysis. Hundreds of bits come at us in pieces never lasting more than a few seconds—for the dumb-downers tell us that average Americans can't assimilate anything more complex or pay attention to anything longer.

This oddly "democratic" procedure makes all bits equal—the cat who fell off a roof in Topeka (and lived) gets the same space as the Soviet withdrawal from Afghanistan. Equality is a magnificent system for human rights and morality in general, but not for the evaluation of information. We are bombarded with too much in our inordinately complex world; if we cannot sort the trivial from the profound, we are lost in terminal overload. The criteria for sorting must involve context and theory—the larger perspective that a good education provides.

In the current dinosaur craze without context, all bits are mined for their superficial news value as items in themselves—a lamentable tendency abetted by the "trivial pursuit" one-upmanship that confers status on people who know (and flaunt) the most bits. (If you play this dangerous game in real life, remember that ignorance of context is the surest mark of a phony. If you approach me in wild lament, claiming that our postal service has mocked the deepest truth of paleontology, I will know that you have only skimmed the surface of my field.)

Consider the four items mentioned earlier in this essay. They are often presented in *USA Today* style as equal factoids. But with a context to sort the trivial from the profound, we may recognize some as statements about words, others as entries to the most general questions we can ask about the history of life. *Apatosaurus* versus *Brontosaurus* is a legalistic quibble about words and rules of naming. Leave the Post Office alone. They take enough flak (much justified of course) as it is. The proper head for *Apatosaurus* is an interesting empirical issue, but of little moment beyond the sauropods. Marsh found no skull associated with either his *Apatosaurus* or his *Brontosaurus* skeleton. He guessed wrong and mounted the head of another sauropod genus called *Camarosaurus*. *Apatosaurus* actually bore a head much more like that of the different genus *Diplodocus*. The head issue (*Camarosaurus*-like versus *Diplodocus*-like) and the name issue (*Apatosaurus* versus *Brontosaurus*) are entirely separate questions, although the press has confused and conflated them.

The question of warm-bloodedness (quite unresolved at the moment) is more general still, as it affects our basic concepts of dinosaur physiology and efficiency. The issue of extinction is the broadest of all—for basic patterns of life's history are set by differential survival of groups through episodes of mass dying. We are here today, arguing about empty issues like *Apatosaurus* versus *Brontosaurus,* because mammals got through the great Cretaceous extinction, while dinosaurs did not.

I hate to be a shill for the Post Office, but I think that they made the right decision this time. Responding to the great *Apatosaurus* flap, Postal Bulletin Number 21744 proclaimed: "Although now recognized by the scientific community as *Apatosaurus,* the name *Brontosaurus* was used for the stamp because it is more familiar to the general population. Similarly, the term "dinosaur" has been used generically to describe all the animals, even though the *Pteranodon* was a flying reptile." Touché and right on; no one bitched about *Pteranodon,* and that's a real error.

The Post Office has been more right than the complainers, for Uncle Sam has worked in the spirit of the plenary powers rule. Names fixed in popular usage may be validated even if older designations have technical priority. But now . . . Oh Lord, why didn't I see it before! Now I suddenly grasp the secret thread behind this overt debate! It's a plot, a dastardly plot sponsored by the apatophiles—that covert society long dedicated to gaining support for Marsh's original name against a potential appeal to the plenary powers. They never had a prayer before. Whatever noise they made, whatever assassinations they attempted, they could never get anyone to pay attention, never disturb the tranquillity and general acceptance of *Brontosaurus.* But now that the Post Office has officially adopted *Brontosaurus,* they have found their opening. Now enough people know about *Apatosaurus* for the first time. Now an appeal to the plenary powers would not lead to the validation of *Brontosaurus,* for *Apatosaurus* has gained precious currency. They have won; we brontophiles have been defeated.

*Apatosaurus* means "deceptive lizard"; *Brontosaurus* means "thunder lizard"—a far, far better name (but appropriateness, alas, as we have seen, counts for nothing). They have deceived us; we brontophiles have been outmaneuvered. Oh well, graciousness in defeat before all (every bit as important as dignity, if not an aspect thereof). I retreat, not with a bang of thunder, but with a whimper of hope that rectification may someday arise from the ashes of my stamp album.

# 6 | The Dinosaur Rip-off

WE LOVE occasional reversals of established order, both to defuse the tension of inequity and to infuse a bit of variety into our lives. Consider the medieval feast of fools (where slaves could be masters, in jest and only for a moment), Sadie Hawkins Day, and the genre of quiz that supplies the answer and asks a contestant to reconstruct the question. I begin this essay in such a spirit by giving my answer to a question that has surpassed all others (except, perhaps, "Where is human evolution going?") in my catalogue of inquiries from people who love natural history. My answer, unfortunately, must be: "Damned if I know"—which won't help you much in trying to guess the question. So I'll reveal the question without further ado: "What's behind the great dinosaur mania that's been sweeping the country during the past few years?"

Readers will scarcely need my words to document the phenomenon, for we are all surrounded by dinosaur tote bags, lunch boxes, pens and pencils, underpants, ties, and T-shirts that say "bossosaurus" or "secretaryosaurus," as the case may be. You can buy dinosaur-egg soap to encourage your kids to take a bath, a rocking stegosaurus for indoor recreation (a mere 800 bucks from F.A.O. Schwarz), a brontosaurus bank to encourage thrift, or a dinosaur growth chart to hang on the wall and measure your tyke's progress toward the N.B.A. In Key West, where dinosaurs have edged out flamingos as icons of kitsch, I even saw dinosaur toilet paper with a different creature on each perforated segment—providing quite a sense of power, I suppose, when used for its customary purpose. (This reminded me of the best attempt

I ever encountered for defusing the Irish situation. I once stayed in a small motel in Eire where the bathrooms had two rolls of toilet paper—one green, the other orange.)

I offer no definitive answer to the cause of this mania, but I can at least document a fact strongly relevant to the solution. Perhaps dinosaur mania is intrinsic and endemic, a necessary and permanent fact of life (once the fossils had been discovered and properly characterized); perhaps dinosaurs act as the trigger for a deep Jungian archetype of the soul; perhaps they rank as incarnations of primal fears and fascinations, programmed into our brains as the dragons of Eden. But these highfalutin suggestions cannot suffice for the simple reason that dinosaurs have been well documented throughout our century, while few people granted them more than passing notice before the recent craze hit.

I can testify to the previous status of dinosaurs among the arcana of our culture, for I was a kiddie dinosaur nut in the late 1940s when nobody gave a damn. I fell in love with the great skeletons at the American Museum of Natural History and then, with all the passion of youth, sought collateral material with thoroughness and avidity. I would pounce on any reinforcement of my greatest interest—a Sinclair Oil logo or a hokey concrete tyrannosaur bestriding (like a colossus) Hole 15 at the local miniature golf course. There sure wasn't much to find—a few overpriced brass figures and a book or two by Roy Chapman Andrews and Ned Colbert, all hard to get anywhere outside the Museum shop. Representations in pop culture were equally scarce, ranging little beyond King Kong versus the pteranodon and Alley Oop riding a brontosaurus.

One story will indicate both the frustration of a young adept in a world of ignorance and the depth of that ignorance itself. At age nine or so, in the Catskills at one of those innumerable summer camps with an Indian name, I got into a furious argument with a bunkmate over the old issue of whether humans and dinosaurs ever inhabited the earth together. We agreed—bad, bad mistake—to abide by the judgment of the first adult claiming to know the answer, and we bet the camp currency, a chocolate bar, on the outcome. We asked all the counselors and staff, but none had ever heard of a brontosaurus. At parents' weekend, his came and mine didn't. We asked his father, who assured us that of course dinosaurs and people lived together; just look at Alley Oop. I paid—and seethed—and still seethe. This could not happen today. Anyone—a few "scientific creationists" excepted—would both know the answer and give you the latest rundown on theories for the extinction of dinosaurs.*

All this I tell for humor, but a part of the story isn't so funny. Kiddie culture can be cruel and fiercely anti-intellectual. I survived because I wasn't hopeless at punchball, and I won some respect for my knowledge of baseball stats. But any kid with a passionate interest in science was a wonk, a square, a dweeb, a doofus, or a geek (I don't remember what word held sway at the time, but one item in that particular litany of cruelty is always in vogue). I was taunted by many classmates as peculiar. I was called "fossil face" on the playground. It hurt.

I once asked my colleague Shep White, a leading child psychologist, why kids were so interested in dinosaurs. He gave an answer both elegant and succinct: "Big, fierce, and extinct." I love this response, but it can't resolve the question that prompted this essay. Dinosaurs were also big, fierce, and extinct twenty years ago, but few kids or adults gave a damn about them. And so I return to the original question: What started the current dinosaur craze?

The optimistic answer for any intellectual must be that public taste follows scientific discovery. The past twenty years have been a heyday for new findings and fundamental revisions in our view of dinosaurs. The drab, lumbering, slow-witted, inefficient beasts of old interpretations have been replaced with smooth, sleek, colorful, well-oiled, and at least adequately intelligent revised versions. The changes have been most significant in three subjects: anatomy, behavior, and extinction. All three have provided a more congenial and more interesting perspective on dinosaurs. For anatomy, a herd of brontosauruses charging through the desert inspires more awe than a few behemoths so encumbered by their own weight that they must live in ponds (see the classic illustration on the dust jacket of this book). For behavior, the

---

*I was too optimistic. Never overestimate the depth of our anti-intellectual traditions! A week after I published this essay, results of a comprehensive survey showed that about 30 percent of American adults accept the probable contemporaneity of humans and dinosaurs. Still, our times are better than before. Seventy percent of that camp could have answered our inquiry before parental arrival.

images of the newly christened *Maiasauria,* the good mother lizard, brooding her young, or a herd of migrating ornithopods, with vulnerable juveniles in the center and strong adults at the peripheries, inspire more sympathy than a dumb stegosaur laying her eggs and immediately abandoning them by instinct and ignorance. For extinction, crashing comets and global dust clouds surely inspire more attention than gradually changing sea levels or solar outputs.

I wish that I could locate the current craze in these exciting intellectual developments. But a moment's thought must convince anyone that this good reason cannot provide the right answer. Dinosaurs might not have been quite so jazzy and sexy twenty years ago, but the brontosaurs weren't any smaller back then, the tyrannosaurs were just as fierce, and the whole clan was every bit as extinct (my camp friend's father notwithstanding). You may accept or reject Shep White's three categories, but choose any alternate criteria and dinosaurs surely had the capacity to inspire a craze at any time—twenty years ago as well as today. (At least two mini-crazes of earlier years—in England after Waterhouse Hawkins displayed his life-sized models at the Crystal Palace in the 1850s, and in America after Sinclair promoted a dinosaur exhibit at the New York World's Fair in 1939—illustrate this permanent potential.) We must conclude, I think, that dinosaurs have never lacked the seeds of appeal, that the missing ingredient must be adequate publicity, and that the key to "why now?" resides in promotion, not new knowledge.

I must therefore assume that the solution lies in that great and dubious driving force of American society—marketing. At some definable point, some smart entrepreneur recognized an enormous and largely unexploited potential for profit. What craze is any different? Did goldfish reach an optimal size and tastiness for swallowing in the early 1940s? Did a breakthrough in yo-yo technology spawn the great passion that swept the streets of New York in my youth? Did hula hoops fit some particular social niche and need uniquely confined to a few months during the 1950s?

I don't doubt that a few more general factors may form part of the story. Perhaps the initial entrepreneurs developed their own interest and insight by reading about new discoveries. Perhaps the vast expansion of museum gift shops—a dubious trend (in my view), with more to lament in skewed priorities than to praise in heightened availability of worthy paraphernalia—gave an essential boost in providing an initial arena for sales. Still, most crazes get started for odd and unpredictable reasons and then propagate by a kind of mass intoxication and social conformity. If I am right in arguing that the current dinosaur craze could have occurred long ago and owes both its origin and initial spread to a marketing opportunity seized by a few diligent entrepreneurs (with later diffusion by odd mechanisms of crowd psychology that engender chain reactions beyond a critical mass), then the source of this phenomenon may not be a social trend or a new discovery, but the cleverness of a person or persons unknown (with a product or products unrecognized). As this craze is no minor item in twentieth-century American cultural history, I would love to identify the instigators and the insights. If anyone knows, please tell me.

I do confess to some cynical dubiety about the inundation of kiddie culture with dinosaurs in every cute, furry, and profitable venue that any marketing agent can devise. I don't, of course, advocate a return to the ignorance and unavailability of information during my youth, but a dinosaur on every T-shirt and milk carton does foreclose any sense of mystery or joy of discovery—and certain forms of marketing do inexorably lead to trivialization. Interest in dinosaurs becomes one of those ephemeral episodes—somewhere between policeman and fireman—in the canonical sequence of childhood interests. Something to burn brightly in its appointed season and then, all too often, to die—utterly and without memory.

As intellectuals, we acknowledge and accept a minority status in our culture (since hope, virtue, and reality rarely coincide). We therefore know that we must seize our advantages by noting popular trends and trying to divert some of their energy into rivulets that might benefit learning and education. The dinosaur craze should be a blessing for us, since the source material is a rip-off of our efforts—the labor of paleontologists, the great skeletons mounted in our museums. Indeed, we have done well—damned well, as things go. Lurking in and around the book covers and shopping bags are a pretty fair number of mighty good books, films, puzzles, games, and other items of—dare I say it—decent intellectual and educational content.

It is now time to segue, via a respectable transition, into the second part of this essay. (But before we do, and while I'm throwing out requests for enlightenment, can anyone tell me how this fairly obscure Italian term from my musical education managed its recent entry into trendy American speech?*)

We all acknowledge the sorry state of primary and secondary education in America, both by contrast with the successes of other nations and by any absolute standard of educational need in an increasingly complex world. We also recognize that the crisis is particularly acute for the teaching of science. Well, being of an optimistic nature, I survey the dinosaur craze and wonder why science suffers so badly within our schools. The dinosaur craze has generated, amidst a supersaurus-sized pile of kitsch and crap, a remarkable range of worthy material that kids seem to like and use. Kids love science so long as fine teaching and good material grace the presentation. If the dinosaur craze of pop culture has been adequately subverted for educational ends, why can't we capitalize on this benevolent spin-off? Why can't we sustain the interest, rather than letting it wither like the flower of grass, as soon as a child moves on to his next stage? Why can't we infuse some of this excitement into our schools and use it to boost and expand interest in all of science? Think of the aggregate mental power vested in 10 million five-year-olds, each with an average of twenty monstrous Latin dinosaur names committed to memory with the effortless joy and awesome talent of human beings at the height of their powers for rote learning. Can't we transfer this skill to all the other domains—arithmetic, spelling, and foreign languages, in particular—that benefit so greatly from rote learning in primary school years? (Let no adult disparage the value of rote because we lose both the ability and the joy in later years.)

Why is the teaching of science in such trouble in our nation's public schools? Why is the shortage of science teachers so desperate that hundreds of high schools have dropped physics entirely, while about half of all science courses still on the books are now being taught by people without formal training in science? To understand this lamentable situation, we must first dispel the silly and hurtful myth that science is simply too hard for pre-adults. (Supporters of this excuse argue that we succeeded in the

---

*See the postscript to this essay for an interesting reaction to this appeal.

past only because science was much simpler before the great explosion of modern knowledge.)

This claim cannot be sustained for two basic reasons. First, science uses and requires no special mental equipment beyond the scope of a standard school curriculum. The subject matter may be different but the cerebral tools are common to all learning. Science probes the factual state of the world; religion and ethics deal with moral reasoning; art and literature treat aesthetic and social judgment.

Second, we may put aside all abstract arguments and rely on the empirical fact that other nations have had great success in science education. If their kids can handle the material, so can ours, with proper motivation and instruction. Korea has made great strides in education, particularly in mathematics and the physical sciences. And if you attempt to take refuge in the cruel and fallacious argument that Orientals are genetically built to excel in such subjects, I simply point out that European nations, filled with people more like most of us, have been just as successful. The sciences are well taught and appreciated in the Soviet Union, for example, where the major popular bookstores on Leninsky Prospekt are stocked with technical books both browsed and purchased in large numbers. Moreover, we proved the point to ourselves in the late 1950s, when the Soviet Sputnik inspired cold war fears of Russian technological takeover, and we responded, for once, with adequate cash, expertise, and enthusiasm, by launching a major effort to improve secondary education in science. But that effort, begun for the wrong reasons, soon petered out into renewed mediocrity (graced, as always, with pinpoints of excellence here and there, whenever a great teacher and adequate resources coincide).

We live in a profoundly nonintellectual culture, made all the worse by a passive hedonism abetted by the spread of wealth and its dissipation into countless electronic devices that impart the latest in entertainment and supposed information—all in short (and loud) doses of "easy listening." The kiddie culture, or playground, version of this nonintellectualism can be even more strident and more one-dimensional, but the fault must lie entirely with adults—for our kids are only enhancing a role model read all too clearly.

I'm beginning to sound like an aging Miniver Cheevy, or like the chief reprobate on Ko-Ko's little list "of society offenders who might well be underground"—and he means dead and buried, not romantically in opposition: "the idiot who praises with enthusiastic tone, all centuries but this and every country but his own." I want to make an opposite and curiously optimistic point about our current mores: We are a profoundly nonintellectural culture, but we are not committed to this attitude; in fact, we are scarcely committed to anything. We may be the most labile culture in all history, capable of rapid and massive shifts of prevailing opinions, all imposed from above by concerted media effort. Passivity and nonintellectual judgment are the greatest spurs to such lability. Everything comes to us in fifteen-second sound bites and photo opportunities. All possibility for ambiguity—the most precious trait of any adequate analysis—is erased. He wins who looks best or shouts loudest. We are so fearful of making judgments ourselves that we must wait until the TV commentators have spoken before deciding whether Bush or Dukakis won the debate.

We are therefore maximally subject to imposition from above. Nonetheless, this dangerous trait can be subverted for good. A few years ago, in the wake of an unparalleled media blitz, drugs rose from insignificance to a strong number one on the list of serious American problems in that most mercurial court of public opinion as revealed by polling. Surely we can provoke the same immediate recognition for poor education. Talk about "wasted minds." Which cause would you pick as the greater enemy, quantitatively speaking, in America: crack or lousy education abetted by conformity and peer pressure in an anti-intellectual culture?

We live in a capitalist economy, and I have no particular objection to honorable self-interest. We cannot hope to make the needed, drastic improvement in primary and secondary education without a dramatic restructuring of salaries. In my opinion, you cannot pay a good teacher enough money to recompense the value of talent applied to the education of young children. I teach an hour or two a day to tolerably well-behaved near-adults—and come home exhausted. By what possible argument are my services worth more in salary than those of a secondary-school teacher with six classes a day, little prestige, less support, massive problems of discipline, and a fundamental role in shaping minds. (In comparison, I only tinker with intellects already largely formed.) Why are salaries so low, and attendant prestige so limited, for the most important job in America? How can our priorities be so skewed that when we wish to raise the status of science teachers, we take the media route and try to place a member of the profession into orbit (with disastrous consequences, as it happened), rather than boosting salaries on earth? (The crisis in science teaching stems directly from this crucial issue of compensation. Science graduates can begin in a variety of industrial jobs at twice the salary of almost any teaching position; potential teachers in the arts and humanities often lack these well-paid alternatives and enter the public schools *faute de mieux.*)

We are now at a crux of opportunity, and the situation may not persist if we fail to exploit it. If I were king, I would believe Gorbachev, realize that the cold war is a happenstance of history—not a necessary and permanent state of world politics—make some agreements, slash the military budget, and use just a fraction of the savings to double the salary of every teacher in American public schools. I suspect that a shift in prestige, and the consequent attractiveness of teaching to those with excellence and talent, would follow.

I don't regard these suggestions as pipe dreams, but having been born before yesterday, I don't expect their immediate implementation either. I also acknowledge, of course, that reforms are not imposed from above without vast and coordinated efforts of lobbying and pressuring from below. Thus, as we work toward a larger and more coordinated solution, and as a small contribution to the people's lobby, could we not immediately subvert more of the dinosaur craze from crass commercialism to educational value?

Dinosaur names can become the model for rote learning. Dinosaur facts and figures can inspire visceral interest and lead to greater wonder about science. Dinosaur theories and reconstructions can illustrate the rudiments of scientific reasoning. But I'd like to end with a more modest suggestion. Nothing makes me sadder than the peer pressure that enforces conformity and erases wonder. Countless Americans have been permanently deprived of the joys of singing because a thoughtless teacher once told them not to sing, but only to mouth the words at the school assembly because they were "off-key." Once told, twice shy and perpetually fearful. Countless others had the light of intellectual wonder extinguished because a thoughtless and swaggering fellow student called them nerds on the playground. Don't point to the obsessives—I was one—who will persist and succeed despite these petty cruelties of youth. For each of us, a hundred are lost—more timid and fearful, but just as capable. We must rage against the dying of the light—and although Dylan Thomas spoke of bodily death in his famous line, we may also apply his words to the extinction of wonder in the mind, by pressures of conformity in an anti-intellectual culture.

The *New York Times,* in an article on science education in Korea, interviewed a nine-year-old girl and inquired after her personal hero. She replied: Stephen Hawking. Believe me, I have absolutely nothing against Larry Bird or Michael Jordan, but wouldn't it be lovely if even one American kid in 10,000 gave such an answer. The article went on to say that science whizzes are class heroes in Korean schools, not isolated and ostracized dweebs.

English wars may have been won on the playing fields of Eton, but American careers in science are destroyed on the playgrounds of Shady Oaks Elementary School. Can we not invoke dinosaur power to alleviate these unspoken tragedies? Can't dinosaurs be the great levelers and integrators—the joint passion of the class rowdy and the class intellectual? I will know that we are on our way when the kid who names *Chasmosaurus* as his personal hero also earns the epithet of Mr. Cool.

---

## Postscript

I had never made an explicit request of readers before, but I was really curious and couldn't find the answer in my etymological books. Hence, my little parenthetical inquiry about *segue:* "Can anyone tell me how this fairly obscure Italian term from my musical education managed its recent entry into trendy American speech?" The question bugged me because two of my students, innocent alas (as most are these days) of classical music, use *segue* all the time, and I longed to know where they found it. Both simply considered *segue* as Ur-English when I asked, perhaps the very next word spoken by our ultimate forefather after his introductory, palindromic "Madam I'm Adam" (as in "segue into the garden with me, won't you").

I am profoundly touched and gratified. The responses came in waves and even yielded, I believe, an interesting resolution. (These letters also produced the salutary effect of reminding me how lamentably ignorant I am about a key element of American culture—pop music and its spin-offs.) I've always said to myself that I write these essays primarily for personal learning; this claim has now passed its own test.

One set of letters (more than two dozen) came from people in their twenties and thirties who had been (or in a case or two, still are) radio deejays for rock stations (a temporary job on a college radio station for most). They all report that *segue* is a standard term for the delicate task (once rather difficult in the days of records and turntables) of making an absolutely smooth transition, without any silence in between or words to cover the change, from one song to the next.

I was quite happy to accept this solution, but I then began to receive letters from old-timers in the radio and film business—all pointing to uses in the 1920s and 1930s (and identifying the lingo of rock deejays as a later transfer). David Emil wrote of his work in television during the mid-1960s:

> The word was in usage as a noun and verb when I worked in the television production industry. . . . It was common for television producers to use the phrase to refer to connections between segments of television shows. . . . Interestingly, although I read a large number of scripts at this time, I never saw the word in writing or knew how it was spelled until the mid-1970s when I came across the word in a more traditional usage.

Bryant Mather, former curator of minerals at the Field Museum in Chicago, sent me an old mimeographed script of his sole appearance on radio—an NBC science show of 1940 entitled *How Do You Know,* and produced "as a public service feature by the Field Museum of Natural History in cooperation with the University Broadcasting Council."

The script, which uses *segue* to describe all transitions between scenes in a dramatization of the history of the Orloff diamond, reminds us by its stereotyping and barely concealed racism (despite the academic credentials of its origin) that some improvements have been made in our attitudes toward human diversity. In one scene, for example, the diamond is bought by Isaacs, described as "a Jewish merchant." His hectoring wife, called "Mama" by Mr. Isaacs, keeps pestering: "Buy it, Isaacs—you hear me—buy it." Isaacs later sells to a shifty Persian, who cheats him by placing lead coins under the surface of gold in his treasure bag. Isaacs, discovering the trick, laments: "Counterfeit—lead—oi, oi, oi—Mama—we are ruined—we are ruined." (Shades of Shylock—my ducats, my daughter.) The script's next line reads "segue to music suggestive of Amsterdam or busy port."

Page Gilman made the earliest link to radio and traced a most sensible transition (dare I say segue) between musical and modern media usages. I will accept his statement as our best resolution to date:

> I think you may find that a bridge between the classical music to which you refer and today's disk jockey use would be the many years of network radio. I began in 1927 and even the earliest scripts would occasionally use "segue" because we had a big staff of professional *working* musicians—people who worked (in those days) in restaurants, theaters (especially), and now radio. Today you'll find a real corps of such folks only in New York and L.A. . . . You'll find me corroborated a little by Pauline Kael of the *New Yorker,* who remembers Horace Heidt's orchestra at the Golden Gate Theater in San Francisco. That was the time when I was dating one of the Downey Sisters in the same orchestra. [May I also report the confirmation of my beloved 92-year-old Uncle Mordie of Rochester, New York, who relished his 1920s daily job in a movie orchestra, playing with the Wurlitzer during the silents and between shows—and never liked nearly as much his forty-year subsequent stint as lead violist in the Rochester Symphony.] I wonder if Bruce Springsteen ever heard of "segue." On such uncultured times have working musicians fallen.

This tracing of origins does not solve the more immediate problem of recent infiltration into general trendy speech. But perhaps this is not even an issue in our media-centered world, where any jargon of the industry stands poised to break out. Among many suggestions for this end of the tale, several readers report that Johnny Carson has prominently used *segue* during the past few years—and I doubt that we would need much more to effect a general spread.

Finally, on my more general inquiry into the sources of our current dinomania, I can't even begin to chronicle the interesting suggestions for fear of composing another book. Just one wistful observation for now. Last year, riding a bus down Haight Street in San Francisco, I approached the junction with Ashbury eager to see what businesses now occupied the former symbolic and actual center of American counterculture. Would you believe that just three or four stores down from the junction itself stands one of those stores that peddles nothing but reptilian paraphernalia and always seems to bear the now-clichéd name "Dino-store." What did Tennyson say in the *Idylls of the King?*

> The old order changeth, yielding
>    place to new;
> And God fulfills himself in
>    many ways,
> Lest one good custom should corrupt
>    the world.

# 3 | Adaptation

# 7 | Of Kiwi Eggs and the Liberty Bell

LIKE OZYMANDIAS, once king of kings but now two legs of a broken statue in Percy Shelley's desert, the great façade of Union Station in Washington, D.C., stands forlorn (but ready to front for a bevy of yuppie emporia now under construction), while Amtrak now operates from a dingy outpost at the side.* Six statues, portraying the greatest of human arts and inventions, grace its parapet. Electricity holds a bar of lightning; his inscription proclaims: "Carrier of light and power. Devourer of time and space. . . . Greatest servant of man. . . . Thou hast put all things under his feet."

Yet I will cast my vote for the Polynesian double canoe, constructed entirely with stone adzes, as the greatest invention for devouring time and space in all human history. These vessels provided sufficient stability for long sea voyages. The Polynesian people, without compass or sextant, but with unparalleled understanding of stars, waves, and currents, navigated these canoes to colonize the greatest emptiness of our earth, the "Polynesian triangle," stretching from New Zealand to Hawaii to Easter Island at its vertices. Polynesians sailed forth into the open Pacific more than a thousand years before Western navigators dared to leave

---

*It is so good and pleasant, in our world of woe and destruction, to report some good news for a change. Union Station has since reopened with a triumphant and vibrant remodeling that fully respects the spirit and architecture of the original. Trains now depart from the heart of this great station, and a renaissance of rational public transportation, with elements of grand style at the termini, may not be a pipe dream.

the coastline of Africa and make a beeline across open water from the Guinea coast to the Cape of Good Hope.

New Zealand, southwestern outpost of Polynesian migrations, is so isolated that not a single mammal (other than bats and seals with their obvious means of transport) managed to intrude. New Zealand was a world of birds, dominated by several species (thirteen to twenty-two by various taxonomic reckonings) of large, flightless moas. Only *Aepyornis,* the extinct elephant bird of Madagascar, ever surpassed the largest moa, *Dinornis maximus,* in weight. Ornithologist Dean Amadon estimated the average weight of *D. maximus* at 520 pounds (although some recent revisions nearly double this bulk), compared with about 220 pounds for ostriches, the largest living birds.

We must cast aside the myths of noble non-Westerners living in ecological harmony with their potential quarries. The ancestors of New Zealand's Maori people based a culture on hunting moas, but soon made short work of them, both by direct removal and by burning of habitat to clear areas for agriculture. Who could resist a 500-pound chicken?

Only one species of New Zealand ratite has survived. (Ratites are a closely related group of flightless ground birds, including moas, African ostriches, South American rheas, and Australian–New Guinean emus and cassowaries. Flying birds have a keeled breastbone, providing sufficient area for attachment of massive flight muscles. The breastbones of ratites lack a keel, and their name honors that most venerable of unkeeled vessels, the raft, or *ratis* in Latin.) We know this curious creature more as an icon on tins of shoe polish or as the moniker for New Zealand's human inhabitants—the kiwi, only hen-sized, but related most closely to moas among birds.

Three species of kiwis inhabit New Zealand today, all members of the genus *Apteryx* (literally, wingless). Kiwis lack an external tail, and their vestigial wings are entirely hidden beneath a curious plumage—shaggy, more like fur than feathers, and similar in structure to the juvenile down of most other birds. (Maori artisans used kiwi feathers to make the beautiful cloaks once worn by chiefs; but the small, secretive, and widely ranging nocturnal kiwis managed to escape the fate of their larger moa relatives.)

The furry bodies, with even contours unbroken by tail or wings, are mounted on stout legs—giving the impression of a double blob (small head and larger body) on sticks. Kiwis eat seeds, berries, and other parts of plants, but they favor earthworms. Their long, thin bills probe the soil continually, suggesting the oddly reversed perspective of a stick leading a blind man. This stick, however, is richly endowed as a sensory device, particularly as an organ of smell. The bill, uniquely among birds, bears long external nostrils, while the olfactory bulb of kiwi brains is second largest among birds relative to size of the forebrain. A peculiar creature indeed.

But the greatest of kiwi oddities centers upon reproduction. Females are larger than males. They lay one to three eggs and may incubate them for a while, but they leave the nest soon thereafter, relegating to males the primary task of incubation, a long seventy to eighty-four days. Males sit athwart the egg, body at a slight angle and bill stretched out along the ground. Females may return occasionally with food, but males must usually fend for

An amazing and famous photo of a female kiwi one day before laying its enormous egg. COURTESY OF THE OTOROHANGA ZOOLOGICAL SOCIETY, NEW ZEALAND.

themselves, covering both eggs and nest entrance with debris and going forth to forage once or twice on most nights.

The kiwi egg is a wonder to behold, and the subject of this essay. It is, by far, the largest of all bird eggs relative to body size. The three species of kiwis just about span the range of domestic poultry: the largest about the size of Rhode Island Reds; the smallest similar to bantams—say five pounds as a rough average (pretty meaningless, given the diversity of species, but setting the general domain). The eggs range to 25 percent of the female's body weight—quite a feat when you consider that she often lays two, and sometimes three, in a clutch, spacing them about thirty-three days apart. A famous X-ray photo of kiwi and egg taken at the kiwi sanctuary of Otorohanga, New Zealand, tells the tale more dramatically than any words I could produce. The egg is so large that females must waddle, legs spread far apart, for several days before laying, as the egg passes down the oviduct toward the cloaca. The incubation patch of male kiwis extends from the top of the chest all the way down to the cloaca—in other words, they need almost all their body to cover the egg.

A study of the general relationship between egg size and body size among birds shows that average birds of kiwi dimensions lay eggs weighing from 55 to 100 grams (as do domestic hens). Eggs of the brown kiwi weigh between 400 and 435 grams (about a pound). Put another way, an egg of this size would be expected from a twenty-eight-pound bird, but brown kiwis are about six times as small.

The obvious question, of course, is why? Evolutionary biologists have a traditional approach to riddles of this sort. They seek some benefit for the feature in question, then argue that natural selection has worked to build these advantages into the animal's way of life. The greatest triumphs of this method center upon odd structures that seem to make no sense or (like the kiwi egg) appear, prima facie, to be out of proportion and probably harmful. After all, anyone can see that a bird's wing (although not a kiwi's) is well designed for flight, so reference to natural selection teaches you little about adaptation that you didn't already know. Thus, the test cases of textbooks are apparently harmful structures that, on closer examination, confer crucial benefits upon organisms in their Darwinian struggle for reproductive success.

This general strategy of research suggests that if you can find out what a structure is good for, you will possess the major ingredient for understanding why it is so big, so colorful, so peculiarly shaped. Kiwi eggs should illustrate this basic method. They seem to be too big, but if we can discover how their large size benefits kiwis, we shall understand why natural selection favored large eggs. Readers who have followed my essays for some time will realize that I wouldn't be writing about this subject if I didn't think that this style of Darwinian reasoning embodied a crucial flaw.

The flaw lies not with the claim of utility. I regard it as proved that kiwis benefit from the unusually large size of their eggs—and for the most obvious reason. Large eggs yield large and well-developed chicks that can fend for themselves with a minimum of parental care after hatching. Kiwi eggs are not only large; they are also the most nutritious of all bird eggs for a reason beyond their maximal bulk: they contain a higher percentage of yolk than any other egg. Brian Reid and G. R. Williams report that kiwi eggs may contain 61 percent yolk and 39 percent albumin (or white). By comparison, eggs of other so-called precocial species (with downy young hatching in an active, advanced, and open-eyed state) contain 35 to 45 percent yolk, while eggs of altricial species (with helpless, blind, and naked hatchlings) carry only 13 to 28 percent yolk.

The lifestyle of kiwi hatchlings demonstrates the benefits of their large, yolky eggs. Kiwis are born fully feathered and usually receive no food from their parents. Before hatching, they consume the unused portion of their massive yolk reserve and do not feed (but live off these egg-based supplies) for their first seventy-two to eighty-four hours alfresco. Newly hatched brown kiwi chicks are often unable to stand because their abdomens are so distended with this reserve of yolk. They rest on the ground, legs splayed out to the side, and only take a first few clumsy steps when they are some sixty hours old. A chick does not leave its burrow until the fifth to ninth day when, accompanied by father, it sallies forth to feed sparingly.

Kiwis thus spend their first two weeks largely living off the yolk supply that their immense egg has provided. After ten to fourteen days, the kiwi chick may weigh one-third less than at hatching—a fasting marked by absorption of ingested yolk from the egg. Brian Reid studied a chick that died a few hours after hatching. Almost half its weight consisted of food reserves—112 grams of yolk and 43 grams of body fat in a 319-gram hatchling. Another chick, killed outside its burrow five to six days after hatching, weighed 281 grams and still held almost 54 grams of enclosed yolk.

I am satisfied that kiwis do very well by and with their large eggs. But can we conclude that the outsized egg was built by natural selection in the light of these benefits? This assumption—the easy slide from current function to reason for origin—is, to my mind, the most serious and widespread fallacy of my profession, for this false inference supports hundreds of conventional tales about pathways of evolution. I like to identify this error of reasoning with a phrase that ought to become a motto: *Current utility may not be equated with historical origin,* or, when you demonstrate that something works well, you have not solved the problem of how, when, or why it arose.

I propose a simple reason for labeling an automatic inference from current utility to historical origin as fallacious: Good function has an alternative interpretation. A structure now useful may have been built by natural selection for its current purpose (I do not deny that the inference often holds), but the structure may also have developed for another reason (or for no particular functional reason at all) and then been co-opted for its present use. The giraffe's neck either got long in order to feed on succulent leaves atop acacia trees or it elongated for a different reason (perhaps unrelated to any adaptation of feeding), and giraffes then discovered that, by virtue of their new height, they could reach some delicious morsels. The simple good fit of form to function—long neck to top leaves—permits, in itself, no conclusion about why giraffes developed long necks. Since Voltaire understood the foibles of human reason so well, he allowed the venerable Dr. Pangloss to illustrate this fallacy in a solemn pronouncement:

> Things cannot be other than they are. . . . Everything is made for the best purpose. Our noses were made to carry spectacles, so we have spectacles. Legs were clearly intended for breeches, and we wear them.

This error of sliding too easily between current use and historical origin is by no means a problem for Darwinian biologists alone, although our faults have been most prominent and unexamined. This procedure of false inference pervades all fields that try to infer history from our present world. My favorite current example is a particularly ludicrous interpretation of the so-called anthropic principle in cosmology. Many physicists have pointed out—and I fully accept their analysis—that life on earth fits intricately with physical laws regulating the universe, in the sense that were various laws even slightly different, molecules of the proper composition and planets with the right properties could never have arisen—and we would not be here. From this analysis, a few thinkers have drawn the wildly invalid inference that human evolution is therefore prefigured in the ancient design of the cosmos—that the universe, in Freeman Dyson's words, must have known we were coming. But the current fit of human life to physical laws permits no conclusion about the reasons and mechanisms of our origin. Since we are here, we have to fit; we wouldn't be here if we didn't—though something else would, probably proclaiming, with all the hubris that a diproton might muster, that the cosmos must have been created with its later appearance in mind. (Diprotons are a prominent candidate for the highest bit of chemistry in another conceivable universe.)

But back to kiwi eggs. Most literature has fallen into the fallacy of equating current use with historical origin, and has defined the problem as explaining why the kiwi's egg should have been actively enlarged from an ancestor with an egg more suited to the expectations of its body size. Yet University of Arizona biologist William A. Calder III, author of several excellent studies on kiwi energetics (see 1978, 1979, and 1984 in the bibliography), has proposed an opposite interpretation that strikes me as much more likely (though I think he has missed two or three good arguments for its support, and I shall try to supply them here).

The alternative interpretation holds that kiwis are phyletic dwarfs, evolved from a lineage of much larger birds. Since these large ancestors laid big eggs appropriate to their body size, kiwis just never (or only slightly) reduced the size of their eggs as their bodies decreased greatly in bulk. In other words, kiwi eggs never became unusually large; kiwi bodies got small—and these statements are not equivalent, just as we know that an obese man is not short for his weight, despite the old jest.

(Such a hypothesis is not anti-adaptationist in the sense that maintenance of a large egg as size decreases—and in the face of energetic and biochemical costs imposed by such a whopping contribution to the next generation—may well require a direct boost from natural selection to prevent an otherwise advantageous decrease more in keeping with life at Colonel Sanders's favorite size. Still, there is a world of difference between retaining something you already have, and first developed for other reasons [in this case simple appropriateness for large body size], and actively evolving such a unique and cumbersome structure for some special benefit.)

Calder's interpretation might seem forced or farfetched but for the outstanding fact of taxonomy and biogeography cited as the introduction to this essay. Moas are the closest cousins of kiwis, and most moas were very large birds. "Is the kiwi perhaps a shrunken moa?" Calder asks. Unfortunately, all moa fossils lie in rocks of a geological yesterday, and kiwi fossils are entirely unknown—so we have no direct evidence about the size of ancestral kiwis. Still, I believe that all the inferential data support Calder's alternative hypothesis for the great size of kiwi eggs—a "structural" or "historical" explanation if you will, not a conventional account based on natural selection for immediate advantages.

Although the best argument for viewing kiwis as much smaller than their ancestors must be the large size of their closest moa cousins, Calder has also developed a quirky and intriguing speculation to support the dwarfed status of kiwis. (I hasten to point out that neither of these arguments amounts to more than a reasonable conjecture. All evidence can be interpreted in other ways. Both moas and kiwis, for example, might have evolved from a kiwi-sized common ancestor, with moas enlarging later. Still, since the kiwi is the smallest of all ratites—a runt among ostriches, rheas, emus, and cassowaries—its decrease seems more probable than moa increase. But we will not know until we have direct evidence of fossil ancestry.)

Calder notes that in many respects, some rather curious, kiwis have adopted forms and lifestyles generally associated with mammals, not birds. Kiwis, for example, are unique among birds in retaining ovaries on both sides (the right ovary degenerates in all other birds)—and eggs alternate between sides, as in mammals. The seventy- to eighty-four-day incubation period matches the eighty-day pregnancy expected for a mammal of kiwi body size, not the forty-four days predicted for birds of this weight. Calder continues: "When one adds to this list, the kiwi's burrow habit, its furlike body feathers, and its nocturnal foraging highly dependent on its sense of smell, the evidence for convergence seems overpowering." Of course, this conjunction of traits could be fortuitous and each might mean something quite unmammalian to a kiwi, but the argument does gain strength when we remember that no terrestrial mammals reached New Zealand, and that the success of many introduced species indicates a hospitable environment for any creature that could exploit a mammalian way of life.

You will be wondering what these similarities with mammals could possibly mean for my key claim that kiwis are probably descendants of much larger birds. After all, mammals are superior, noble, and large. But they aren't. The original and quintessential mammalian way of life (still exploited by a majority of species) is secretive, furtive, nocturnal, smell-oriented in a nonvisual world—and, above all, small. Remember that for two-thirds of their geological history, all mammals were little creatures living in the interstices of a world ruled by dinosaurs. If a large bird converged upon a basically mammalian lifestyle in the absence of "proper" inhabitants as a result of geographic isolation, *decrease* in size would probably be a first and best step.

Perhaps I have convinced you that kiwis probably decreased in size during their evolution. But why should this dwarfing help to explain their large eggs? Why didn't egg size just keep pace with body size as kiwis scaled down? We now come to the strong evidence of the case.

The study of changes in form and proportion as organisms increase or decrease in size is called allometry. It has been a popular and fruitful subject in evolutionary research since Julian Huxley's pioneering work of the 1920s. One of Huxley's own classic studies (*Journal of the Linnaean Society of London*, 1927) bore the title: "On the Relation between Egg-weight and Body-weight in Birds." Huxley found that if you plot one point for each species on the hummingbird-to-moa curve for egg weight versus body weight, relative egg size decreases in an even and predictable way. The eggs of large birds, he found, are absolutely larger, but relatively smaller in proportion to body weight, than those of small birds.

Huxley's work has since been extended several times with more voluminous and consistent data. In the two best studies that I know, Samuel Brody (in his masterful compendium, *Bioenergetics and Growth,* 1945) calculated a slope of 0.73, while H. Rahn, C. V. Paganelli, and A. Ar (1975), with even more data from some 800 species, derived a similar value of 0.67. This means that as birds increase in body weight, egg weight enlarges only about two-thirds as fast. Conversely, as birds decrease in size, egg weight diminishes more slowly—so little birds have relatively heavy eggs.

This promising datum will not, however, explain the kiwi's out-sized egg, for the two-thirds slope represents the general standard for all birds. Kiwi eggs are huge compared with the *expected* egg weight for a bird of kiwi body weight along this standard curve.

But the literature of allometry has also yielded a generality that will, I think, explain the kiwi's massive egg. The two-thirds slope of the egg weight/body weight curve represents a type of allometry technically called interspecific scaling—that is, you plot one point for each species in a related group of organisms and attempt to establish the characteristic change of proportion along a gradient of increasing size. (These curves are popularly called mouse-to-elephant for relationships among mammals—hence my designation hummingbird-to-moa for birds.) Allometricians have established hundreds of interspecific curves for birds and mammals.

Another kind of allometry is called intraspecific scaling. Here you plot one point for each individual among adults of varying body weights within a single species—the Tom Thumb–to–Manute Bol curve for human males, if you will. Since the similarity of these technical terms—interspecific and intraspecific—is so confusing, I shall call them, instead, among-species (for mouse-to-elephant) and within-species (for Thumb-to-Bol).

As an important generality in allometric studies, within-species curves usually have a substantially lower slope than among-species curves for the same property. For example (and in our best-studied case), the mouse-to-elephant curve for brain weight versus body weight in mammals has a slope of about two-thirds (as does the egg weight/body weight curve for birds). But the within-species curve from small to large adults of a single species, while varying from one group to another, almost always has a much lower slope in the range of 0.2 to 0.4. In other words, while brains increase about two-thirds as fast as bodies among species (implying that large mammals have relatively small brains), brains only increase about one-fifth to two-fifths as fast as bodies when we move from small to large adults within a single mammalian species.

Such a regularity, if it applied to egg weight as well, could resolve the kiwi paradox—if kiwis evolved from larger ancestors. Suppose that kiwi forebears start at moa size. By the humming-bird-to-moa among-species standard, egg size should decrease along the two-thirds slope. But suppose that natural selection is operating to favor small adults within a population. If the within-species curve for egg weight had a slope much lower than two-thirds, then size decrease by continued selection of small adults might produce a new species with outsized eggs well above the two-thirds slope, and therefore well above the expected weight for a bird of this reduced size. (Quantitative arguments like this are always easier to grasp by picture than by words—and a glance at the accompanying graph should resolve any confusion.)

But what is the expected within-species relationship for egg weight? Is the shape of the curve low, as for brain weight, thus affirming my conjecture? I reached for my well-worn copy of Brody's unparalleled compendium and found that for adults of domestic fowl, egg weight increases not two-thirds as fast, but only 15 percent as fast as body weight! (Brody uses this fact to argue that small hens are usually better than large, so long as egg production remains the same—for egg size diminishes very little with a large decrease in body mass, and the small loss in egg volume is more than compensated by large decreases in feeding costs.)

The same argument might apply to kiwis. As a poultryman might choose small hens for minimal decrease in egg size with maximal decline in body weight, natural selection for smaller adults might markedly decrease the average body weight within a species with very little accompanying reduction in egg weight.

I believe that this general argument, applied to kiwis, may be

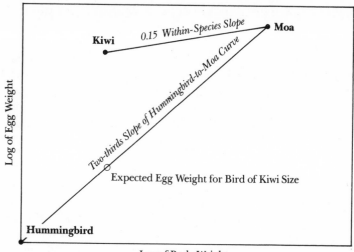

Proposed allometric explanation for the large egg of the kiwi. The kiwi probably evolved from a much larger bird by backing down the very shallow within-species slope (upper line). Most birds arrange themselves on the standard hummingbird-to-moa curve with its steeper slope (lower line). Therefore, a kiwi has a much heavier egg than predicted for a bird of its body size. BEN GAMIT. ADAPTED FROM JOE LEMONNIER. COURTESY OF *NATURAL HISTORY*.

defended on three strong grounds. First, as stated above, a general finding in allometric studies teaches us that within-species slopes for adults of one species are usually much lower than among-species slopes along mouse-to-elephant curves. Thus, any evolution of decreasing size along the within-species curve should produce a dwarfed descendant with more of the particular item being measured than an average nondwarfed species at the same body weight. Second, we have actual data, for domestic poultry at least, indicating that the within-species curve does have a substantially lower slope than the hummingbird-to-moa curve for our crucial measure of egg weight.

Third, I have studied many cases of dwarfism, and I believe we can state as a general phenomenon—rooted in the first point above—that decline in body size often far outstrips decrease in many particular features. Dwarfs, in several respects, always seem to have much more of certain body parts than related non-dwarfed species of the same body size. For example, I once studied tooth size in three species of dwarfed hippos (two fossil and the modern Liberian pygmy)—and found their molar teeth substantially larger, for each of three separate evolutionary events, than expected values for related hoofed mammals at their body size (*American Zoologist*, 1975).

In another example, the talapoin, a dwarfed relative of the rhesus monkey, has the largest relative brain weight among monkeys. Since within-species brain curves have substantially lower slopes than the two-thirds value for the marmoset-to-baboon curve, evolution to smaller size by backing down the within-species curve would yield a dwarf with a far larger brain than an ordinary monkey at the same body size.

Put all this together and a resolution fairly jumps at you for kiwis. Their enormous eggs require no special explanation if kiwis have evolved by marked decrease in size. Kiwi eggs exhibit the weight expected for backing down the within-species curve if natural selection operates only to decrease body size and no other factor intervenes to favor an active reduction in egg size—as we might anticipate in New Zealand, this easy land of no natural predators, where a female might waddle without fear as an enormous egg distends her abdomen during passage down the oviduct.

In this interpretation, if you ask me why kiwi eggs are so large, I reply, "Because kiwis are dwarfed descendants of larger birds, and just followed ordinary principles of scaling in their evolution." This answer differs sharply from the conventional form of evolutionary explanation: "Because these big eggs are good for something now, and natural selection favored them."

My answer will also strike many people as deeply unsatisfactory. It provides a reason rooted in history, pure and simple (with a bit of scaling theory thrown in)—kiwis are as they are because their ancestors were as they were. Don't we want answers that invoke general laws of nature rather than particular contingencies of history?

I would reply that my resolution is quite satisfactory, that evolutionary arguments are often properly resolved by such historical statements, and that we would do well to understand this important and neglected principle of reasoning—for we might save ourselves many a stumble in trying to apply preferred, but inappropriate, styles of explanation to situations encountered again and again in our daily lives.

To cite just one example where I learned, to my deep chagrin, that a peculiarity of history, rather than a harmonious generality, resolved an old personal puzzle: I had been troubled for a long time by something I didn't understand in the inscription on the Liberty Bell—not losing any sleep to be sure, but troubled nonetheless, for little things count. This national symbol bears, like most bells, an appropriate quotation: "Proclaim liberty throughout all the land unto all the inhabitants thereof" (Lev. 25:10). But the bell also says, "Pass and Stow." I assumed that this line must also be a quotation, fit to the purpose of the bell (as selection fits the features of organisms to their needs)—part of the general harmony and chosen plan. I pondered these cryptic words quite a bit because I didn't recognize the source. I consulted Bartlett's and found nothing. I constructed various possibilities: This too will pass, as we stow courage for the coming conflict; oh ye who pass by, remember, they prosper that stow and do not waste; pass the grass and stow the dough. Finally I asked the attendant on duty in Philadelphia. Of course, I should have figured it out, but I was too busy trying to make intrinsic sense of the inscription. The bell was cast by Messrs. John Pass and John Stow. Pass and Stow is a statement about the particular history of the bell; nothing more.

My odd juxtapositions sometimes cause consternation; some readers might view this particular comparison as outright sacrilege. Some may claim that the only conceivable similarity between kiwi eggs and the Liberty Bell is that both are cracked, but I reply that they stand united in owing their peculiarity and meaning to pathways of history.

The Liberty Bell on display in Philadelphia, advertising its makers, Mr. Pass and Mr. Stow. THE BETTMANN ARCHIVE.

# 8 | Male Nipples and Clitoral Ripples*

THE MARQUIS DE CONDORCET, enthusiast of the French Revolution but not radical enough for the Jacobins—and therefore forced into hiding from a government that had decreed, and would eventually precipitate, his death—wrote in 1793 that "the perfectibility of man is really boundless. . . . It has no other limit than the duration of the globe where nature has set us." As Dickens so aptly remarked, "It was the best of times, it was the worst of times."

The very next year, as Condorcet lay dying in prison, a famous voice from across the channel published another paean to progress in a world that many judged on the brink of ruin. This treatise, called *Zoonomia, or the Laws of Organic Life,* was written by Erasmus Darwin, grandfather of Charles.

*Zoonomia* is primarily a dissertation on the mechanisms of human physiology. Yet, in the anachronistic tradition that judges biological works by their attitude to the great watershed of evolution, established by grandson Charles in 1859, *Zoonomia* owes its modern reputation to a few fleeting passages that look upon organic transmutation with favor.

---

*The proper and most accurate title of this piece should be "Tits and Clits"—but such a label would be misread as sexist because people would not recognize the reference point as *male* tits. My wife, a master at titles, suggested this alternative. (During the short heyday of that most unnecessary of all commercially touted products—vaginal deodorants—she wanted to market a male counterpart to be known as "cocksure.") *Natural History* magazine, published by a group of fine but slightly overcautious folks, first brought out this essay under their imposed title: "Freudian Slip." Not terrible; but not really descriptive either.

The evolutionary passages of *Zoonomia* occur in Item 8, Part 4, of Section 39, entitled, "Of Generation," Erasmus Darwin's thoughts on reproduction and embryology. He viewed embryology as a tale of continuous progress to greater size and complexity. Since his evolutionary speculations are strictly analogous to his concept of embryology, organic transformation also follows a single pathway to more and better:

> Would it be too bold to imagine that in the great length of time, since the earth began to exist . . . all warm-blooded animals have arisen from one living filament . . . possessing the faculty of continuing to improve by its own inherent activity, and of delivering down those improvements by generation to its posterity, world without end?

As the last sentence states, Erasmus Darwin's proposed mechanism of evolution lay in the inheritance of *useful* characters acquired by organisms during their lifetimes. This false theory of heredity has passed through later history under the label of Lamarckism, but the citation by Erasmus (a contemporary of Lamarck) illustrates the extent of this misnomer. Inheritance of acquired characters was the standard folk wisdom of the time, used by Lamarck to be sure, but by no means original or distinctive with him. For Erasmus, this mechanism of evolution required a concept of pervasive utility. New structures arose only when needed and by direct organic striving for an evident purpose. Erasmus discusses adaptations in three great categories: reproduction, protection and defense, and food. Of the last, he writes:

> All . . . seem to have been gradually produced during many generations by the perpetual endeavor of the creatures to supply the want of food, and to have been delivered to their posterity with constant improvement of them for the purposes required.

In this long section, Erasmus considers only one potential exception to the principle of pervasive utility: "the breasts and teats of all male quadrupeds, to which no use can be now assigned." He also suggests two exits from this potential dilemma: first, that male nipples are vestiges of a previous utility if, as Plato had suggested, "mankind with all other animals were originally hermaphrodites during the infancy of the world, and were in process of time separated into male and female"; and second, that some males may lactate and therefore help to feed their babies (in the absence of any direct evidence, Erasmus cites the milky-colored feeding fluids, produced in the crops of both male and female pigeons, as a possible analogue).

The tenacity of anomalies through centuries of changing beliefs can be truly astounding. As a consequence of writing these essays for so many years, I receive hundreds of letters from readers puzzled about one or another apparent oddity of nature. With so large a sample, I have obtained a pretty good feel for the issues and particulars of evolution that pose conundrums for well-informed nonscientific readers. I have been fascinated (and, I confess, surprised) over the years to discover that no single item has evoked more puzzlement than the very issue that Erasmus Darwin chose as a primary challenge to his concept of pervasive utility—male nipples. I have received more than a dozen requests to explain how evolution could possibly produce such a useless structure.

Consider my latest example from a troubled librarian. "I have a question that no one can answer for me, and I don't know where or how to look up the answer. Why do men have nipples? . . . This question nags at me whenever I see a man's bare chest!"

I was fascinated to note that her two suggestions paralleled exactly the explanations floated by Erasmus Darwin. First, she reports, she asked a doctor. "He told me that men in primitive societies used to nurse babies." Finding this incredible, she tried Darwin's first proposal for nipples as a vestige of previous utility: "Can you tell me—was there once only one sex?"

If you are committed—as Erasmus was, and as a distressingly common version of "pop," or "cardboard," Darwinism still is— to a principle of pervasive utility for all parts of all creatures, then male nipples do raise an insoluble dilemma, hence (I assume) my voluminous correspondence. But as with so many persistent puzzles, the resolution does not lie in more research within an established framework but rather in identifying the framework itself as a flawed view of life.

Suppose we begin from a different point of view, focusing on rules of growth and development. The external differences between male and female develop gradually from an early embryo so generalized that its sex cannot be easily determined. The clitoris and penis are one and the same organ, identical in early form, but later enlarged in male fetuses through the action of testosterone. Similarly, the labia majora of women and the scrotal sacs of men are the same structure, indistinguishable in young embryos, but later enlarged, folded over, and fused along the midline in male fetuses.

I do not doubt that the large size and sensitivity of the female breast should count as an adaptation in mammals, but the smaller male version needs no adaptive explanation at all. Males and females are not separate entities, shaped independently by natural selection. Both sexes are variants upon a single ground plan, elaborated in later embryology. Male mammals have nipples because females need them—and the embryonic pathway to their development builds precursors in all mammalian fetuses, enlarging the breasts later in females but leaving them small (and without evident function) in males.

In a similar case that illuminates the general principle, the panda develops a highly functional false "thumb" from the radial sesamoid bone of its wrist. Interestingly, the corresponding bone of the foot, the tibial sesamoid, is also enlarged in the same manner (but not nearly so much), although increase of the tibial sesamoid has no apparent function.

As D. Dwight Davis argued in his great monograph on the giant panda (1964), evolution works on growth fields. Radial and tibial sesamoids are homologous structures, probably affected in concert by the same genetic factors. If natural selection operates for an enlarged radial sesamoid, a bigger tibial sesamoid will probably "come along for the ride." Davis drew a profound message from this case: Organisms are integral and constrained structures, "pushing back" against the force of selection to channel changes along permitted paths; complex animals are not a dissociable collection of independent, optimal parts. Davis wrote that "the effect seen in the sympathetic enlargement of the tibial sesamoid . . . strongly suggests that a very simple mechanism, perhaps involving a single factor, lies behind the hypertrophy of the radial sesamoid."

In my view of life, akin to Davis's concept of constraint and integration, male nipples are an expectation based on pathways of sexual differentiation in mammalian embryology.

At this point, readers might demur with the most crushing of all rejoinders: "Who cares?" Why worry about little items that ride piggyback on primary adaptations? Let's concentrate on the important thing—the adaptive value of the female breast—and leave aside the insignificant male ornament that arises as its consequence. Adaptations are preeminent; their side effects are nooks and crannies of organic design, meaningless bits and pieces. This argument is, I think, the standard position of strict Darwinian adaptationists.

I could defend the importance of structural nonadaptation with a long and abstruse general argument (I have done so in several technical papers). Let me proceed instead by the most compelling route I know by presenting a second example based on human sexuality, a case entirely comparable in concept with the origin of male nipples but differing in importance for human culture—a case, moreover, where the bias of utility has brought needless pain and anxiety into the lives of millions (where, indeed, one might argue that Freudian traditions have provided a manifestly false but potent weapon, however unintentional, for the subjugation of women). Consider the anatomical site of orgasm in human females.

As women have known since the dawn of our time, the primary site for stimulation to orgasm centers upon the clitoris. The revolution unleashed by the Kinsey report of 1953 has, by now, made this information available to men who, for whatever reason, had not figured it out for themselves by the more obvious routes of experience and sensitivity.

The data are unambiguous. Consider only the three most widely read of extensive surveys—the Kinsey report of 1953, Masters and Johnson's book of 1966, and *The Hite Report* of 1976. In his study of genital anatomy, Kinsey reports that the female clitoris is as richly supplied with sensory nerves as the male penis—and therefore as capable of excitation. The walls of the vagina, on the other hand, "are devoid of end organs of touch and are quite insensitive when they are gently stroked or lightly pressed. For most individuals the insensitivity extends to every part of the vagina."

The data on masturbation are particularly convincing. Kinsey reports from his sample of 8,000 women that 84 percent of individuals who have ever masturbated depend "primarily on labial and/or clitoral techniques." *The Hite Report* on 3,000 individuals found that 79 percent of women who masturbate do so by directly stimulating the clitoris and surrounding vulva, while only 1.5 percent use vaginal entry.

The data on intercourse affirm this pattern. Shere Hite reports a frequency of orgasm with intercourse at 30 percent and often attained only with simultaneous stimulation of the clitoris by hand. She concludes: *"not* to have orgasm from intercourse is the experience of the majority of women." Masters and Johnson only included women who experienced orgasm with intercourse in their study. But they concluded that all orgasms are identical in physiology and clitoral in origin. These findings led Hite to comment that human copulation "sounds more like a Rube Goldberg scheme than a reliable way to orgasm. . . . Intercourse was never meant to stimulate women to orgasm." As Kinsey had said earlier with his characteristic economy and candor: "The techniques of masturbation and of petting are more specifically calculated to effect orgasm than the techniques of coitus itself."

This conclusion should be utterly unsurprising—once we grasp the proper role and limitation of adaptationist argument in evolutionary biology. I don't believe in the mystery style of writing essays: build up suspense but save the resolution until the end—for then readers miss the significance of details along the way for want of proper context. The reason for a clitoral site of orgasm is simple—and exactly comparable with the nonpuzzle of male nipples. The clitoris is the homologue of the penis—it is the same organ, endowed with the same anatomical organization and capacity of response.

Anatomy, physiology, and observed responses all agree. Why then do we identify an issue at all? Why, in particular, does the existence of clitoral orgasm seem so problematic? Why, for example, did Freud label clitoral orgasm as infantile and define feminine maturity as the shifting to an unattainable vaginal site?

Part of the reason, of course, must reside in simple male vanity. We (and I mean those of my sex, not the vague editorial pronoun) simply cannot abide the idea—though it flows from obvious biology—that a woman's sexual pleasure might not arise most reliably as a direct result of our own coital efforts. But the issue extends further. Clitoral orgasm is a paradox not only for the traditions of Darwinian biology but also for the bias of utility that underlies all functionally based theories of evolution (including Lamarck's and Darwin's) and, in addition, the much older tradition of natural theology that saw God's handiwork in the exquisite fit of organic form to function.

Consider the paradox of clitoral orgasm in any world of strict functionalism (I present a Darwinian version, but parallel arguments can be made for the entire range of functionalist thinking, from Paley's natural theology to Cuvier's creationism): Evolution arises from a struggle among organisms for differential reproductive success. Sexual pleasure, in short, must evolve as a stimulus for reproduction.

This formulation works for men since the peak of sexual excitement occurs during ejaculation—a primary and direct adjunct of intercourse. For men, maximal pleasure is linked with the greatest possibility of fathering offspring. In this perspective, the sexual pleasure of women should also be centered upon the act that causes impregnation—on intercourse itself. But how can our world be functional and Darwinian if the site of orgasm is divorced from the place of intercourse? How can sexual pleasure be so separated from its functional significance in the Darwinian game of life? (For the most divergent, but equally functionalist, view of some conservative Christians, sex was made by God to foster procreation; any use in any other context is blasphemy.)

Elisabeth Lloyd, a philosopher of science at Berkeley, has just completed a critical study of explanations recently proposed by evolutionary biologists for the origin and significance of female orgasm. Nearly all these proposals follow the lamentable tradition of speculative storytelling in the a priori adaptationist mode. In all the recent Darwinian literature, I believe that Donald Symons is the only scientist who presented what I consider the proper answer—that female orgasm is not an adaptation at all. (See his book, *The Evolution of Human Sexuality,* 1979.)

Many of these scientists don't even know the simple facts of the matter; they assume that female orgasms are triggered by intercourse and draw the obvious Darwinian conclusion. A second group recognizes the supposed paradox of nonassociation between orgasm and intercourse and then proposes another sort of adaptive explanation, usually based on maintenance of their pair bond by fostering close relationships through sexual pleasure. Desmond Morris (*The Naked Ape*, 1969), the most widely read promoter of this view, writes that female orgasm evolved for its role in promoting the pair bond by "the immense behavioral reward it brings to the act of sexual cooperation with the mated partner." Perhaps no popular speculation has been more androcentric than George Pugh's (*Biological Origin of Human Values*, 1977), who speaks about "the development of a female orgasm, which makes it easier for a female to be satisfied by one male, and which also operates psychologically to produce a stronger emotional bond in the female." Or Eibl-Eibesfeldt, who argues (1975) that the evolution of female orgasm "increases her readiness to submit and, in addition, strengthens her emotional bond to the partner."

This popular speculation about pair bonding usually rests upon an additional biological assumption—almost surely false—that capacity for female orgasm is an especially human trait. Yet Symons shows, in his admirable review of the literature, that whereas most female mammals do not experience orgasm during ordinary copulation, prolonged clitoral stimulation—either artificially in the laboratory (however unpleasant a context from the human point of view) or in nature by rubbing against another animal (often a female)—does produce orgasm in a wide range of mammals, including many primates. Symons concludes that "orgasm is most parsimoniously interpreted as a potential all female mammals possess."

Adaptive stories for female orgasm run the full gamut—leaving only the assumption of adaptation itself unquestioned. Sarah Hrdy (1981), for example, has taken up the cudgels against androcentrism in evolutionary speculation, not by branding the entire enterprise as bankrupt, but by showing that she can tell just as good an adaptive story from a female-centered point of view. She argues—turning the old pair-bond theory on its head—that the dissociation between orgasm and intercourse is an adaptation for promiscuous behavior, permitting females to enlist the support of several males to prevent any one from harming her babies. (In many species, a male that displaces a female's previous partner may kill her offspring, presumably to foster his own reproductive success by immediate remating.)

Indeed, no one surpasses Hrdy in commitment to the adaptationist assumption that orgasm must have evolved for Darwinian utility in promoting reproductive success. Chosen language so often gives away an underlying bias; note Hrdy's equation of nonadaptation both with despair in general and with the denigration of women's sexuality in particular.

> Are we to assume, then, that [the clitoris] is irrelevant? . . . It would be safer to suspect that, like most organs . . . it serves a purpose, or once did. . . . The lack of obvious purpose has left the way open for both orgasm, and female sexuality in general, to be dismissed as "nonadaptive."

But why are adaptationist arguments "safer," and why is nonadaptation a "dismissal"? I do not feel degraded because my nipples are concomitants of a general pattern in human development and not a sign that ancestors of my sex once lactated. In fact, I find this nonadaptationist explanation particularly fascinating, both because it teaches me something important about structural rules of development and because it counters a pervasive and constraining bias that has harmed evolutionary biology by restricting the range of permitted hypotheses. Why should the dissociation of orgasm from intercourse degrade women when it merely records a basic (if unappreciated) fact of human anatomy that happens to unite both sexes as variations of a common pattern in development? (Such an argument would only hold if adaptations were "good" and all other aspects of anatomy "irrelevant." I, for one, am quite attached to all my body parts and do not make such invidious rankings and distinctions among them.)

I could go on but will stop here for the obvious reason that this discussion, however amusing, might be deemed devoid of social importance. After all, these biologists may be enjoying themselves and promoting their view of life, but isn't all this strictly *entre nous?* I mean, after all, who cares about speculative ideas if they impose no palpable harm upon people's lives? But unfortunately, the history of psychology shows that one of the most influential theories of our century—a notion that had a direct and deeply negative effect upon millions of women—rested upon the false assumption that clitoral orgasm cannot be the natural way of a mature female. I speak, of course, about Sigmund Freud's theory of transfer from clitoral to vaginal orgasm.

In Freud's landmark and most influential book *Three Essays on the Theory of Sexuality* (1905, but first published in complete form in 1915), the third essay on "transformations of puberty" argues that "the leading erotogenic zone in female children is located at the clitoris." He also, as a scientist originally trained in anatomy, knows the reason—that the clitoris "is homologous to the masculine genital zone of the glans penis."

Freud continues: "All my experience concerning masturbation in little girls has related to the clitoris and not the regions of the external genitalia that are important in later sexual functioning." So far so good; Freud recognizes the phenomenon, knows its anatomical basis, and should therefore identify clitoral orgasm as a proper biological expression of female sexuality. Not at all, for Freud then describes a supposed transformation in puberty that defines the sexuality of mature women.

Puberty enhances the libido of boys but produces an opposite effect in girls—"a fresh wave of repression." Later, sexuality resumes in a new way. Freud writes:

> When at last the sexual act is permitted and the clitoris itself becomes excited, it still retains a function: the task, namely, of transmitting the excitation to the adjacent female sexual parts, just as—to use a simile—pine shavings can be kindled in order to set a log of harder wood on fire.

Thus, we encounter Freud's famous theory of female sexual maturity as a transfer from clitoral to vaginal orgasm:

> When erotogenic susceptibility to stimulation has been successfully transferred by a woman from the clitoris to the vaginal orifice, it implies that she has adopted a new leading zone for the purposes of her later sexual activity.

This dogma of transfer from clitoral to vaginal orgasm became a shibboleth of pop culture during the heady days of pervasive

Freudianism. It shaped the expectations (and therefore the frustration and often misery) of millions of educated and "enlightened" women told by a brigade of psychoanalysts and by hundreds of articles in magazines and "marriage manuals" that they must make this biologically impossible transition as a definition of maturity.

Freud's unbiological theory did further harm in two additional ways. First, Freud did not define frigidity only as an inability to perform sexually or as inefficacy in performance, but proposed as his primary definition a failure to produce this key transfer from clitoris to vagina. Thus, a woman who greatly enjoys sex, but only by clitoral stimulation, is frigid by Freud's terminology. "This anaesthesia," Freud writes, "may become permanent if the clitoridal zone refuses to abandon its excitability."

Second, Freud attributed a supposedly greater incidence of neurosis and hysteria in women to the difficulty of this transfer— for men simply retain their sexual zone intact from childhood, while women must undergo the hazardous switch from clitoris to vagina. Freud continues:

> The fact that women change their leading erotogenic zone in this way, together with the wave of repression at puberty . . . are the chief determinants of the greater proneness of women to neurosis and especially to hysteria. These determinants, therefore, are intimately related to the essence of femininity.

In short, Freud's error may be encapsulated by stating that he defined the ordinary biology of female sexuality as an aberration based on failure to abandon an infantile tendency.

The sources of Freud's peculiar theory are complex and involve many issues not treated in this essay (in particular his androcentric biases in interpreting the act of intercourse from a man's point of view and in defining both clitoral and penile stimulation in childhood as a fundamentally masculine form of sexuality that must be shunned by a mature woman). But another important source resides in the perspective underlying all the fanciful theories that I have discussed throughout this essay, from male nipples as sources of milk to clitoral orgasm as a clever invention to cement pair bonds—the bias of utility, or the exclusive commitment to functionalist explanations.

The more I read Kinsey, the more he wins my respect for his humane sensibility, and for his simple courage. (His 1953 report on *Sexual Behavior in the Human Female* appeared during the height of McCarthyism in America and led to a withdrawal of funding for his research and the effective end, during his lifetime, of his programs—see the essay "Of Wasps and WASPs" in my previous book, *The Flamingo's Smile.*) Kinsey was a measured man. He wrote in a dry and clinical fashion (probably more for reasons of necessity than inclinations of temperament). Yet, every once in a while, his passion spills forth and his rage erupts in a single, well-controlled phrase. Nowhere does Kinsey express more agitation than in his commentary on Freud's theory of the shift from clitoral to vaginal orgasm.

Kinsey locates his discussion of Freud in the proper context—in his section on sexual anatomy (Chapter 14, "Anatomy of Sexual Response and Orgasm"). He reports the hard data on adult masturbation and on the continuing clitoral site of orgasm in mature women. He locates the reason for clitoral orgasm not in any speculative theory about function but in the basic structure of sexual anatomy.

> In any consideration of the functions of the adult genitalia, and especially of their liability to sensory stimulation, it is important and imperative that one take into account the homologous origins of the structures in the two sexes.

Kinsey then provides a long and beautifully clear discussion of anatomical homologies, particularly the key unity of penis and clitoris. He concludes that "the vaginal walls are quite insensitive in the great majority of females. . . . There is no evidence that the vagina is ever the sole source of arousal, or even the primary source of erotic arousal in any female." Kinsey has now laid the foundation for a swift demolition of Freud's hurtful theory. He cites (in a long footnote, for his text is not contentious) a compendium of psychoanalytical proclamations from the Freudian heyday of the 1920s to 1940s. Consider just three items on his list:

1. (from 1936): "If this transition [from clitoris to vagina] is not successful, then the woman cannot experience satisfaction in the sexual act. . . . The first and decisive requisite of a normal orgasm is vaginal sensitivity."

2. (again from 1936): "The sole criterion of frigidity is the absence of the vaginal orgasm."

3. (from 1927): "In frigidity the pleasurable sensation is as a rule situated in the clitoris and the vaginal zone has none."

Kinsey's sole paragraph of evaluation ranks as the finest dismissal by understatement (and by incisive phrase at the end) that I have ever read.

> This question is one of considerable importance because much of the literature and many of the clinicians, including psychoanalysts and some of the clinical psychologists and marriage counselors, have expended considerable effort trying to teach their patients to transfer "clitoral responses" into "vaginal responses." Some hundreds of women in our own study and many thousands of the patients of certain clinicians have consequently been much disturbed by their failure to accomplish this biological impossibility.

I then must ask myself, why could Kinsey be so direct and sensible in 1953, while virtually all evolutionary discussion of female orgasm during the past twenty years has been not only biologically erroneous but also obtuse and purely speculative? I'm sorry to convert this essay into something of a broken record in contentious repetition, but the same point pervades the discussion all the way from Erasmus Darwin on male nipples to Sarah Hrdy on clitoral orgasm. The fault lies in a severely restrictive (and often false) functionalist view of life. Most functionalists have not misinterpreted male nipples, for their unobtrusive existence poses no challenge. But clitoral orgasm is too central to the essence of life for any explanation that does not focus upon the role of sexuality in reproductive success. And yet the obvious, nonadaptive structural alternative stares us in the face as the most elementary fact of sexual anatomy—the homology of penis and clitoris.

Kinsey's ability to cut through this morass right to the core of the strong developmental argument has interesting roots. Kinsey began his career by devoting twenty years to the taxonomy of gall-forming wasps. He pursued this work in the 1920s and 1930s before American evolutionary biology congealed around Darwinian functionalism. In Kinsey's day, many (probably most) taxonomists accepted the nonadaptive nature of much small-scale geographic variability within species. Kinsey followed this structuralist tradition and never absorbed the bias of utility. He was therefore able to grasp the meaning of this elemental fact of homology between penis and clitoris—a fact that stares everyone in the face, but becomes invisible if the bias of utility be strong enough.

I well remember something that Francis Crick said to me many years ago, when my own functionalist biases were strong. He remarked, in response to an adaptive story I had invented with alacrity and agility to explain the meaning of repetitive DNA: "Why do you evolutionists always try to identify the value of something before you know how it is made?" At the time, I dismissed this comment as the unthinking response of a hidebound molecular reductionist who did not understand that evolutionists must always seek the "why" as well as the "how"—the final as well as the efficient causes of structures.

Now, having wrestled with the question of adaptation for many years, I understand the wisdom of Crick's remark. If all structures had a "why" framed in terms of adaptation, then my original dismissal would be justified for we would know that "whys" exist whether or not we had elucidated the "how." But I am now convinced that many structures (including male nipples and clitoral orgasm) have no direct adaptational "why." And we discover this by studying pathways of genetics and development—or, as Crick so rightly said to me, by first understanding how a structure is built. In other words, we must first establish "how" in order to know whether or not we should be asking "why" at all.

I began with Charles Darwin's grandpa Erasmus and end with his namesake, Desiderius Erasmus, the greatest of all Renaissance scholars. Of more than 3,000 proverbs from antiquity collected in his *Adagia* of 1508, perhaps two are best known and wonderfully apt for the point of this essay (which is not a diatribe against adaptation but a plea for expansion by alternative hypotheses and for fruitful competition and synthesis between functional and structural perspectives). First a comment on limitations of outlook: "No one is injured save by himself." Second, probably the most famous of zoological metaphors about human temperament: "The fox has many tricks, and the hedgehog only one, but that is the best of all." Some have taken the hedgehog's part in this dichotomy, but I will cast my lot for a diversity of options—for our complex world may offer many paths to salvation, and the hounds of hell press continually upon us.

# 9 | Not Necessarily a Wing

FROM *Flesh Gordon* to *Alex in Wonderland,* title parodies have been a stock-in-trade of low comedy. We may not anticipate a tactical similarity between the mayhem of *Mad* magazine's movie reviews and the titles of major scientific works, yet two important nineteenth-century critiques of Darwin parodied his most famous phrases in their headings.

In 1887, E. D. Cope, the American paleontologist known best for his fossil feud with O. C. Marsh (see Essay 5) but a celebrated evolutionary theorist in his own right, published *The Origin of the Fittest*—a takeoff on Herbert Spencer's phrase, borrowed by Darwin as the epigram for natural selection: survival of the fittest. (Natural selection, Cope argued, could only preserve favorable traits that must arise in some other manner, unknown to Darwin. The fundamental issue of evolution cannot be the differential survival of adaptive traits, but their unexplained origin—hence the title parody.)

St. George Mivart (1817–1900), a fine British zoologist, tried to reconcile his unconventional views on religion and biology but ended his life in tragedy, rejected by both camps. At age seventeen, he abandoned his Anglican upbringing, became a Roman Catholic, and consequently (in a less tolerant age of state religion) lost his opportunity for training in natural history at Oxford or Cambridge. He became a lawyer but managed to carve out a distinguished career as an anatomist nonetheless. He embraced evolution and won firm support from the powerful T. H. Huxley, but his strongly expressed and idiosyncratic anti-Darwinian views led to his rejection by the biological establishment of Britain. He tried to unite his biology with his religion in a series of books and essays, and ended up excommunicated for his trouble six weeks before his death.

Cope and Mivart shared the same major criticism of Darwin—that natural selection could explain the preservation and increase of favored traits but not their origin. Mivart, however, went gunning for a higher target than Darwin's epigram. He shot for the title itself, naming his major book (1871) *On the Genesis of Species.* (Darwin, of course, had called his classic *On the Origin of Species.*)

Mivart's life may have ended in sadness and rejection thirty years later, but his *Genesis of Species* had a major impact in its time. Darwin himself offered strong, if grudging, praise and took Mivart far more seriously than any other critic, even adding a chapter to later editions of the *Origin of Species* primarily to counter Mivart's attack.

Mivart gathered, and illustrated "with admirable art and force" (Darwin's words), all objections to the theory of natural selection—"a formidable array" (Darwin's words again). Yet one particular theme, urged with special attention by Mivart, stood out as the centerpiece of his criticism. This argument continues to rank as the primary stumbling block among thoughtful and friendly scrutinizers of Darwinism today. No other criticism seems so troubling, so obviously and evidently "right" (against a Darwinian claim that seems intuitively paradoxical and improbable).

Mivart awarded this argument a separate chapter in his book, right after the introduction. He also gave it a name, remembered ever since. He called his objection "The Incompetency of 'Natural Selection' to Account for the Incipient Stages of Useful Structures." If this phrase sounds like a mouthful, consider the easy translation: We can readily understand how complex and fully developed structures work and how their maintenance and preservation may rely upon natural selection—a wing, an eye, the resemblance of a bittern to a branch or of an insect to a stick or dead leaf. But how do you get from nothing to such an elaborate something if evolution must proceed through a long sequence of intermediate stages, each favored by natural selection? You can't fly with 2 percent of a wing or gain much protection from an iota's similarity with a potentially concealing piece of vegetation.

How, in other words, can natural selection explain the incipient stages of structures that can only be used in much more elaborated form?

I take up this old subject for two reasons. First, I believe that Darwinism has, and has long had, an adequate and interesting resolution to Mivart's challenge (although we have obviously been mightily unsuccessful in getting it across). Second, a paper recently published in the technical journal *Evolution* has provided compelling experimental evidence for this resolution applied to its most famous case—the origin of wings.

The dilemma of wings—*the* standard illustration of Mivart's telling point about incipient stages—is set forth particularly well in a perceptive letter that I recently received from a reader, a medical doctor in California. He writes:

> How does evolutionary theory as understood by Darwin explain the emergence of items such as wings, since a small move toward a wing could hardly promote survival? I seem to be stuck with the idea that a significant quality of wing would have to spring forth all at once to have any survival value.

Interestingly, my reader's proposal that much or most of the wing must arise all at once (because incipient stages could have no adaptive value) follows Mivart's own resolution. Mivart first enunciated the general dilemma (1871, p. 23):

> Natural selection utterly fails to account for the conservation and development of the minute and rudimentary beginnings, the slight and infinitesimal commencements of structures, however useful those structures may afterwards become.

After fifty pages of illustration, he concludes: "Arguments may yet be advanced in favor of the view that new species have from time to time manifested themselves with suddenness, and by modifications appearing at once." Advocating this general solution for wings in particular, he concludes (p. 107): "It is difficult, then, to believe that the Avian limb was developed in any other way than by a comparatively sudden modification of a marked and important kind."

Darwin's theory is rooted in the proposition that natural selection acts as the primary creative force in evolutionary change. This creativity will be expressed only if the fortuitous variation forming the raw material of evolutionary change can be accumulated sequentially in tiny doses, with natural selection acting as the sieve of acceptance. If new species arise all at once in an occasional lucky gulp, then selection has no creative role. Selection, at best, becomes an executioner, eliminating the unfit following this burst of good fortune. Thus, Mivart's solution—bypassing incipient stages entirely in a grand evolutionary leap—has always been viewed, quite rightly, as an anti-Darwinian version of evolutionary theory.

Darwin well appreciated the force, and potentially devastating extent, of Mivart's critique about incipient stages. He counterattacked with gusto, invoking the standard example of wings and arguing that Mivart's solution of sudden change presented more problems than it solved—for how can we believe that so complex a structure as a wing, made of so many coordinated and co-adapted parts, could arise all at once:

> He who believes that some ancient form was transformed suddenly through an internal force or tendency into, for instance, one furnished with wings, will be . . . compelled to believe that many structures beautifully adapted to all the other parts of the same creature and to the surrounding conditions, have been suddenly produced; and of such complex and wonderful co-adaptations, he will not be able to assign a shadow of an explanation. . . . To admit all this is, as it seems to me, to enter into the realms of miracle, and to leave those of Science.

(This essay must now go in other directions but not without a small, tangential word in Mivart's defense. Mivart did appreciate the problem of complexity and coordination in sudden origins. He did not think that any old complex set of changes could arise all at once when needed—*that* would be tantamount to miracle. Most of Mivart's book studies the regularities of embryology and comparative anatomy to learn which kinds of complex changes might be possible as expressions and elaborations of developmental programs already present in ancestors. He advocates these changes as possible and eliminates others as fanciful.)

Darwin then faced his dilemma and developed the interestingly paradoxical resolution that has been orthodox ever since (but more poorly understood and appreciated than any other principle in evolutionary theory). If complexity precludes sudden origin, and the dilemma of incipient stages forbids gradual development in functional continuity, then how can we ever get from here to there? Darwin replies that we must reject an unnecessary hidden assumption in this argument—the notion of functional continuity. We will all freely grant that no creature can fly with 2 percent of a wing, but why must the incipient stages be used for flight? If incipient stages originally performed a different function suited to their small size and minimal development, natural selection might superintend their increase as adaptations for this original role until they reached a stage suitable for their current use. In other words, the problem of incipient stages disappears because these early steps were not inadequate wings but well-adapted something-elses. This principle of *functional change in structural continuity* represents Darwin's elegant solution to the dilemma of incipient stages.

Darwin, in a *beau geste* of argument, even thanked Mivart for characterizing the dilemma so well—all the better to grant Darwin a chance to elaborate his solution. Darwin writes: "A good opportunity has thus been afforded [by Mivart] for enlarging a little on gradations of structure, often associated with changed functions—an important subject, which was not treated at sufficient length in the former editions of this work." Darwin, who rarely added intensifiers to his prose, felt so strongly about this principle of functional shift that he wrote: "In considering transitions of organs, it is so important to bear in mind the probability of conversion from one function to another."

Darwin presented numerous examples in Chapters 5 and 7 of the final edition of the *Origin of Species.* He discussed organs that perform two functions, one primary, the other subsidiary, then relinquish the main use and elaborate the formerly inconspicuous operation. He then examined the flip side of this phenomenon—functions performed by two separate organs (fishes breathing with both lungs and gills). He argues that one organ may assume the entire function, leaving the other free for evolution to some other role (lungs for conversion to air bladders, for example, with respiration maintained entirely by gills). He does not, of course, neglect the classic example of wings, arguing that insects evolved their organs of flight from tracheae (or breathing organs—a minority theory today, but not without supporters). He writes: "It is therefore highly probable that in this great class organs which once served for respiration have been actually converted into organs of flight."

Darwin's critical theory of functional shift, usually (and most unfortunately) called the principle of "preadaptation,"* has been with us for a century. I believe that this principle has made so little headway not only because the basic formulation seems paradoxical and difficult, but mainly because we have so little firm, direct evidence for such functional shifts. Our technical literature contains many facile verbal arguments—little more than plausible "just-so" stories. The fossil record also presents some excellent examples of sequential development through intermediary stages that could not work as modern organs do—but we lack a rigorous mechanical analysis of function at the various stages.

Let us return, as we must, to the classic case of wings. *Archaeopteryx,* the first bird, is as pretty an intermediate as paleontology

---

*This dreadful name has made a difficult principle even harder to grasp and understand. Preadaptation seems to imply that the proto-wing, while doing something else in its incipient stages, knew where it was going—predestined for a later conversion to flight. Textbooks usually introduce the word and then quickly disclaim any odor of foreordination. (But a name is obviously ill-chosen if it cannot be used without denying its literal meaning.) Of course, by "preadaptation" we only mean that some structures are fortuitously suited to other roles if elaborated, not that they arise with a different future use in view—now there I go with the standard disclaimer. As another important limitation, preadaptation does not cover the important class of features that arise without functions (as developmental consequences of other primary adaptations, for example) but remain available for later co-optation. I suspect, for example, that many important functions of the human brain are co-opted consequences of building such a large computer for a limited set of adaptive uses. For these reasons, Elizabeth Vrba and I have proposed that the restrictive and confusing word "preadaptation" be dropped in favor of the more inclusive term "exaptation"—for any organ not evolved under natural selection for its current use—either because it performed a different function in ancestors (classical preadaptation) or because it represented a nonfunctional part available for later co-optation. See our technical article, "Exaptation: A Missing Term in the Science of Form," *Paleobiology,* 1981.

could ever hope to find—a complex mélange of reptilian and avian features. Scientists are still debating whether or not it could fly. If so, *Archaeopteryx* worked like the Wrights' biplane to a modern eagle's Concorde. But what did the undiscovered ancestors of *Archaeopteryx* do with wing rudiments that surely could not produce flight? Evolutionists have been invoking Darwin's principle of functional shift for more than 100 years, and the list of proposals is long. Proto-wings have been reconstructed as stabilizers, sexual attractors, or insect catchers. But the most popular hypothesis identifies thermoregulation as the original function of incipient stages that later evolved into feathered wings. Feathers are modified reptilian scales, and they work very well as insulating devices. Moreover, if birds evolved from dinosaurs (as most paleontologists now believe), they arose from a lineage particularly subject to problems with temperature control. *Archaeopteryx* is smaller than any dinosaur and probably arose from the tiniest of dinosaur lineages. Small animals, with high ratios of surface area to volume, lose heat rapidly and may require supplementary devices for thermoregulation. Most dinosaurs could probably keep warm enough just by being large. Surface area (length × length, or length squared) increases more slowly than volume (length × length × length, or length cubed) as objects grow. Since animals generate heat over their volumes and lose it through their surfaces, small animals (with their relatively large surface areas) have most trouble keeping warm.

There I go again—doing what I just criticized. I have presented a plausible story about thermoregulation as the original function of organs that later evolved into wings. But science is tested evidence, not tall tales. This lamentable mode of storytelling has been used to illustrate Darwin's principle of functional shift only *faute de mieux*—because we didn't have the goods so ardently desired. At least until recently, when my colleagues Joel G. Kingsolver and M. A. R. Koehl published the first hard evidence to support a shift from thermoregulation to flight as a scenario for the evolution of wings. They studied insects, not birds—but the same argument has long been favored for nature's smaller and far more abundant wings (see their article, "Aerodynamics, Thermoregulation, and the Evolution of Insect Wings: Differential Scaling and Evolutionary Change," in *Evolution,* 1985).

In preparing this essay, I spent several days reading the classical literature on the evolution of insect flight—and emerged with a deeper understanding of just how difficult Darwin's principle of functional shift can be, even for professionals. Most of the literature hasn't even made the first step of applying functional shift at all, not to mention the later reform of substituting direct evidence for verbal speculation. Most reconstructions are still trying to explain the incipient stages of insect wings as somehow involved in airborne performance from the start—not for flapping flight, of course, but still for some aspect of motion aloft rather than, as Darwin's principle would suggest, for some quite different function.

To appreciate the dilemma of such a position (so well grasped by Mivart more than 100 years ago), consider just one recent study (probably the best and most widely cited) and the logical quandaries that a claim of functional continuity entails. In 1964, J. W. Flower presented aerodynamic arguments for wings evolved from tiniest rudiment to elaborate final form in the interest of airborne motion. Flower argues, supporting an orthodox view, that wings evolved from tiny outgrowths of the body used for gliding prior to elaboration for sustained flight. But Flower recognizes that these incipient structures must themselves evolve from antecedents too small to function as gliding planes. What could these very first, slight outgrowths of the body be for? Ignoring Darwin's principle of functional shift, Flower searches for an aerodynamic meaning even at this very outset. He tries to test two suggestions: E. H. Hinton's argument that initial outgrowths served for "attitude control," permitting a falling insect to land in a suitable position for quick escape from predators; and a proposal of the great British entomologist Sir Vincent Wigglesworth (wonderful name for an insect man, I always thought) that such first stages might act as stabilizing or controlling devices during takeoff in small, passively aerial insects.

Flower proceeded by performing aerodynamic calculations on consequences of incipient wings for simple body shapes when dropped—and he quickly argued himself into an inextricable logical corner. He found, first of all, that tiny outgrowths might help, as Wigglesworth, Hinton, and others had suggested. But the argument foundered on another observation: The same advantages could be gained far more easily and effectively by another, readily available alternative route—evolution to small size (where increased surface/volume ratios retard falling and enhance the probability of takeoff). Flower then realized that he would have to specify a reasonably large body size for incipient wings to have any aerodynamic effect. But he then encountered another problem: At such sizes, legs work just as well as, if not better than, proto-wings for any suggested aerodynamic function. Flower admitted:

> The first conclusion to be drawn from these calculations is that the selective pressure in small insects is towards smaller insects, which would have no reason to evolve wings.

I would have stopped and searched elsewhere (in Darwin's principle of functional shift) at this point, but Flower bravely continued along an improbable path:

> The main conclusions, however, are that attitude control of insects would be by the use of legs or by very small changes in body shape [*i.e.*, by evolving small outgrowths, or proto-wings].

Flower, in short, never considered an alternative to his assumption of functional continuity based upon some aspect of aerial locomotion. He concluded:

> At first they [proto-wings] would affect attitude; later they could increase to a larger size and act as a true wing, providing lift in their own right. Eventually they could move, giving the insect greater maneuverability during descent, and finally they could "flap," achieving sustained flight.

As an alternative to such speculative reconstructions that work, in their own terms, only by uncomfortable special pleading, may I suggest Darwin's old principle of functional shift (preadaptation—ugh—for something else).

The physiological literature contains voluminous testimony to the thermodynamic efficiency of modern insect wings: in presenting, for example, a large surface area to the sun for quick heating (see B. Heinrich, 1981). If wings can perform this subsidiary function now, why not suspect thermoregulation as a primary role at the outset? M. M. Douglas (1981), for example, showed that, in *Colias* butterflies, only the basal one-third of the wing operates in thermoregulation—an area approximately equal to the thoracic lobes (proto-wings) of fossil insects considered ancestral to modern forms.

Douglas then cut down some *Colias* wings to the actual size of these fossil ancestral lobes and found that insects so bedecked showed a 55 percent greater increase in body temperature than bodies deprived of wings entirely. These manufactured proto-wings measured 5 by 3 millimeters on a body 15 millimeters long. Finally, Douglas determined that no further thermoregulatory advantage could be gained by wings longer than 10 millimeters on a 15-millimeter body.

Kingsolver and Koehl performed a host of elaborate and elegant experiments to support a thermoregulatory origin of insect proto-wings. As with so many examples of excellent science producing clear and interesting outcomes, the results can be summarized briefly and cleanly.

Kingsolver and Koehl begin by tabulating all the aerodynamic hypotheses usually presented in the literature as purely verbal speculations. They arrange these proposals of functional continuity (the explanations that do not follow Darwin's solution of Mivart's dilemma) into three basic categories: proto-wings for gliding (aerofoils for steady-state motion), for parachuting (slowing the rate of descent in a falling insect), and attitude stability (helping an insect to land right side up). They then transcended the purely verbal tradition by developing aerodynamic equations for exactly how proto-wings should help an insect under these three hypotheses of continuity in adaptation (increasing the lift/drag ratio as the major boost to gliding, increasing drag to slow the descent rate in parachuting, measuring the moment about the body axis produced by wings for the hypothesis of attitude stability).

They then constructed insect models made of wire, epoxy, and other appropriate materials to match the sizes and body shapes of flying and nonflying forms among early insect fossils. To these models, they attached wings (made of copper wire enclosing thin, plastic membranes) of various lengths and measured the actual aerodynamic effects for properties predicted by various hypotheses of functional continuity. The results of many experiments in wind tunnels are consistent and consonant: Aerodynamic benefits begin for wings above a certain size, and they increase as wings get larger. But at the small sizes of insect proto-wings, aerodynamic advantages are absent or insignificant and do not increase with growing wing length. These results are independent of body shape, wind velocity, presence or placement of legs, and mounting position of wings. In other words, large wings work well and larger wings work better—but small wings (at the undoubted sizes of Mivart's troubling incipient stages) provide no aerodynamic edge.

Kingsolver and Koehl then tested their models for thermoregulatory effects, constructing wings from two materials with different thermal conductivities (construction paper and aluminum foil) and measuring the increased temperature of bodies supplied with wings of various lengths versus wingless models. They achieved results symmetrically opposite to the aerodynamic experiments. For thermoregulation, wings work well at the smallest sizes, with benefits increasing as the wing grows. However, beyond a measured length, further increase of the wing confers no additional effect. Kingsolver and Koehl conclude:

> At any body size, there is a relative wing length above which there is no additional thermal effect, and below which there is no significant aerodynamic effect.

The accompanying chart illustrates these combined results. Note how the thermoregulatory effect of excess body temperature due to wings (solid line) increases rapidly at small wing sizes but not at all above an intermediate wing length. Conversely, the aerodynamic effect of lift/drag ratio does not increase at all until intermediate wing length, but grows rapidly thereafter.

We could not hope for a more elegant experimental confirmation of Darwin's solution to Mivart's challenge. Kingsolver and Koehl have actually measured the functional shift by showing that incipient wings aid thermoregulation but provide no aerodynamic benefit—while larger wings provide no further thermoregulatory oomph but initiate aerodynamic advantage and increase the benefits steadily thereafter. The crucial intermediate

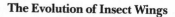

**The Evolution of Insect Wings**

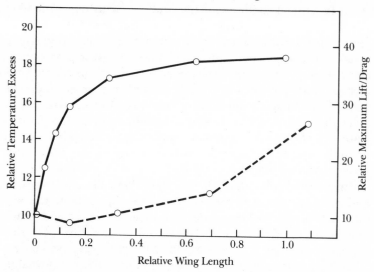

The thermoregulatory (upper curve) and aerodynamic (lower curve) advantages for increasing wing length in insects. Note that thermodynamic benefits accrue rapidly when the wing is very small (too small for flight), but scarcely increase at all for wings of larger size. Aerodynamic advantages, on the other hand, are insignificant for small size, but increase rapidly at larger wing dimensions, just as the thermodynamic benefits cease. BEN GAMIT. ADAPTED FROM JOE LEMONNIER. COURTESY OF *NATURAL HISTORY.*

wing length, where thermoregulatory gain ceases and aerodynamic benefits begin, represents a domain of functional shift, as aerodynamic advantages pick up the relay from waning thermoregulation to continue the evolutionary race to increasing wing size.

But what might push an insect across the transition? Why reach this crucial domain at all? If wings originally worked primarily for thermoregulation, why not just stop as the length of maximum benefit approached? Here, Kingsolver and Koehl present an interesting speculation based on another aspect of their data. They found that the domain of transition between thermal and aerial effects varied systematically with body size: The larger the body, the sooner the transition (in terms of relative wing length). For a body 2 centimeters long, the transition occurred with wings 40 to 60 percent of body length; but a 10-centimeter body switches to aerodynamic advantage at only 10 percent of body length.

Now suppose that incipient ancestral wings worked primarily for thermoregulation and had reached a stable, optimum size for greatest benefit. Natural selection would not favor larger wings and a transition to the available domain of aerodynamic advantage. But if body size increased for other reasons, an insect might reach the realm of aerial effects simply by growing larger, without any accompanying change of body shape or relative wing length.

We often think, naively, that size itself should make no profound difference. Why should just more of the same have any major effect beyond simple accumulation? Surely, any major improvement or alteration must require an extensive and explicit redesign, a complex reordering of parts with invention of new items.

Nature does not always match our faulty intuitions. Complex objects often display the interesting and paradoxical property of major effect for apparently trifling input. Internal complexity can translate a simple quantitative change into a wondrous alteration of quality. Perhaps that greatest and most effective of all evolutionary inventions, the origin of human consciousness, required little more than an increase of brain power to a level where internal connections became rich and varied enough to force this seminal transition. The story may be much more complex, but we have no proof that it must be.

Voltaire quipped that "God is always for the big battalions." More is not always better, but more can be very different.

# 4 | Fads and Fallacies

# 10 | The Case of the Creeping Fox Terrier Clone

WHEN ASTA the fox terrier exhumed the body of the Thin Man, his delightfully tipsy detective master, Nick Charles, exclaimed, "You're not a terrier; you're a police dog" (*The Thin Man*, MGM 1934 original with William Powell and Myrna Loy). May I now generalize for Asta's breed in the case of the telltale textbook.

The wisdom of our culture abounds with mottoes that instruct us to acknowledge the faults within ourselves before we criticize the failings of others. These words range from clichés about pots and kettles to various sayings of Jesus: "And why beholdest thou the mote that is in thy brother's eye, but perceivest not the beam that is in thine own eye?" (Luke 6:41); "He that is without sin among you, let him first cast a stone at her" (John 8:7). I shall follow this wisdom by exposing my own profession in trying to express what I find so desperately wrong about the basic tool of American teaching, the textbook.

In March 1987, I spent several hours in the exhibit hall of the National Science Teachers Association convention in Washington, D.C. There I made an informal, but reasonably complete, survey of evolution as treated (if at all) in major high-school science textbooks. I did find some evidence of adulteration, pussyfooting, and other forms of capitulation to creationist pressure. One book, *Life Science*, by L. K. Bierer, K. F. Liem, and E. P. Silberstein (Heath, 1987), in an accommodation that at least makes you laugh while you weep for lost integrity in education, qualifies every statement about the ages of fossils—usually in the most barbarous of English constructions, the passive infinitive.

We discover that trilobites are "believed to have lived 500–600 million years ago," while frozen mammoths are "thought to have roamed the tundra 22,000 years ago." But of one poor bird, we learn with terrible finality, "There are no more dodoes living today." Their extinction occurred within the bounds of biblical literalism and need not be hedged.

But I was surprised and pleased to note that most books contained material at reasonable length about evolution, and with no explicit signs of tampering to appease creationists. Sins imposed by others were minimal. But I then found the beam in our own eye and became, if anything, more distressed than by any capitulation to the yahoos. The problem does not lie in what others are doing to us, but in what we are doing to ourselves. In book after book, the evolution section is virtually cloned. Almost all authors treat the same topics, usually in the same sequence, and often with illustrations changed only enough to avoid suits for plagiarism. Obviously, authors of textbooks are copying material on a massive scale and passing along to students an ill-considered and virtually Xeroxed version with a rationale lost in the mists of time.

Just two months after making this depressing observation, I read Diane B. Paul's fascinating article "The Nine Lives of Discredited Data" (*The Sciences,* May 1987). Paul analyzed the sections on heritability of IQ from twenty-eight textbooks on introductory genetics published between 1978 and 1984. She paid particular attention to their treatment of Sir Cyril Burt's data on identical twins raised separately. We now know that these "studies" represent one of the most striking cases of fraud in twentieth-century science—for Burt invented both data and co-workers. His sad story had been well publicized, and all authors of texts published since 1978 surely knew that Burt's data had been discredited and could not be used. Several texts even included discussions of the Burt scandal as a warning about caution and scrutiny in science.

But Paul then found that nearly half these books continued to cite and use Burt's data, probably unconsciously. Of nineteen textbooks that devoted more than a paragraph to the subject of genetics and IQ, eleven based their conclusions about high heritability on a review article published in *Science* in 1963. This review featured a figure that ten of these textbooks reproduced either directly or in slightly altered and simplified form. This figure includes, as a prominent feature, the results of Sir Cyril Burt (not yet suspect in 1963). We must conclude that the authors of these texts either had not read the 1963 article carefully or had not consulted it at all. Paul infers (correctly, I am sure) that this carelessness arises because authors of textbooks copy from other texts and often do not read original sources. How else to explain the several books that discussed the Burt scandal explicitly and then, unbeknownst to their authors, used the same discredited data in a figure?

Paul argues that the increasing commercialization of textbooks has engendered this virtual cloning of contents. Textbook publishing is a big business, replete with market surveys, fancy art programs, and subsidiary materials in the form of slide sets, teachers' guides, even test-making and grading services. The actual text of the book can become secondary and standardized; any departure from a conventional set of topics could derail an entire industry of supporting materials. Teachers are also locked into a largely set curriculum based on this flood of accoutrements. Paul concludes: "Today's textbooks are thicker, slicker, more elaborate, and more expensive than they used to be. They are also more alike. Indeed, many are virtual clones, both stylistic and substantive, of a market leader."

The marketplace rules. Most publishing houses are now owned by conglomerates—CBS, Raytheon, and Coca-Cola among them—with managers who never raise their eyes from the financial bottom line, know little or nothing about books, and view the publishing arm of their diversified empire as but one more item for the ultimate balance. I received a dramatic reminder of this trend last week when I looked at the back cover of my score for Mozart's *Coronation Mass,* now under rehearsal in my chorus. It read: "Kalmus Score. Belwin Mills Publishing Company, distributed by Columbia Pictures Publication, a unit of the Coca-Cola Company." I don't say that Bill Cosby or Michael Jackson or whoever advertises the stuff doesn't like Mozart; I merely suspect that Don Giovanni can't be high on the executive agenda when the big boys must worry about such really important issues as whether or not to market Cherry Coke (a resounding "yes" vote from this old New York soda fountain junkie).

Paul quotes a leading industry analyst from the 1984 *Book Publishing Annual.* Future textbooks, the analyst argues, will have "more elaborate designs and greater use of color. . . . The ancillary packages will become more comprehensive. . . . New, more aggressive marketing plans will be needed just to maintain a company's position. The quality of marketing will make the difference." Do note the conspicuous absence of any mention whatsoever about the quality of the text itself.

Paul is obviously correct in arguing that this tendency to cloning has accelerated remarkably as concerns of the market overwhelm scholarly criteria in the composition of textbooks. But I believe that the basic tendency has always been present and has a human as well as a corporate face. Independent thought has always been more difficult than borrowing, and authors of textbooks have almost always taken the easier way out. Of course I have no objection to the similar recording of information by textbooks. No author can know all the byways of a profession, and all must therefore rely on written sources for areas not enlightened by personal expertise. I speak instead of the thoughtless, senseless, and often false copying of phrase, anecdote, style of argument, and sequence of topics that perpetuates itself by degraded repetition from text to text and thereby loses its anchor in nature.

I present an example that may seem tiny and peripheral in import. Nevertheless, and perhaps paradoxically, such cases provide our best evidence for thoughtless copying. When a truly important and well-known fact graces several texts in the same form, we cannot know whether it has been copied from previous sources or independently extracted from any expert's general knowledge. But when a quirky little senseless item attains the frequency of the proverbial bad penny, copying from text to text is the only reasonable interpretation. There is no other source. My method is no different from the standard technique of bibliographic scholars, who establish lineages of texts by tracing errors (particularly for documents spread by copyists before the invention of printing).

When textbooks choose to illustrate evolution with an example from the fossil record, they almost invariably trot out that greatest warhorse among case studies—the history of horses themselves (see the next essay in this section for fallacies of the usual tale). The standard story begins with an animal informally called *Eohippus* (the dawn horse), or more properly, *Hyracotherium*. Since evolutionary increase in size is a major component of the traditional tale, all texts report the diminutive stature of ancestral *Hyracotherium*. A few give actual estimates or measurements, but most rely upon a simile with some modern organism. For years, I have been much amused (and mildly bothered) that the great majority of texts report *Hyracotherium* as "like a fox-terrier" in size. I was jolted into action when I found myself writing the same line, and then stopped. "Wait a minute," said my inner voice, "beyond some vague memories of Asta last time I watched a Thin Man movie, I haven't the slightest idea what a fox terrier is. I can't believe that the community of textbook authors includes only dog fanciers—so if I don't know, I'll bet most of them don't either." Clearly, the classic line has been copied from text to text. Where did it begin? What has been its history? Is the statement even correct?

My immediate spur to action came from a most welcome and unexpected source. I published a parenthetical remark about the fox terrier issue (see Essay 11), ending with a serious point: "I also wonder what the textbook tradition of endless and thoughtless copying has done to retard the spread of original ideas."

I have, over the years, maintained a correspondence about our favorite common subject with Roger Angell of the *New Yorker*, who is, among other things, the greatest baseball writer ever. I assumed that his letter of early April would be a scouting report for the beginning of a new season. But I found that Roger Angell is a man of even more dimensions than I had realized; he is also a fox terrier fancier. He had read my parenthetical comment and wrote, "I am filled with excitement and trepidation at the prospect of writing you a letter about science instead of baseball."

Angell went on to suggest a fascinating and plausible explanation for the origin of the fox terrier simile (no excuse, of course, for its later cloning). Fox terriers were bred "to dig out foxes from their burrows, when a fox had gone to earth during a traditional British hunt." Apparently, generations of fox-hunting gentlemen selected fox terriers not only for their functional role in the hunt but also under a breeder's artifice to make them look as much like horses as possible. Angell continues, "The dogs rode up on the saddle during the hunt, and it was a pretty conceit for the owner-horseman to appear to put down a little simulacrum of a horse when the pack of hounds and the pink-coated throng had arrived at an earth where the animal was to do his work." He also pointed out that fox terriers tend to develop varied patches of color on a basically white coat and that a "saddle" along the back is "considered desirable and handsome." Thus, Angell proposed his solution: "Wouldn't it seem possible that some early horse geologist, in casting about for the right size animal to fit his cliché-to-be, might have settled, quite unconsciously, on a breed of dog that fitted the specifications in looks as well as size?"

This interesting conjecture led me to devise the following, loosely controlled experiment. I asked David Backus, my research assistant, to record every simile for *Hyracotherium* that he could find in the secondary literature of texts and popular books during more than a century since O. C. Marsh first recognized this animal as a "dawn horse." We would then use these patterns in attempting to locate original sources for favored similes in the primary literature of vertebrate paleontology. We consulted the books in my personal library as a sample, and compiled a total of eighty-six descriptions. The story turns out to be much more ascertainable and revealing than I had imagined.

The tradition of simile begins at the very beginning. Richard Owen, the great British anatomist and paleontologist, described the genus *Hyracotherium* in 1841. He did not recognize its relationship with horses (he considered this animal, as his chosen name implies, to be a possible relative of hyraxes, a small group of Afro-Asian mammals, the "coneys" of the Bible). In this original article, Owen likened his fossil to a hare in one passage and to something between a hog and a hyrax in another. Owen's simile plays no role in later history because other traditions of comparison had been long established before scientists realized that Owen's older discovery represented the same animal that Marsh later named *Eohippus*. (Hence, under the rules of taxonomy, Owen's inappropriate and uneuphonious name takes unfortunate precedence over Marsh's lovely *Eohippus*—see Essay 5 on the rules of naming.)

The modern story begins with Marsh's description of the earliest horses in 1874. Marsh pressed "go" on the simile machine by writing, "This species was about as large as a fox." He also described the larger descendant *Miohippus* as sheeplike in size.

Throughout the nineteenth century all sources that we have found (eight references, including such major figures as Joseph Le Conte, Archibald Geikie, and even Marsh's bitter enemy E. D.

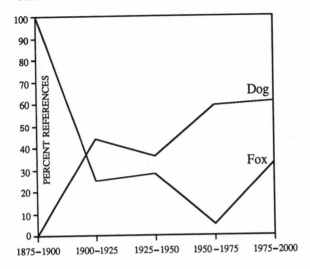

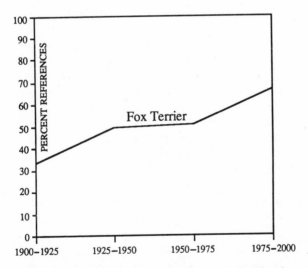

The rise to dominance of fox terriers as similes for the size of the earliest horses. Top graph: Increasing domination of dogs over foxes through time. Lower graph: Increase in percentage of fox terrier references among sources citing dogs as their simile. IROMIE WEERAMANTRY. COURTESY OF *NATURAL HISTORY*.

Cope) copy Marsh's favored simile—they all describe *Eohippus* as fox-sized. We are confident that Marsh's original description is the source because most references also repeat his statement that *Miohippus* is the size of a sheep. How, then, did fox terriers replace their prey?

The first decade of our century ushered in a mighty Darwinian competition among three alternatives and led to the final triumph of fox terriers. By 1910, three similes were battling for survival. Marsh's original fox suffered greatly from competition, but managed to retain a share of the market at about 25 percent (five of twenty citations between 1900 and 1925 in our sample)—a frequency that has been maintained ever since (see accompanying figure). Competition came from two stiff sources, however—both from the American Museum of Natural History in New York.

First, in 1903, W. D. Matthew, vertebrate paleontologist at the Museum, published his famous pamphlet *The Evolution of the Horse* (it remained in print for fifty years, and was still being sold at the Museum shop when I was a child). Matthew wrote: "The earliest known ancestors of the horse were small animals not larger than the domestic cat." Several secondary sources picked up Matthew's simile during this quarter century (also five of twenty references between 1900 and 1925), but felines have since faded (only one of fifteen references since 1975), and I do not know why.

Second, the three-way carnivorous competition of vulpine, feline, and canine began in earnest when man's best friend made his belated appearance in 1904 under the sponsorship of Matthew's boss, American Museum president and eminent vertebrate paleontologist Henry Fairfield Osborn. Remember that no nineteenth-century source (known to us) had advocated a canine simile, so Osborn's late entry suffered a temporal handicap. But Osborn was as commanding (and enigmatic) a figure as American natural history has ever produced (see Essay 29)—a powerful patrician in science and politics, imperious but kind, prolific and pompous, crusader for natural history and for other causes of opposite merit (Osborn wrote, for example, a glowing preface to the most influential tract of American scientific racism, *The Passing of the Great Race,* by his friend Madison Grant).

In the *Century Magazine* for November 1904, Osborn published a popular article, "The Evolution of the Horse in America."

(Given Osborn's almost obsessively prolific spate of publications, we would not be surprised if we have missed an earlier citation.) His first statement about *Eohippus* introduces the comparison that would later win the competition:

> We may imagine the earliest herds of horses in the Lower Eocene (*Eohippus,* or "dawn horse" stage) as resembling a lot of small fox-terriers in size. . . . As in the terrier, the wrist (knee) was near the ground, the hand was still short, terminating in four hoofs, with a part of the fifth toe (thumb) dangling at the side.

Osborn provides no rationale for his choice of breeds. Perhaps he simply carried Marsh's old fox comparison unconsciously in his head and chose the dog most similar in name to the former standard. Perhaps Roger Angell's conjecture is correct. Osborn certainly came from a social set that knew about fox hunting. Moreover, as the quotation indicates, Osborn extended the similarity of *Eohippus* and fox terrier beyond mere size to other horselike attributes of this canine breed (although, in other sources, Osborn treated the whippet as even more horselike, and even mounted a whippet's skeleton for an explicit comparison with *Eohippus*). Roger Angell described his fox terrier to me: "The back is long and straight, the tail is held jauntily upward like a trotter's, the nose is elongated and equine, and the forelegs are strikingly thin and straight. In motion, the dog comes down on these forelegs in a rapid and distinctive, stiff, flashy style, and the dog appears to walk on his tiptoes—on hooves, that is."

In any case, we can trace the steady rise to domination of dog similes in general, and fox terriers in particular, ever since. Dogs reached nearly 50 percent of citations (nine of twenty) between 1900 and 1925, but have now risen to 60 percent (nine of fifteen) since 1975. Meanwhile, the percentage of fox terrier citations among dog similes had also climbed steadily, from one-third (three of nine) between 1900 and 1925 to one-half (eight of sixteen) between 1925 and 1975, to two-thirds (six of nine) since 1975. Osborn's simile has been victorious.

Copying is the only credible source for these shifts of popularity—first from experts; then from other secondary sources. Shifts in fashion cannot be recording independent insights based on observation of specimens. *Eohippus* could not, by itself, say "fox" to every nineteenth-century observer and "dog" to most twentieth-century writers. Nor can I believe that two-thirds of all dog-inclined modern writers would independently say, "Aha, fox terrier" when contemplating the dawn horse. The breed is no longer so popular, and I suspect that most writers, like me, have only the vaguest impression about fox terriers when they copy the venerable simile.

In fact, we can trace the rise to dominance of fox terriers in our references. The first post-Osborn citation that we can find (Ernest Ingersoll, *The Life of Animals,* MacMillan, 1906) credits Osborn explicitly as author of the comparison with fox terriers. Thereafter, no one cites the original, and I assume that the process of text copying text had begun.

Two processes combined to secure the domination of fox terriers. First, experts began to line up behind Osborn's choice. The great vertebrate paleontologist W. B. Scott, for example, stood in loyal opposition in 1913, 1919, and 1929 when he cited both alternatives of fox and cat. But by 1937, he had switched: *"Hyracotherium* was a little animal about the size of a fox-terrier, but horse-like in all parts." Second, dogs became firmly ensconced in major textbooks. Both leading American geology textbooks of the early twentieth century (Chamberlin and Salisbury, 1909 edition, and Pirsson and Schuchert, 1924 edition) opt for canines, as does Hegner's zoology text (1912) and W. Maxwell Read's fine children's book (a mainstay of my youth) *The Earth for Sam* (1930 edition).

Fox terriers have only firmed up their position ever since. Experts cite this simile, as in A. S. Romer's leading text, *Vertebrate Paleontology* (3d edition, 1966): " 'Eohippus' was a small form, some specimens no larger than a fox terrier." They have also entered the two leading high-school texts: (1) Otto and Towle (descendant of Moon, Mann, and Otto, the dominant text for most of the past fifty years): "This horse is called *Eohippus*. It had four toes and was about the size of a fox-terrier" (1977 edition); (2) the *Biological Sciences Curriculum Study, Blue Edition* (1968): "The fossil of a small four-toed animal about the size of a fox-terrier was found preserved in layers of rock." College texts also comply. W. T. Keeton, in his *Biological Science,* the Hertz of the profession, writes (1980 edition): "It was a small animal, only about the size of a fox-terrier." Baker and Allen's *The Study of Biology,* a strong Avis, agrees (1982 edition): "This small animal *Eohippus* was not much bigger than a fox-terrier."

You may care little for dawn horses or fox terriers and might feel that I have made much of nothing in this essay. But I cite the case of the creeping fox terrier clone not for itself, but rather as a particularly clear example of a pervasive and serious disease—the debasement of our textbooks, the basic tool of written education, by endless, thoughtless copying.

My younger son started high school last month. For a biology text, he is using the 4th edition of *Biology: Living Systems,* by R. F. Oram, with consultants P. J. Hummer and R. C. Smoot (Charles E. Merrill, 1983, but listed on the title page, following our modern reality of conglomeration, as a Bell and Howell Company). I was sad and angered to find several disgraceful passages of capitulation to creationist pressure. Page one of the chapter on evolution proclaims in a blue sidebar: "The theory of evolution is the most widely accepted scientific explanation of the origin of life and changes in living things. You may wish to investigate other theories." Similar invitations are not issued for any other well-established theory. Students are not told that "most folks accept gravitation, but you might want to check out levitation" or that "most people view the earth as a sphere, but you might want to consider the possibility of a plane." When the text reaches human history, it doesn't even grant majority status to our evolutionary consensus: "Humans are indeed unique, but because they are also organisms, many scientists believe that humans have an evolutionary history."

Yet, as I argued at the outset, I find these compromises to outside pressure, disgraceful though they be, less serious than the internal disease of cloning from text to text. There is virtually only one chapter on evolution in all high-school biology texts, copied and degraded, then copied and degraded again. My son's book is no exception. This chapter begins with a discussion of Lamarck and the inheritance of acquired characters. It then moves to Darwin and natural selection and follows this basic contrast with a picture of a giraffe and a disquisition of Lamarckian and Darwinian explanations for long necks. A bit later, we reach industrial melanism in moths and dawn horses of you-know-what size.

What is the point of all this? I could understand this development if Lamarckism were a folk notion that must be dispelled before introducing Darwin, or if Lamarck were a household name. But I will lay 100 to 1 that few high-school students have ever heard of Lamarck. Why begin teaching evolution by explicating a false theory that is causing no confusion? False notions are often wonderful tools in pedagogy, but not when they are unknown, are provoking no trouble, and make the grasp of an accepted theory more difficult. I would not teach more sophisticated college students this way; I simply can't believe that this sequence works in high school. I can only conclude that someone once wrote the material this way for a reason lost in the mists of time, and that authors of textbooks have been dutifully copying "Lamarck . . . Darwin . . . giraffe necks" ever since.

(The giraffe necks, by the way, make even less sense. This venerable example rests upon no data at all for the superiority of Darwinian explanation. Lamarck offered no evidence for his interpretation and only introduced the case in a few lines of speculation. We have no proof that the long neck evolved by natural selection for eating leaves at the tops of acacia trees. We only prefer this explanation because it matches current orthodoxy. Giraffes do munch the topmost leaves, and this habit obviously helps them to thrive, but who knows how or why their necks elongated? They may have lengthened for other reasons and then been fortuitously suited for acacia leaves.)

If textbook cloning represented the discovery of a true educational optimum, and its further honing and propagation, then I would not object. But all evidence—from my little story of fox terriers to the larger issue of a senseless but nearly universal sequence of Lamarck, Darwin, and giraffe necks—indicates that cloning bears an opposite and discouraging message. It is the easy way out, a substitute for thinking and striving to improve. Somehow I must believe—for it is essential to my notion of scholarship—that good teaching requires fresh thought and genuine excitement, and that rote copying can only indicate boredom and slipshod practice. A carelessly cloned work will not excite students, however pretty the pictures. As an antidote, we need only the most basic virtue of integrity—not only the usual, figurative meaning of honorable practice but the less familiar, literal definition of wholeness. We will not have great texts if authors cannot shape content but must serve a commercial master as one cog in an ultimately powerless consortium with other packagers.

To end with a simpler point amid all this tendentiousness and generality: Thoughtlessly cloned "eternal verities" are often false. The latest estimate I have seen for the body size of *Hyracotherium* (MacFadden, 1986), challenging previous reconstructions congenial with the standard simile of much smaller fox-terriers, cites a weight of some twenty-five kilograms, or fifty-five pounds.

Lassie come home!

# 11 | Life's Little Joke

I STILL DON'T UNDERSTAND why a raven is like a writing desk, but I do know what binds Hernando Cortés and Thomas Henry Huxley together.

On February 18, 1519, Cortés set sail for Mexico with about 600 men and, perhaps more important, 16 horses. Two years later, the Aztec capital of Tenochtitlán lay in ruins, and one of the world's great civilizations had perished.

Cortés's victory has always seemed puzzling, even to historians of an earlier age who did not doubt the intrinsic superiority of Spanish blood and Christian convictions. William H. Prescott, master of this tradition, continually emphasizes Cortés's diplomatic skill in making alliances to divide and conquer—and his good fortune in despoiling Mexico during a period of marked internal dissension among the Aztecs and their vassals. (Prescott published his *History of the Conquest of Mexico* in 1843; it remains among the most exciting and literate books ever written.)

Prescott also recognized Cortés's two "obvious advantages on the score of weapons"—one inanimate and one animate. A gun is formidable enough against an obsidian blade, but consider the additional impact of surprise when your opponent has never seen a firearm. Cortés's cavalry, a mere handful of horses and their riders, caused even more terror and despair, for the Aztecs, as Prescott wrote,

> had no large domesticated animals, and were unacquainted with any beast of burden. Their imaginations were bewildered when they beheld the strange apparition of the horse

168

and his rider moving in unison and obedient to one impulse, as if possessed of a common nature; and as they saw the terrible animal, with "his neck clothed in thunder," bearing down their squadrons and trampling them in the dust, no wonder they should have regarded him with the mysterious terror felt for a supernatural being.

On the same date, February 18, in 1870, Thomas Henry Huxley gave his annual address as president of the Geological Society of London and staked his celebrated claim that Darwin's ideal evidence for evolution had finally been uncovered in the fossil record of horses—a sequence of continuous transformation, properly arrayed in temporal order:

> It is easy to accumulate probabilities—hard to make out some particular case, in such a way that it will stand rigorous criticism. After much search, however, I think that such a case is to be made out in favor of the pedigree of horses.

Huxley delineated the famous trends to fewer toes and higher-crowned teeth that we all recognize in this enduring classic among evolutionary case histories. Huxley viewed this lineage as a European affair, proceeding from fully three-toed *Anchitherium,* to *Hipparion* with side toes "reduced to mere dew-claws [that] do not touch the ground," to modern *Equus,* where, "finally, the crowns of the grinding-teeth become longer. . . . The phalanges of the two outer toes in each foot disappear, their metacarpal and metatarsal bones being left as the 'splints.' "

In *Cat's Cradle,* Kurt Vonnegut speaks of the subtle ties that can bind people across worlds and centuries into aggregations forged by commonalities so strange that they must be meaningful. Cortés and Huxley must belong to the same karass (Vonnegut's excellent word for these associations)—for they both, on the same date, unfairly debased America with the noblest of animals. Huxley was wrong and Cortés, by consequence, was ever so lucky.

Horses evolved in America, through a continuity that extends unbroken across 60 million years. Several times during this history, different branches migrated to Europe, where Huxley arranged three (and later four) separate incursions as a false continuity. But horses then died in America at the dawn of human history in our hemisphere, leaving the last European migration as a source of recolonization by conquest. Huxley's error became Montezuma's sorrow, as an animal more American than Babe Ruth or apple pie came home to destroy her greatest civilization. (Montezuma's revenge would come later, and by another route.)

During our centennial year of 1876, Huxley visited America to deliver the principal address for the founding of Johns Hopkins University. He stopped first at Yale to consult the eminent paleontologist Othniel C. Marsh. Marsh, ever gracious, offered Huxley an architectural tour of the campus, but Huxley had come for a purpose and would not be delayed. He pointed to the buildings and said to Marsh: "Show me what you have got inside them; I can see plenty of bricks and mortar in my own country." Huxley was neither philistine nor troglodyte; he was simply eager to study some particular fossils: Marsh's collection of horses.

Two years earlier, Marsh had published his phylogeny of American horses and identified our continent as the center stage, while relegating Huxley's European sequence to a periphery of discontinuous migration. Marsh began with a veiled and modest criticism (*American Journal of Science*, 1874):

> Huxley has traced successfully the later genealogy of the horse through European extinct forms, but the line in America was probably a more direct one, and the record is more complete.

Later, he stated more baldly (p. 258): "The line of descent appears to have been direct, and the remains now known supply every important intermediate form."

Marsh had assembled an immense collection from the American West (prompted largely by a race for priority in his bitter feud with Edwin D. Cope—see Essay 5 for another consequence of this feud!). For every query, every objection that Huxley raised, Marsh produced a specimen. Leonard Huxley describes the scene in his biography of his father:

> At each inquiry, whether he had a specimen to illustrate such and such a point or to exemplify a transition from earlier and less specialized forms to later and more special-

ized ones, Professor Marsh would simply turn to his assistant and bid him fetch box number so and so, until Huxley turned upon him and said, "I believe you are a magician; whatever I want, you just conjure it up."

Years before, T. H. Huxley had coined a motto; now he meant to live by it: "Sit down before fact as a little child, be prepared to give up every preconceived notion." He capitulated to Marsh's theory of an American venue. Marsh, with growing pleasure and retreating modesty, reported his impression of personal triumph:

> He [Huxley] then informed me that this was new to him, and that my facts demonstrated the evolution of the horse beyond question, and for the first time indicated the direct line of descent of an existing animal. With the generosity of true greatness, he gave up his own opinions in the face of new truth and took my conclusions.

A few days later, Huxley was, if anything, more convinced. He wrote to Marsh from Newport, his next stop: "The more I think of it the more clear it is that your great work is the settlement of the pedigree of the horse." But Huxley was scheduled to lecture on the evolution of horses less than a month later in New York. As he traveled about eastern America, Huxley rewrote his lecture from scratch. He also enlisted Marsh's aid in preparing a chart that would show the new evidence to his New York audience in pictorial form. Marsh responded with one of the most famous illustrations in the history of paleontology—the first pictorial pedigree of the horse.

Scholars are trained to analyze words. But primates are visual animals, and the key to concepts and their history often lies in iconography. Scientific illustrations are not frills or summaries; they are foci for modes of thought. The evolution of the horse—both in textbook charts and museum exhibits—has a standard iconography. Marsh began this traditional display in his illustration for Huxley. In so doing, he also initiated an error that captures pictorially the most common of all misconceptions about the shape and pattern of evolutionary change.

Errors in science are diverse enough to demand a taxonomy of

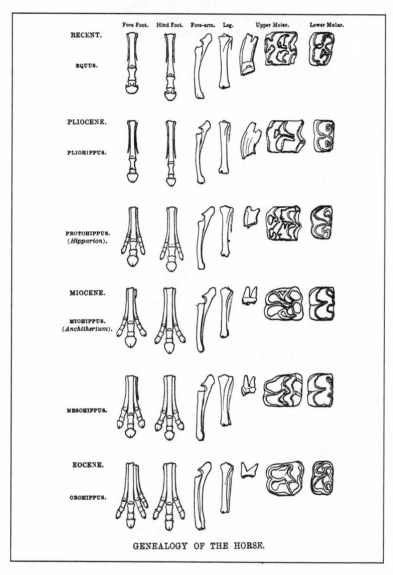

GENEALOGY OF THE HORSE.

The celebrated original figure drawn by O.C. Marsh for T.H. Huxley's New York lecture on the evolution of horses. This version appeared in an article by Marsh in the *American Journal of Science* for 1879. NEG. NO. 123823. COURTESY DEPARTMENT OF LIBRARY SERVICES, AMERICAN MUSEUM OF NATURAL HISTORY.

categories. Some make me angry, particularly those that arise from social prejudice, masquerade as objectively determined truth, and directly limit the lives of those caught in their thrall (scientific justifications for racism and sexism, as obvious examples). Others make me sad because honest effort ran headlong into unresolvable complexities of nature. Still others, as errors of logic that should not have occurred, bloat my already extended ego when I discover them. But I reserve a special place in perverse affection for a small class of precious ironies—errors that pass nature through a filter of expectation and reach a particular conclusion only because nature really works in precisely the opposite way. This result, I know, sounds both peculiar and unlikely, but bear with me for the premier example of life's little joke—as displayed in conventional iconography (and interpretation) for the most famous case study of all, the evolution of the horse.

In his original 1874 article, Marsh recognized the three trends that define our traditional view of old dobbin's genealogy: increase in size, decrease in the number of toes (with the hoof of modern horses made from a single digit, surrounded by two vestigial splints as remnants of side toes), and increase in the height and complexity of grinding teeth. (I am not treating the adaptive significance of these changes here, but wish to record the conventional explanation for the major environmental impetus behind trends in locomotion and dentition: a shift from browsing on lush lowland vegetation to grazing of newly evolved grasses upon drier plains. Tough grasses with less food value require considerably more dental effort.)

Marsh's famous chart, drawn for Huxley, depicts these trends as an ascending series—a ladder of uninterrupted progress toward one toe and tall, corrugated teeth (by scaling all his specimens to the same size, Marsh does not show the third "classic" trend toward increasing bulk).

We are all familiar with this traditional picture—the parade of horses from little eohippus (properly called *Hyracotherium*), with four toes in front and three behind, to Man o' War. (*Hyracotherium* is always described as "fox terrier" in size. Such traditions disturb and captivate me. I know nothing about fox terriers but have dutifully copied this description. I wonder who said it first, and why this simile has become so canonical. I also wonder what the textbook tradition of endless and thoughtless copying has done to retard the spread of original ideas.\*)

In conventional charts and museum displays, the evolution of horses looks like a line of schoolchildren all pointed in one direction and arrayed in what my primary-school drill instructors called "size place" (also stratigraphic order in this case). The most familiar of all illustrations, first drawn early in the century for the American Museum of Natural History's pamphlet on the evolution of horses, by W. D. Matthew, but reproduced hundreds of times since then, shows the whole story: size, toes, and teeth arranged in a row by order of appearance in the fossil record. To cite just one example of this figure's influence, George W. Hunter reproduced Matthew's chart as the primary illustration of

THE EVOLUTION OF THE HORSE.

Most widely reproduced of all illustrations showing the evolution of horses as a ladder towards progress. Note increase in skull size, decrease in the number of toes, and increase in the height of teeth. The skulls are also arranged in stratigraphic order. W.D. Matthew used this illustration in several publications. This version comes from an article in the *Quarterly Review of Biology* for 1926. NEG. NO. 37969. COURTESY DEPARTMENT OF LIBRARY SERVICES, AMERICAN MUSEUM OF NATURAL HISTORY.

---

\*This parenthetical comment inspired Roger Angell's letter and led directly to research and writing of the essay preceding this piece.

evolution in his high-school textbook of 1914, *A Civic Biology*. John Scopes assigned this book to his classes in Tennessee and was convicted for teaching its chapters on evolution, as William Jennings Bryan issued his last hurrah (see Essay 28): "No more repulsive doctrine was ever proclaimed by man . . . may heaven defend the youth of our land from [these] impious babblings."

But what is so wrong with these evolutionary ladders? Surely we can trace an unbroken continuity from *Hyracotherium* to modern horses. Yes, but continuity comes in many more potential modes than the lock step of the ladder. Evolutionary genealogies are copiously branching bushes—and the history of horses is more lush and labyrinthine than most. To be sure, *Hyracotherium* is the base of the trunk (as now known), and *Equus* is the surviving twig. We can, therefore, draw a pathway of connection from a common beginning to a lone result. But the lineage of modern horses is a twisted and tortuous excursion from one branch to another, a path more devious than the road marked by Ariadne's thread from the Minotaur at the center to the edge of our culture's most famous labyrinth. Most important, the path proceeds not by continuous transformation but by lateral stepping (with geological suddenness when punctuated equilibrium applies, as in this lineage, at least as read by yours truly, who must confess his bias as coauthor of the theory).

Each lateral step to a new species follows one path among several alternatives. Each extended lineage becomes a set of decisions at branching points—only one among hundreds of potential routes through the labyrinth of the bush. There is no central direction, no preferred exit to this maze—just a series of indirect pathways to every twig that ever graced the periphery of the bush.

As an example of distortions imposed by converting tortuous paths through bushes into directed ladders, consider the men associated with the two classical iconographies reproduced here. When Huxley made his formal capitulation to Marsh's interpretation in print (1880), he extended the ladder of horses as a metaphor for all vertebrates. Speaking of modern reptiles and teleost fishes, Huxley wrote (1880, p. 661): "They appear to me to be off the main line of evolution—to represent, as it were, side tracks starting from certain points of that line." But teleosts (modern bony fishes) are an enormously successful group. They stock the world's oceans, lakes, and rivers and maintain nearly 100 times as many species as primates (and more than all mammals combined). How can we call them "off the main line" just because we

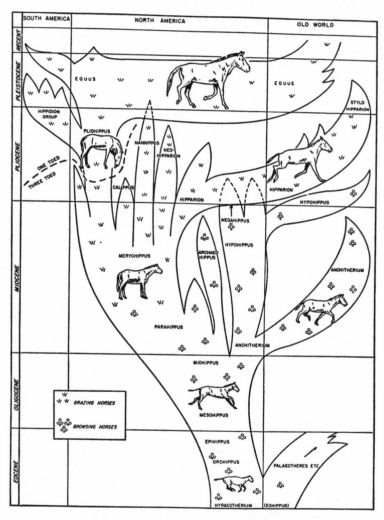

The evolution of horses depicted as at least a modest bush by G.G. Simpson in 1951. NEG. NO. 328907. COURTESY DEPARTMENT OF LIBRARY SERVICES, AMERICAN MUSEUM OF NATURAL HISTORY.

can trace our own pathway back to a common ancestry with theirs more than 300 million years ago?

W. D. Matthew slipped into an equally biased assessment of value because his designation of one pathway as a ladder forced an interpretation of all others as diversions. Matthew (1926, p. 164) designated his ladder as the "direct line of succession," but acknowledged that "there are also a number of side branches, more or less closely related." Three pages later, Matthew adds the opprobrium of near indecency to his previous charge of mere laterality, as he describes (p. 167) "a number of side branches leading up in a similar manner to aberrant specialized Equidae now extinct." But in what way are extinct lineages more specialized than a modern horse or in any sense more peculiar? Their historical death is the only possible rationale for a designation of aberrancy, but more than 99 percent of all species that ever lived are extinct—and disappearance cannot be the biological equivalent of a scarlet letter. We might as well call modern horses aberrant because, much to Montezuma's later sorrow, they became extinct in the land of their birth.

Yet we have recognized the bushiness of horse evolution from the very beginning. How else did Marsh forestall Huxley but by convincing him that his European "genealogy" of horses formed a stratigraphic sequence of discontinuous stages, falsely linking several side branches that had disappeared without issue?

As an example of bushiness, and a plug for the value of appropriate metaphors in general, consider the finest book on the evolution of horses ever written for popular audiences—G. G. Simpson's *Horses* (1951). Simpson redrew the genealogy of horses as a modest bush with no preferred main line. He also criticized the conceptual lock imposed by the bias of the ladder when he noted that modern one-toed horses are a side branch and extinct three-toed creatures the main line (if any center can be designated at all).

> As nearly as there is a straight line in horse evolution, it culminated and ended with these animals [the three-toed anchitheres], which, like their ancestors, were multiple-toed browsers. From this point of view, it is the line leading to modern horses that was the side branch, even though it outlasted the straighter line of horse evolution [p. 130].

Yet Simpson, who held a lifelong commitment to the predominant role of evolution by transformational change within populations rather than by accumulation across numerous events of discrete, branching speciation, could not entirely let go of biases imposed by the metaphor of the ladder. In one revealing passage, he accepts bushiness, but bemoans the complexities thus introduced, as though they clouded evolution's essence of transformational change:

> *Miohippus* . . . intergraded with several different descendant groups. It is sad that this introduces possible confusion into the story, but there is not much point in criticizing nature for something that happened some millions of years ago. It would also be foolish to try to ignore the complications, which did occur and which are a very important part of the record.

But these "complications" are not a veil upon the essence of lineal descent; they are the primary stuff of evolution itself.

Moreover, Simpson restricted his bushiness as much as possible and retained linearity wherever he could avoid an inference of branching. In particular, he proposes the specific and testable hypothesis (see his illustration) that the early part of the record—the sequence of *Hyracotherium—Orohippus—Epihippus—Mesohippus—Miohippus—Hypohippus*—tells a story of linear descent, only later interrupted by copious branching among three-toed browsers: "The line from *Eohippus* to *Hypohippus,* for example, exemplifies a fairly continuous phyletic evolution" (p. 217). Simpson especially emphasizes the supposedly gradual and continuous transformation from *Mesohippus* to *Miohippus* near the top of this sequence:

> The more progressive horses of the middle Oligocene and all the horses of the late Oligocene are placed by convention in a separate genus, *Miohippus.* In fact *Mesohippus* and *Miohippus* intergrade so perfectly and the differences between them are so slight and variable that even experts find it difficult, at times nearly impossible, to distinguish them clearly.

The enormous expansion of collections since Simpson proposed this hypothesis has permitted a test by vertebrate paleontologists Don Prothero and Neil Shubin. Their results falsify Simpson's gradual and linear sequence for the early stages of horse evolution and introduce extensive bushiness into this last stronghold of the ladder.

Prothero and Shubin have made four major discoveries in the crucial segment of history that Simpson designated as the strongest case for a gradualistic sequence of lineal transformation—the transition from *Mesohippus* to *Miohippus*.

1. Previous experts were so convinced about the imperceptibly gradual transition between these two genera that they declared any search for distinguishing characters as vain, and arbitrarily drew the division between *Mesohippus* and *Miohippus* at a stratigraphic boundary. But far richer material available to Prothero and Shubin has permitted the identification of characters that cleanly distinguish the two genera. (Teeth are the hardest part of a vertebrate skeleton and the fossil record of mammals often contains little else. A technical course in the evolution of mammals is largely an exercise in the identification of teeth, and an old professional quip holds that mammalian evolution is the interbreeding of two sets of teeth to produce some slightly modified descendant choppers. *Miohippus* and *Mesohippus* do not have distinctive dentitions, and previous failure to find a clear separation should not surprise us. The new material is rich in skull and limb bones.) In particular, Prothero and Shubin found that *Miohippus* develops a distinctive articulation, absent in ancestral *Mesohippus*, between the enlarging third metatarsal (the foot bone of the digit that will become the entire hoof of modern horses) and the cuboid bone of the tarsus (ankle) above.

2. *Mesohippus* does not turn into *Miohippus* by insensible degrees of gradual transition. Rather, *Miohippus* arises by branching from a *Mesohippus* stock that continues to survive long afterward. The two genera overlap in time by at least 4 million years.

3. Each genus is itself a bush of several related species, not a rung on a ladder of progress. These species often lived and interacted in the same area at the same time (as different species of zebra do in Africa today). One set of strata in Wyoming, for example, has yielded three species of *Mesohippus* and two of *Miohippus,* all contemporaries.

4. The species of these bushes tend to arise with geological suddenness, and then to persist with little change for long periods. Evolutionary change occurs at the branch points themselves, and trends are not continuous marches up ladders, but concatenations of increments achieved at nodes of branching on evolutionary bushes. Of this phenomenon Prothero and Shubin write:

> There is no evidence of long-term changes within these well-defined species [of *Mesohippus* and *Miohippus*] through time. Instead, they are strikingly static through millions of years. Such stasis is apparent in most Neogene [later] horses as well, and in *Hyracotherium*. This is contrary to the widely-held myth about horse species as gradualistically-varying parts of a continuum, with no real distinctions between species. Throughout the history of horses, the species are well-marked and static over millions of years. At high resolution, the gradualistic picture of horse evolution becomes a complex bush of overlapping, closely related species.

Bushiness now pervades the entire phylogeny of horses.

We can appreciate this fundamental shift in iconography and meaning, but where is the "precious irony" that I promised? What is "life's little joke" of my title? Simply this. The model of the ladder is much more than merely wrong. It never could provide the promised illustration of evolution progressive and triumphant— for *it could only be applied to unsuccessful lineages.*

Bushes represent the proper topology of evolution. Ladders are false abstractions, made by running a steamroller over a labyrinthine pathway that hops from branch to branch through a phylogenetic bush. We cannot force a successful bush of evolution into a ladder because we may follow a thousand pathways through the maze of twigs, and we cannot find a criterion for preferring one route over another. Who ever heard of the evolutionary trend of rodents or of bats or of antelopes? Yet these are the greatest success stories in the history of mammals. Our proudest cases do not become our classic illustrations because we can draw no ladder of progress through a vigorous bush with hundreds of surviving twigs.

But consider the poor horses. Theirs was once a luxuriant bush, yet they barely survive today. Only one twig (the genus *Equus,* with horses, zebras, and asses) now carries all the heritage of a group that once dominated the history of hoofed mammals—and with fragility at that, for *Equus* died in the land of its birth and had to be salvaged from a stock that had migrated elsewhere. (In a larger sense, horses form one of three dwindling lines—tapirs and rhinos are the others—that now represent all the diversity of the formerly dominant order Perissodactyla, or odd-toed ungulates, among hoofed mammals. This mighty group once included the giant titanotheres, the clawed chalicotheres, and *Baluchitherium,* the largest land mammal that ever lived. It now hangs on as a remnant in a world increasingly dominated by the Artiodactyla, or even-toed ungulates—cows, deer, antelope, camels, hippos, giraffes, pigs, and their relatives.)

This is life's little joke. By imposing the model of the ladder upon the reality of bushes, we have guaranteed that our classic examples of evolutionary progress can only apply to unsuccessful lineages on the very brink of extermination—for we can linearize a bush only if it maintains but one surviving twig that we can falsely place at the summit of a ladder. I need hardly remind everybody that at least one other mammalian lineage, preeminent among all in our attention and concern, shares with horses the sorry state of reduction from a formerly luxuriant bush to a single surviving twig—the very property of extreme tenuousness that permits us to build a ladder reaching only to the heart of our own folly and hubris.

# 12 | The Chain of Reason versus the Chain of Thumbs

THE *Weekly World News,* most lurid entry in the dubious genre of shopping mall tabloids, shattered all previous records for implausibility with a recent headline: "Siamese Twins Make Themselves Pregnant." The story recounted the sad tale of a conjoined brother-sister pair from a remote Indian village (such folks never hail from Peoria, where their non-existence might be confirmed). They knew that their act was immoral, but after years of hoping in vain for ordinary partners, and in the depths of loneliness and frustration, they finally succumbed to an ever-present temptation. The story is heart-rending, but faces one major obstacle to belief: All Siamese twins are monozygotic, formed from a single fertilized egg that failed to split completely in the act of twinning. Thus, Siamese twins are either both male or both female.

I will, however, praise the good people at *Weekly World News* for one slight scruple. They did realize that they had created a problem with this ludicrous tale, and they did not shrink from the difficulty. The story acknowledged that, indeed, Siamese twins generally share the same sex, but held that this Indian pair had been formed, uniquely and differently, from two eggs that had fused! Usually, however, *Weekly World News* doesn't even bother with minimal cover-ups. Recently, for example, they ran a screaming headline about a monster from Mars, just sighted in a telescope and now on its way to earth. The accompanying photo of the monster showed a perfectly ordinary chambered nautilus (an odd-looking and unfamiliar creature to be sure). I mean, they

182

didn't even bother to retouch the photo or to hide in any way their absurd transmogrification of a marine mollusk into an extra-terrestrial marauder!

The sad moral of this tale lies not with the practices of *Weekly World News,* but with the nature of a readership that permits such a publication to prosper—for if *Weekly World News* could not rely, with complete confidence, on the ignorance of its consumers, the paper would be exposed and discredited. The Siamese twin story at least showed a modicum of respect for the credulity of readers; the tale of the Martian monster records utter contempt both for the consuming public and for truth in general.

We like to cite an old motto of our culture on the factual and ethical value of veracity: "And ye shall know the truth, and the truth shall make you free" (John 8:32). But ignorance has always prospered, serving the purposes of demagogues and profit-mongers. An overly optimistic account might try to link our in-creasing factual knowledge with the suppression of cruelties and abuses ranging from execution for witchcraft to human sacrifice for propitiating deities. But this hope cannot be sustained, for no century has exceeded our own in quantity of imposed cruelty (as "improvements" in the technology of genocide and warfare more than balance any overall gains in sensibility). Moreover, despite a great spread in the availability of education, the favored irration-alisms of the ages show no signs of abatement. Presidential calen-dars are still set by astrologers, while charlatans do a brisk business in necklaces made of colored glass masquerading as crystals that supposedly bathe believers in a salutary and intangi-ble "energy." An astounding percentage of "educated" Ameri-cans think that the earth might be less than 10,000 years old, even while their own kids delight in dinosaurs at the local museum.

The champions of beleaguered rationalism—all heroes in my book—have been uncovering charlatans throughout the ages: from Elijah denouncing the prophets of Baal to Houdini expos-ing the tricks of mediums to James Randi on the trail of modern hoaxers and hucksters. Obviously, we have not won the war, but we have developed effective battle strategies—and would have triumphed long ago were our foe not able, like the Lernean Hydra, to grow several new heads every time we lop one off. Still, tales of past victories—including the story of this essay—are not only useful as spurs of encouragement; they also teach us effective methods of attack. For reason is timeless, and its application to unfamiliar contexts can be particularly instructive.

How many of us realize that we are invoking a verbal remnant of "the greatest vogue of the 1780s" (according to historian Robert Darnton) when we claim to be "mesmerized" by a wonderful concert or a beautiful sunset? Franz Anton Mesmer was a German physician who had acquired wealth through marriage to a well-endowed widow; connections by assiduous cultivation (Mozart,* a valued friend, had staged the first performance of his comic opera *Bastien und Bastienne* at Mesmer's private theater); and renown with a bizarre, if fascinating, theory of "animal magnetism" and its role in human health. In 1778, Mesmer transferred to Paris, then the most "open" and vibrant capital of Europe, a city embracing the odd mixture so often spawned by liberty—intellectual ferment of the highest order combined with quackery at its most abject: Voltaire among the fortune tellers; Benjamin Franklin surrounded by astrologers; Antoine Lavoisier amidst the spiritualists.

Mesmer, insofar as one can find coherence in his ideas at all, claimed that a single (and subtle) fluid pervaded the universe, uniting and connecting all bodies. We give different names to this fluid according to its various manifestations: gravity for planets in their courses; electricity in a thunderstorm; magnetism for navigation by compass. The same fluid flows through organisms and

---

*I thank Gerald A. Le Boff and Ernest F. Marmorek for informing me, after reading this essay at its initial publication, of another explicit link between Mozart (and his great librettist, DaPonte) and Mesmer. In *Cosi Fan Tutte,* the maid Despina, disguised as a physician, "cures" Ferrando and Guglielmo of their feigned illness by touching their foreheads with a large magnet and then gently stroking the length of their bodies. An orchestral tremolo recalls the curing mesmeric crisis, while Despina describes her magnet as:

> pietra Mesmerica
> ch'ebbe l'origine
> nell' Alemagna
> che poi si celebre
> lá in Francia fù.

—a mesmeric stone that had its origin in Germany and then was so famous in France (a fine epitome of Mesmer's tactic and its geographic history).

may be called animal magnetism. A blockage of this flow causes disease, and cure requires a reestablishment of the flux and a restoration of equilibrium. (Mesmer himself never went so far as to ascribe all bodily ills to blocked magnetism, but several disciples held this extreme view, and such a motto came to characterize the mesmeric movement: "There is only one illness and one healing.")

Cure of illness requires the intervention of an "adept," a person with unusually strong magnetism who can locate the "poles" of magnetic flow on the exterior of a human body and, by massaging these areas, break the blockage within to reestablish the normal flux. When working one on one, Mesmer would sit directly opposite his patient, establishing the proper contact and flow by holding the sufferer's knees within his own, touching fingers, and staring directly into her face (most patients were women, thus adding another dimension to charges of exploitation). Mesmer, by all accounts, was a most charismatic man—and we need no great psychological sophistication to suspect that he might have produced effects more by power of suggestion than by flow of any fluid.

In any case, the effects could be dramatic. Within a few minutes of mesmerizing, sensitive patients would fall into a characteristic "crisis" taken by Mesmer as proof of his method. Bodies would begin to shake, arms and legs move violently and involuntarily, teeth chatter loudly. Patients would grimace, groan, babble, scream, faint, and fall unconscious. Several repetitions of these treatments would reestablish magnetic equilibrium and produce cures. Mesmer carried sheaves of testimonials claiming recovery from a variety of complaints. Even his most determined critics did not deny all cures, but held that Mesmer had only relieved certain psychosomatic illnesses by the power of suggestion and had produced no physical effects with his putative universal fluid.

Mesmer's popularity required the development of methods for treating large numbers of patients simultaneously (such a procedure didn't hurt profits either), and Mesmer imposed high charges, in two senses, upon his mostly aristocratic crowd. Moreover, as a master of manipulation, Mesmer surely recognized the social value of treatment in groups—both the reinforcing effect of numerous crises and the simple value of conviviality in spread-

A patient falls into a Mesmeric crisis as the eponymous hero himself performs a cure. THE BETTMANN ARCHIVE.

ing any vogue as a joint social event and medical cure. Mesmer therefore began to magnetize inanimate objects and to use these charged bodies as instruments of unblocking and cure.

Many contemporary descriptions and drawings of Mesmer's sessions depict the same basic scene. Mesmer placed a large vat, called a *baquet,* in the center of a room. He then filled the *baquet* with "magnetized" water and, sometimes, a layer of iron filings as well. Some twenty thin metal rods protruded from the *baquet.* A patient would grab hold of a rod and apply it to the mesmeric poles of his body. To treat more than twenty, Mesmer would loop a rope from those who surrounded the *baquet* (and held the iron rods) to others in the room, taking care that the rope contained no knots, for such constrictions impeded the flux. Patients would then form a "mesmeric chain" by holding a neighbor's left thumb between their own right thumb and forefinger, while extending their own left thumb to the next patient down the line. By squeezing a neighbor's left thumb, magnetic impulses could be sent all the way down the chain.

Mesmer, whether consciously or not, surely exploited both the art and politics of psychosomatic healing. Everything in his curing room was carefully arranged to maximize results, efficiency, and profit. He installed mirrors to reflect the action and encourage mass response; he heightened the effect with music played on the ethereal tones of a glass harmonica, the instrument that Benjamin Franklin had developed; he employed assistants to carry convulsive patients into a "crisis room" lined with mattresses, lest they should hurt themselves in their frenzy. To avoid the charge of profitmongering among the rich alone, Mesmer provided a poor man's cure by magnetizing trees and inviting the indigent to take their relief gratis and alfresco.

I don't want to commit the worst historical error of wrenching a person from his own time and judging him by modern standards and categories. Thus, Franz Mesmer was not Uri Geller teleported to 1780. For one thing, historical records of Mesmer are scanty, and we do not even know whether he was a simple charlatan, purveying conscious fakery for fame and profit, or a sincere believer, deluded no less than his patients in mistaking the power of suggestion for the physical effects of an actual substance. For another, the lines between science and pseudoscience were not so clearly drawn in Mesmer's time. A strong group of rationalists was laboring to free science from speculation, system building, and untestable claims about universal harmonies. But their campaign also demonstrates that all-embracing and speculative systems were still viewed by many scholars as legitimate parts of science in the eighteenth century. Robert Darnton, who has written the best modern book on mesmerism, describes the French intellectual world of the 1780s (*Mesmerism and the End of the Enlightenment in France,* 1968):

They looked out on a world so different from our own that we can hardly perceive it; for our view is blocked by our own cosmologies assimilated, knowingly or not, from the scientists and philosophers of the 19th and 20th centuries. In the 18th century, the view of literate Frenchmen opened upon a splendid, baroque universe, where their gaze rode on waves of invisible fluid into realms of infinite speculation.

Still, whatever the differing boundaries and cultural assumptions, the fact remains that Mesmer based his system on specific claims about fluids, their modes of flow, and their role in causing and curing human disease—claims subject to test by the ordinary procedures of experimental science. The logic of argument has a universality that transcends culture, and late eighteenth century debunking differs in no substantial way from the modern efforts. Indeed, I write this essay because the most celebrated analysis of mesmerism, the report of the Royal Commission of 1784, is a masterpiece of the genre, an enduring testimony to the power and beauty of reason.

Mesmerism became such a craze in the 1780s that many institutions began to worry and retaliate. Conventional medicine, which offered so little in the way of effective treatment, was running scared. Empirical and experimental scientists viewed Mesmer as a throwback to the worst excesses of speculation. People in power feared the irrationalism, the potential for sexual license, the possibility that Mesmer's mass sessions might rupture boundaries between social classes. Moreover, Mesmer had many powerful friends in high circles, and his disturbing ideas might spread by export. (Mesmer counted Lafayette among his most ardent disciples. King Louis XVI asked Lafayette before he departed for America in 1784: "What will Washington think when he learns that you have become Mesmer's chief journeyman apothecary?" Lafayette did proselytize for Mesmer on our shores, although Thomas Jefferson actively opposed him. Lafayette even visited a group of Shakers, thinking that they had discovered a form of mesmerism in their religious dances.)

The mesmeric vogue became sufficiently serious that Louis XVI was persuaded to establish a Royal Commission in 1784 to evaluate the claims of animal magnetism. The commission was surely stacked against Mesmer, but it proceeded with scrupulous fairness and thoroughness. Never in history has such an extraordinary and luminous group been gathered together in the service of rational inquiry by the methods of experimental science. For this reason alone, the *Rapport des commissaires chargés par le roi de l'examen du magnétisme animal* (Report of the Commissioners Charged by the King to Examine Animal Magnetism) is a key document in the history of human reason. It should be rescued from its current obscurity, translated into all languages, and reprinted by organizations dedicated to the unmasking of quackery and the defense of rational thought.

The commissioners included several of France's leading physicians and scientists, but two names stand out: Benjamin Franklin and Antoine Lavoisier. (Franklin served as titular head of the commission, signed the report first, and designed and performed several of the experiments; Lavoisier was the commission's guiding spirit and probably wrote the final report.) The conjunction may strike some readers as odd, but no two men could have been more appropriate or more available. Franklin lived in Paris, as official representative of our newborn nation, from 1776 to 1785. American intellectuals sometimes underestimate Franklin's status, assuming perhaps that we revere him *faute de mieux* and for parochial reasons—and that he was really a pipsqueak and amateur among the big boys of Europe. Not at all. Franklin was a universally respected scholar and a great, world-class scientist in an age when nearly all practitioners were technically amateurs. As the world's leading expert on electricity—a supposed manifestation of Mesmer's universal fluid—Franklin was an obvious choice for the commission. His interest also extended to smaller details, in particular to Mesmer's use of the glass harmonica (Franklin's own invention) as an auxiliary in the precipitation of crises. As for Lavoisier, he ranks as one of the half-dozen greatest scientific geniuses of all time: He wrote with chilling clarity, and he thought with commanding rigor. If the membership contains any odd or ironic conjunction, I would point rather to the inclusion of Dr. Guillotin among the physicians—for Lavoisier would die, ten years later, under the knife that bore the good doctor's name (see Essay 24).

The experimental method is often oversold or promulgated as

the canonical, or even the only, mode of science. As a natural historian, I have often stressed and reported the different approaches used in explaining unique and complex historical events—aspects of the world that cannot be simulated in laboratories or predicted from laws of nature (see my book *Wonderful Life*, 1989). Moreover, the experimental method is fundamentally conservative, not innovative—a set of procedures for evaluating and testing ideas that originate in other ways. Yet, despite these caveats about nonexclusivity and limited range, the experimental method is a tool of unparalleled power in its appropriate (and large) domain.

Lavoisier, Franklin, and colleagues conclusively debunked Mesmer by applying the tools of their experimental craft, tried and true: standardization of complex situations to delineate possible causal factors, repetition of experiments with control and variation, and separation and independent testing of proposed causes. The mesmerists never recovered, and their leader and namesake soon hightailed it out of Paris for good, although he continued to live in adequate luxury, if with reduced fame and prestige, until 1815. Just a year after the commission's report, Thomas Jefferson, replacing Franklin as American representative in Paris, noted in his journal: "animal magnetism dead, ridiculed." (Jefferson was overly optimistic, for irrationalism born of hope never dies; still, the report of Franklin and Lavoisier was probably the key incident that turned the tide of opinion—a subtle fluid far more palpable and powerful than animal magnetism—against Mesmer.)

The commissioners began with a basic proposition to guide their testing: "Animal magnetism might well exist without being useful, but it cannot be useful if it doesn't exist." Yet, any attempt to affirm the existence of animal magnetism faced an intense and immediate frustration: The mesmerists insisted that their subtle fluid had no tangible or measurable attributes. Imagine the chagrin of a group of eminent physical scientists trying to test the existence of a fluid without physical properties! They wrote, with the barely concealed contempt that makes Lavoisier's report both a masterpiece of rhetoric and an exemplar of experimental method (the two are not inconsistent because fair and scrupulous procedures do not demand neutrality, but only strict adherence to the rules of the craft):

It didn't take the Commissioners long to recognize that this fluid escapes all sensation. It is not at all luminous and visible like electricity [the reference, of course, is to lightning before the days of "invisible" flow through modern wires]. Its action is not clearly evident, as the attraction of a magnet. It has no taste, no odor. It works without sound, and surrounds or penetrates you without warning you of its presence. If it exists in us and around us, it does so in an absolutely insensible manner. [All quotations from the commissioners' report are my translations from an original copy in Harvard's Houghton Library.]

The commissioners therefore recognized that they would have to test for the existence of animal magnetism through its effects, not its physical properties. This procedure suggested a focus either on cures or on the immediate (and dramatic) crises supposedly provoked by the flow of magnetism during Mesmer's sessions. The commissioners rejected a test of cures for three obvious and excellent reasons: Cures take too long and time was awasting as the mesmeric craze spread; cures can be caused by many factors, and the supposed effects of magnetism could not be separated from other reasons for recovery; nature, left to her own devices, relieves many ills without any human intervention. (Franklin wryly suspected that an unintended boost to nature lay at the root of Mesmer's successes. His fluid didn't exist, and his sessions produced no physical effect. But patients in his care stayed away from conventional physicians and therefore didn't take the ordinary pills and potions that undoubtedly did more harm than good and impeded natural recovery.) Mesmer, on the other hand, wanted to focus upon cures, and he refused to cooperate with the commission when they would not take his advice. The commission therefore worked in close collaboration with Mesmer's chief disciple, Charles Deslon, who attended the tests and attempted to magnetize objects and people. (Deslon's cooperation indicates that the chief mesmerists were not frauds, but misguided believers in their own system. Mesmer tried to dissociate himself from the commission's findings, arguing that Deslon was a blunderer unable to control the magnetic flux— but all to no avail, and the entire movement suffered from the exposé.)

The commissioners began by trying to magnetize themselves. Once a week, and then for three days in a row (to test a claim that such concentrated time boosted the efficiency of magnetism), they sat for two and a half hours around Deslon's *baquet* in his Paris curing room, faithfully following all the mesmeric rituals. Nobody felt a thing beyond boredom and discomfort. (I am, somehow, greatly taken by the image of these enormously talented and intensely skeptical men sitting around a *baquet*, presumably under their perukes, joined by a rope, each holding an iron rod, and "making from time to time," to quote Lavoisier, "the chain of thumbs." I can picture the scene, as Lavoisier says—Okay boys, ready? One, two, squeeze those thumbs now.)

The commissioners recognized that their own failure scarcely settled the issue, for none was seriously ill (despite Franklin's gout), and Mesmer's technique might only work on sick people with magnetic blockages. Moreover, they acknowledged that their own skepticism might be impeding a receptive state of mind. They therefore tested seven "common" people with assorted complaints and then, in a procedure tied to the social assumptions of the *ancien régime*, seven sufferers from the upper classes, reasoning that people of higher status would be less subject, by their refinement and general superiority, to the power of suggestion. The results supported power of suggestion as the cause of crises, rather than physical effects of a fluid. Only five of fourteen subjects noted any results, and only three—all from the lower classes—experienced anything severe enough to label as a crisis. "Those who belong to a more elevated class, endowed with more light, and more capable of recognizing their sensations, experienced nothing." Interestingly, two commoners who felt nothing—a child and a young retarded woman—might be judged less subject to the power of suggestion, but not less able to experience the flow of a fluid, if it existed.

These preliminaries brought the commissioners to the crux of their experiments. They had proceeded by progressive elimination and concentration on a key remaining issue. They had hoped to test for physical evidence of the fluid itself, but could not and chose instead to concentrate on its supposed effects. They had decided that immediate reactions rather than long-term cures must form the focus of experiments. They had tried the standard techniques on themselves, without result. They had given mesmerists the benefit of all doubt by using the same methods on people with illnesses and inclined to accept the mesmeric system—still without positive results. The investigation now came down to a single question, admirably suited for experimental resolution: The undoubted crises that mesmerists could induce might be caused by one of two factors (or perhaps both)—the psychological power of suggestion or the physical action of a fluid.

The experimental method demands that the two possible causes be separated in controlled situations. People must be subjected to the power of suggestion but not magnetized, and then magnetized but not subject to suggestion. These separations demanded a bit of honorable duplicity from the commissioners—for they needed to tell people that nonmagnetized objects were really full of mesmeric fluid (suggestion without physical cause), and then magnetize people without letting them know (physical cause without suggestion).

In a clever series of experiments, designed mainly by Lavoisier and carried out at Franklin's home in Passy, the commissioners made the necessary separations and achieved a result as clear as any in the history of debunking: Crises are caused by suggestion; not a shred of evidence exists for any fluid, and animal magnetism, as a physical force, must be firmly rejected.

For the separation of suggestion from magnetism, Franklin asked Deslon to magnetize one of five trees in his garden. A young man, certified by Deslon as particularly sensitive to magnetism, was led to embrace each tree in turn, but not told about the smoking gun. He reported increasing strength of magnetization in each successive tree and finally fell unconscious in a classic mesmeric crisis before the fourth tree. Only the fifth, however, had been magnetized by Deslon! Mesmerists rejected the result, arguing that all trees have some natural magnetization anyway, and that Deslon's presence in the garden might have enhanced the effect. But Lavoisier replied scornfully:

> But then, a person sensitive to magnetization would not be able to chance a walk in a garden without the risk of suffering convulsions, and such an assertion is therefore denied by ordinary, everyday experience.

Nevertheless, the commissioners persisted with several other experiments, all leading to the same conclusion—that suggestion without magnetism could easily produce full-scale mesmeric crises. They blindfolded a woman and told her that Deslon was in the room, filling her with magnetism. He was nowhere near, but the woman had a classic crisis. They then tested the patient without a blindfold, telling her that Deslon was in the next room directing the fluid at her. He was not, but she had a crisis. In both cases, the woman was not magnetized or even touched, but her crises were intense.

Lavoisier conducted another experiment at his home in the Arsenal (where he worked as Commissioner of Gunpowder, having helped America's revolution with matériel, as much as Lafayette had aided with men). Several porcelain cups were filled with water, one supposedly strongly magnetized. A particularly sensitive woman who, in anticipation, had already experienced a crisis in Lavoisier's antechamber, received each cup in turn. She began to quiver after touching the second cup and fell into a full crisis upon receiving the fourth. When she recovered and asked for a cup of water, the foxy Lavoisier finally passed her the magnetized liquid. This time, she not only held, but actually imbibed, although "she drank tranquilly and said that she felt relieved."

The commissioners then proceeded to the reverse test of magnetizing without unleashing the power of suggestion. They removed the door between two rooms at Franklin's home and replaced it with a paper partition (offering no bar at all, according to Deslon, to the flow of mesmeric fluid). They induced a young seamstress, a woman with particularly acute sensitivity to magnetism, to sit next to the partition. From the other side, but unknown to the seamstress, an adept magnetizer tried for half an hour to fill her with fluid and induce a crisis, but "during all this time, Miss B . . . made gay conversation; asked about her health, she freely answered that she felt very well." Yet, when the magnetizer entered the room, and his presence became known (while acting from an equal or greater distance), the seamstress began to convulse after three minutes and fell into a full crisis in twelve minutes.

The evident finding, after so many conclusive experiments—that no evidence exists for Mesmer's fluid and that all noted effects may be attributed to the power of imagination—seems almost anticlimactic, and the commissioners offered their result with clarity and brevity: "The practice of magnetization is the art of increasing the imagination by degrees." Lavoisier then ended the report with a brilliant analysis of the reasons for such frequent vogues of irrationalism throughout human history. He cited two major causes, or predisposing factors of the human mind and heart. First, our brains just don't seem to be well equipped for reasoning by probability. Fads find their most fertile ground in subjects, like the curing of disease, that require a separation of many potential causes and an assessment of probability in judging the value of a result:

> The art of concluding from experience and observation consists in evaluating probabilities, in estimating if they are high or numerous enough to constitute proof. This type of calculation is more complicated and more difficult than one might think. It demands a great sagacity generally above the power of common people. The success of charlatans, sorcerers, and alchemists—and all those who abuse public credulity—is founded on errors in this type of calculation.

I would alter only Lavoisier's patrician assumption that ordinary folks cannot master this mode of reasoning—and write instead that most people surely can but, thanks to poor education and lack of encouragement from general culture, do not. The end result is the same—riches for Las Vegas and disappointment for Pete Rose. But at least the modern view does not condemn us to a permanent and inevitable status as saps, dupes, and dunces.

Second, whatever our powers of abstract reasoning, we are also prisoners of our hopes. So long as life remains disappointing and cruel for so many people, we shall be prey to irrationalisms that promise relief. Lavoisier regarded his countrymen as more sophisticated than previous suckers of centuries past, but still victims of increasingly sly manipulators (nothing has changed today, as the Gellers and von Danikens remain one step ahead of their ever-gullible disciples):

> This theory [mesmerism] is presented today with the more imposing apparatus [I presume that Lavoisier means both ideas and contraptions] necessary in our more enlightened

century—but it is no less false. Man seizes, abandons, but then commits again the errors that flatter him.

Since hope is an ever-present temptress in a world of woe, mesmerism "attracts people by the two hopes that touch them the most: that of knowing the future and that of prolonging their days."

Lavoisier then drew an apt parallel between the communal crises of mesmeric sessions and the mass emotionalism so often exploited by demagogues and conquerors throughout history—*"l'enthousiasme du courage"* (enthusiasm of courage) or *"l'unité d'ivresse"* (unity of intoxication). Generals elicit this behavior by sounding drums and playing bugles; promoters by hiring a claque to begin and direct the applause after performances; demagogues by manipulating the mob.

Lavoisier's social theory offered no solution to the destructive force of irrationalism beyond a firm and continuing hegemony of the educated elite. (As my one criticism of the commissioners' report, Lavoisier and colleagues could see absolutely nothing salutary, in any conceivable form, in the strong emotionalism of a mesmeric crisis. They did not doubt the power of the psyche to cure, but as sons of the Enlightenment, children of the Age of Reason, they proclaimed that only a state of calm and cheerfulness could convey any emotional benefit to the afflicted. In this restriction, they missed an important theme of human complexity and failed to grasp the potential healing effect of many phenomena that call upon the wilder emotions—from speaking in tongues to catharsis in theatrical performance to aspects of Freudian psychoanalysis. In this sense, some Freudians view Mesmer as a worthy precursor with a key insight into human nature. I hesitate to confer such status upon a man who attained great wealth from something close to quackery—but I see the point.)

I envision no easy solution either, but I adopt a less pessimistic attitude than Lavoisier. Human nature is flexible enough to avert the baleful effects of intoxicated unity, and history shows that revolutionary enthusiasm need not devolve into hatred and mass murder. Consider Franklin and Lavoisier one last time. Our revolution remained in the rational hands of numerous Franklins, Jeffersons, and Washingtons; France descended from the Declaration of the Rights of Man into the Reign of Terror. (I do recognize the different situations, particularly the greater debt of hatred, based on longer and deeper oppression, necessarily discharged by the new rulers of France. Still, no inevitability attended the excesses fanned by mass emotionalism.) In other words:

> Antoine Lavoisier
> Lost his head
> Benjamin Franklin
> Died in bed.

From which, I think, we can only conclude that Mr. Franklin understood a thing or two when he remarked, speaking of his fellow patriots, but extended here to all devotees of reason, that we must either hang together or hang separately.

# 5 | Art and Science

# 13 | Madame Jeanette

THIRTY YEARS AGO, on April 30, 1958, to be exact, I sat with 250 students facing one of the most formidable men of our generation—Peter J. Wilhousky, director of music in the New York City schools and conductor of the New York All-City High School Chorus. As the warm, and primarily parental, applause receded at the concert's end, Wilhousky returned to the podium of Carnegie Hall, gestured for silence, and raised his baton to conduct the traditional encore, "Madame Jeanette." Halfway through, he turned and, without missing a beat (to invoke a cliché in its appropriate, literal sense), smiled to acknowledge the chorus alumni who stood at their seats or surrounded the podium, singing with their current counterparts. These former members seemed so ancient to me—though none had passed forty, for the chorus itself was then only twenty years old—and their solidarity moved me to a rare fit of tears at a time when teenage boys did not cry in public.

"Madame Jeanette" is a dangerous little piece, for it ventures so near the edge of cloying sentimentality. It tells the tale, in close four-part a cappella harmony, of a French widow who sits at her door by day and at her window by night. There she thinks only of her husband, killed so many years before on the battlefield of St. Pierre, and dreams of the day that they will be reunited at the cemetery of Père Lachaise. With 250 teenagers and sloppy conducting, "Madame Jeanette" becomes a maudlin and embarrassing tearfest. Wilhousky, ever the perfectionist, ever the rationalist, somehow steered to the right side of musicality, and ended each concert with integrity and control.

"Madame Jeanette" was our symbol of continuity. For a very insecure boy, singing second bass on the brink of manhood, "Madame Jeanette" offered another wonderful solace. It ends, for the basses, on a low D-flat, just about as far down the scale as any composer would dare ask a singer to venture. Yes, I knew even then that low did not mean masculine, or capable, or mature, or virile—but that fundament resonated with hope and possibility, even in pianissimo.

Len and I met at the bus stop every Saturday morning at 7:30, took the Q17 to 169th Street and the subway to Lexington Avenue, walked uptown along the line of the old Third Avenue El, and arrived at Julia Richman High School just in time for the 9 A.M. rehearsal.

We lived, thirty years ago, in an age of readier obedience, but I still marvel at the discipline that Wilhousky could maintain with his mixture of awe (inspired) and terror (promulgated). He forged our group of blacks from Harlem, Puerto Ricans from the great migration then in progress, Jews from Queens, and Italians from Staten Island into a responsive singing machine. He worked, in part, through intimidation by public ridicule. One day, he stopped the rehearsal and pointed to the tenor section, saying: "You, third row, fourth seat, stand up. You're singing flat. Ten years ago, Julius La Rosa sat in that same seat—and sang flat. And he's still singing flat." (Memory is a curious trickster. La Rosa, in a recent *New Yorker* profile, states that Wilhousky praised him in the same forum for singing so true to pitch. But I know what I heard. Or is the joke on me?) Each year, he cashiered a member or two for talking or giggling—in public, and with no hope of mercy or reinstatement.

But Peter Wilhousky had another side that inspired us all and conveyed the most important lesson of intellectual life. He was one of the finest choral conductors in America, yet he chose to spend every Saturday morning with high school kids. His only rule, tacit but pervasive, proclaimed: "No compromises." We could sing, with proper training and practice, as well as any group in America—nothing else would be tolerated or even conceptualized. Anything less would not be worth doing at all. I had encountered friendliness, grace, kindness, animation, clarity, and dedication among my teachers, but I had never even considered the notion that unqualified excellence could emerge from anything touched or made by students. The idea, however, is infectious. As I worked with Wilhousky, I slowly personalized the dream that excellence in one activity might be extended to become the pattern, or at least the goal, of an actual life.

Len phoned me a few months ago and suggested that we attend this year's concert, the thirtieth since our valedictory. I hesitated for two reasons. I feared that my memory of excellence would not be supported by reality, and I didn't relish the role of a graybeard from springs long past, standing and singing "Madame Jeanette" from the audience, should that peculiar tradition still be honored. But sentiment and curiosity prevailed, and we went.

Yes, Heraclitus, you cannot step twice into the same river. The raw material remains—talented kids of all colors, shapes, backgrounds. But the goal has been inverted. Wilhousky tried to mold all this diversity into the uncompromising, single standard of elite culture as expressed in the classical repertory for chorus and orchestra. In the auditorium of Julia Richman High School, before his arrival, we used to form small pickup groups to sing the latest rock-and-roll numbers. But when our sentinels spotted the maestro, they quickly spread the alarm and dead silence descended. Wilhousky claimed that rock-and-roll encouraged poor habits of voice and pitch, and he would expel anyone caught singing the stuff in his bailiwick.

Diversity has now triumphed, and the forbidden fruit of our era has become the entire first part of the program. The concert began with the All-City marching band, complete with drum major, baton twirlers, and flag carriers. Then the All-City jazz ensemble.

A full concert and two hours later, the orchestra and chorus finally received their turn. Not only has the number of ensembles expanded to respect the diversity of tastes and inclinations in our polyglot city, but each group has also retained a distinctive signature. Blacks predominate in the chorus; the string sections of the orchestra are overwhelmingly Asian. The chorus is now led by Edith Del Valle, a tall, stunning woman who heads the vocal department at Fiorello H. La Guardia High School of the Arts. (As a single sign of continuity, Anna Ext still coaches the sopranos, as she did in our day and has for thirty-two years. How can we convey adequate praise to a woman who has devoted so much, for so long, to a voluntary, weekend organization—except to say that our language contains no word more noble than "teacher"?)

The chorus still sings the same basic repertory—Randall Thompson's "Alleluia," Wilhousky's own arrangement of "The Battle Hymn of the Republic," some Bach and Beethoven, and an Irving Berlin medley for the season of his centennial.

How good are they, and how good were we? Was Wilhousky's insistence on full professionalism just a vain conceit? They sing by memory, and therefore (since eyes can be fixed on the conductor), with uncanny precision and unanimity. But I demur for two reasons. First, the sound, though lovely in raw quality, is so emotionless, as though text and style of composition have no influence upon interpretation. Perhaps we sang in the same manner. The soul of these classics may not be accessible before the legal age of drinking, driving, and voting.

But my second reservation troubles me more. The chorus is terribly unbalanced, with 129 women and only 31 men. The tenors are reduced to astringent shouting as the evening wears on. This cannot be by design, and can only mean that the chorus is not attracting anywhere near the requisite number of male applicants. Thirteen of the 31 men hail from the conductor's own specialty school, La Guardia High. Have they been pressed into desperate service? In our chorus, all sections were balanced. We clamored in our local high schools for the strictly limited right to audition, and fewer than half the applicants succeeded.

I mused upon these inadequacies as the evening wore on (and the tenors tired). The expanded diversity of bands and jazz is both exciting and a proper testimony to cultural pluralism. The relaxed attitude of performers contrasts pleasantly with the rigid formalism and nervousness of our era (I could have died in a spectacular backward plunge off the top riser of Carnegie Hall when I felt the chair's rearward creep, but didn't dare stop to fidget and readjust).

But has the evening's diversity and spontaneous joy pushed aside Wilhousky's uncompromising excellence? Can the two ideals, each so important in itself, coexist at all? And if not, whatever shall we do to keep alive that harsh vision of the best of the greatest?

But if I felt this single trouble amidst my pleasure, at least I wouldn't have to worry about "Madame Jeanette" in this new river. Surely, that tradition had evaporated, and I would not have to face brightness and acne from the depths of advancing middle age in the fifteenth row. After all, "Madame Jeanette" is a quiet classical piece for chorus alone—and the chorus no longer holds pride of place among the various ensembles.

I applauded warmly after the finale, pleasure only slightly tinged with a conceptual sort of sadness, and then turned to leave. But Edith Del Valle strode out from the wings and, with a presence fully equal to Wilhousky's, stepped onto the podium—to conduct "Madame Jeanette." Old members scurried to the front. Len and I looked at each other and, without exchanging a word, rose in unison.

No tears. We are both still terrified of Wilhousky's wrath, and his ghost surely stood on that stage, watching carefully for any sign of inattention or departure from pitch. This time, the chorus sang exquisitely, for "Madame Jeanette" succeeds by precision or fails by overinvolvement. The imbalance of sections does not affect such a quiet song, while its honest, but simple, sentimentality can be encompassed by the high-school soul.

Edith Del Valle, the black woman from La Guardia High, blended with her absolute opposite, the silver-haired Slavic aristocrat, Peter J. Wilhousky. The discipline and precision of her chorus—their species of excellence—had triumphed to convert the potentially maudlin into thoughtful dignity for tradition's sake. It was a pleasure to make music with her. If youth and age can produce such harmony, there must be hope for pluralism *and* excellence—but only if we can recover, and fully embrace, Wilhousky's dictum: No compromises.

I learned something else at this final celebration of continuity, something every bit as important to me, if only parochially: I can still hit that low D-flat. Father Lachaise may be beckoning, but "Madame Jeanette" and I are still hanging tough and young in our separate ways.

---

## Postscript

This essay, which first appeared in the *New York Times Magazine*, unleashed a flood of reminiscence by correspondence, mostly from former chorus members and others who knew Peter Wilhousky. I was regaled with many sweet memories, particularly of our custom in jamming subway cars after leaving the rehearsals *en masse* and singing (generally to the keen surprise and enjoyment of passengers) until the accumulating departures of homebound choristers reduced our ranks to less than four-part harmony. But one theme, in its several guises, pervaded all the letters and reinforced the serious, and decidedly nonsentimental, *raison d'être* of this essay—Wilhousky's commitment to excellence and its impact upon us. One woman wrote from a generation before mine:

> Mr. Wilhousky was my music teacher and mentor 55 years ago when I was a student at New Utrecht High in Brooklyn. We had an outstanding choir that won every competition in my four years at the school. How we adored and esteemed this wonderful man who by the way we were sure was a prince: so handsome and aristocratic. He was then, as well as you say later, a stickler for seriousness, discipline, and dedication to our work. He encouraged those of us with some talent to continue our studies and many of us did.

Another who sang in the chorus five years before me said:

> What memories you stirred for me, and brought forth some tears too. Only another choir member could share how special those rehearsals and concerts were. Just to be chosen to audition was an honor. . . . Madame Jeanette is turning around in my head now. I recall teaching the bass part to my kid brother so that we could sing. I've taught it to my husband and kids too. I was in awe of Peter J. Wilhousky. Discipline was never a problem in this group. How we loved to sing!

And from ten years after my watch:

> Today I am a professional singer in Philadelphia, having sung with umpteen college groups, choruses, community theaters, opera workshops, etc., but nothing will ever match that full-bodied enthusiastic blend of voices I remember now so well. My children poked gentle fun at me today as I

waxed enthusiastic over your story and they listened to
INXS on their Walkmans as I hummed Madame Jeanette
over and over again.

And finally, from a Wilhousky counterpart in Portland, Oregon:
"The taste of excellence is the hook. The kids never forget—as
you obviously have not."

This accumulated weight of testimony made me reassess the
tone of the essay itself. I now think that I was a bit too ecumeni-
cally forgiving of the chorus's present insufficiencies. We probably were very good (if not quite so subtle and professional as
clouds of memory suggest); in any case, the ideal of uncompromising excellence certainly pervaded our concepts and did pass
down into our subsequent lives. I don't see how the present
chorus can be engendering such an attitude with an appeal so
feeble that male singers must be dredged up rather than turned
away after dreaming, scheming, and begging for a chance (as we
did). This is simply too great a loss for any gain in diversity or
relaxation. Islands of excellence are too rare and precious in our
world of mediocrity; any erosion and foundering is tragic.

Finally, for I really do not wish to end a sweet story on a sour
note, I report Julius La Rosa's version of his incident with
Wilhousky. He writes in a letter of November 17, 1988, that the
chorus was rehearsing "Begin the Beguine" (during his tenure in
the late 1940s). Wilhousky wanted the men to sing with a cello-
like tone. La Rosa writes:

> I swear to you, I can still see him holding an imaginary cello,
> his left hand on the neck, fingers pressing down on the
> strings and vibrating to achieve the desired tremolo. But we
> weren't getting it so he told us to stand up, *individually,* and
> sing the phrase. My turn. I sang it. He asked me to do it
> again, then exclaimed, "That's it!" And all I remember after
> that was walking back to the subway with Jeanette feeling
> seven feet tall.

La Rosa was also gracious enough to add: "And yes, though time
does distort the memory, I wouldn't be surprised if I was flat the
day he, Mr. Wilhousky, singled me out. I was terrified—and probably didn't take a good deep breath!!" (Actually, I don't doubt for

a moment that both La Rosa's version and what I heard ten years later from Wilhousky are entirely accurate. We scarcely need *Rashomon* to teach us that rich events are remembered for different parts and different emphases—so that equally accurate, but partial versions yield almost contradictory impressions.)

And if La Rosa's reference to Jeanette (*not* Madame) puzzled you, let me close with his ultimate touché from earlier in his letter:

> All-City Chorus was an enchantment. . . . And lucky, too, I was 'cause I could walk from the subway along Third Avenue with Jeanette—Yes! Jeanette Caponegro, second alto—while you were stuck with Len!

# 14 | Red Wings in the Sunset

TEDDY ROOSEVELT borrowed an African proverb to construct his motto: Speak softly, but carry a big stick. In 1912, a critic turned Roosevelt's phrase against him, castigating the old Roughrider for trying to demolish an opponent by rhetoric alone: "Ridicule is a powerful weapon and the temptation to use it unsparingly is a strong one. . . . Even if we don't agree with him [Roosevelt's opponent], it is not necessary either to cut him into little pieces or to break every bone in his body with the 'big stick.' "

This criticism appeared in the midst of Roosevelt's presidential campaign (when he split the Republican party by trying to wrest the nomination from William Howard Taft, then formed his own Progressive, or Bull Moose, party to contest the election, thereby scattering the Republican vote and bringing victory to Democrat Woodrow Wilson). Surely, therefore, the statement must record one of Roosevelt's innumerable squabbles during a tough political year. It does not. Francis H. Allen published these words in an ornithological journal, *The Auk.* He was writing about flamingos.

When, as a cynical and posturing teenager, I visited Mount Rushmore, I gazed with some approval at the giant busts of Washington, Jefferson, and Lincoln, and then asked as so many others have—what in hell is Teddy Roosevelt doing up there? Never again shall I question his inclusion, for I have just discovered something sufficiently remarkable to warrant a sixty-foot stone likeness all by itself. In 1911, an ex-president of the United States, after seven exhausting years in office, and in the throes of preparing his political comeback, found time to write and publish a technical scientific article, more than one hundred pages long: "Revealing and Concealing Coloration in Birds and Mammals."

Roosevelt wrote his article to demolish a theory proposed by the artist-naturalist Abbott H. Thayer (and defended by Mr. Allen, who castigated Roosevelt for bringing the rough language of politics into a scientific debate). In 1896 Thayer, as I shall document in a moment, correctly elucidated the important principle of countershading (a common adaptation that confers near invisibility upon predators or prey). But he then followed a common path to perdition by slowly extending his valid theory to a doctrine of exclusivity. By 1909, Thayer was claiming that *all* animal colors, from the peacock's tail to the baboon's rump, worked primarily for concealment. As a backbreaking straw that sealed his fate and inspired Roosevelt's wrath, Thayer actually argued that natural selection made flamingos red, all the better to mimic the sunset. In the book that will stand forever as a monument to folly, to cockeyed genius, and to inspiration gone askew, Thayer stated in 1909 (in *Concealing-Coloration in the Animal Kingdom,* written largely by his son Gerald H. Thayer and published by Macmillan):

> These traditionally "showy" birds are, at their most critical moments, perfectly "obliterated" by their coloration. Conspicuous in most cases, when looked at from above, as man is apt to see them, they are wonderfully fitted for "vanishment" against the flushed, rich-colored skies of early morning and evening.

Roosevelt responded with characteristic vigor in his 1911 article:

> Among all the wild absurdities to which Mr. Thayer has committed himself, probably the wildest is his theory that flamingos are concealingly colored because their foes mistake them for sunsets. He has never studied flamingos in their haunts, he knows nothing personally of their habits or their enemies or their ways of avoiding their enemies . . . and certainly has never read anything to justify his suppositions; these suppositions represent nothing but pure guesswork, and even to call them guesswork is a little over-

conservative, for they come nearer to the obscure mental processes which are responsible for dreams.

Roosevelt's critique (and many others equally trenchant) sealed poor Thayer's fate. In 1896, Thayer had begun his campaign with praise, promise, and panache (his outdoor demonstrations of disappearing decoys became legendary). He faced the dawn of World War I in despair and dejection (though the war itself brought limited vindication as our armies used his valid ideas in theories of camouflage). He lamented to a friend that his avocation (defending his theory of concealing coloration) had sapped his career:

> Never . . . have I felt less a painter . . . I am like a man to whom is born, willy nilly, a child whose growth demanded his energies, he the while always dreaming that this growing offspring would soon go forth to seek his fortune and leave him to his profession, but the offspring again and again either unfolding some new faculties that must be nurtured and watched, or coming home and bursting into his parent's studio, bleeding and bruised by an insulted world, continued to need attention so that there was nothing for it but to lay down the brush and take him once more into one's lap.

I must end this preface to my essay with a confession. I have known about Thayer's "crazy" flamingo theory all my professional life—and for a particular reason. It is the standard example always used by professors in introductory courses to illustrate illogic and unreason, and dismissed in a sentence with the ultimate weapon of intellectual nastiness—ridicule that forecloses understanding. When I began my research for this essay, I thought that I would write about absurdity, another comment on unthinking adaptationism. But my reading unleashed a cascade of discovery, leading me to Roosevelt and, more importantly, to the real Abbott Thayer, shorn of his symbolic burden. The flamingo theory is, of course, absurd—that will not change. But how and why did Thayer get there from an excellent start that the standard dismissive anecdote, Thayer's unfortunate historical legacy, never acknowledges? The full story, if we try to understand Abbott Thayer aright, contains lessons that will more than compensate for laughter lost.

Who was Abbott Handerson Thayer anyway? I had always assumed, from the name alone, that he was an eccentric Yankee who used wealth and social postion to gain a hearing for his absurd ideas. I could find nothing about him in the several scientific books that cite the flamingo story. I was about to give up when I located his name in the *Encyclopedia Britannica.* I found, to my astonishment, that Abbott Thayer was one of the most famous painters of late nineteenth century America (and an old Yankee to be sure, but not of the wealthy line of Thayers—see the biography by Nelson G. White, *Abbott H. Thayer: Painter and Naturalist*). He specialized in ethereal women, crowned with suggestions of halos and accompanied by quintessentially innocent children. Art and science are both beset by fleeting tastes that wear poorly—far be it for me to judge. I had begun to uncover a human drama under the old pedagogical caricature.

But let us begin, as they say, at the beginning. Standard accounts of the adaptive value of animal colors use three categories to classify nature's useful patterns (no one has substantially improved upon the fine classic by Hugh B. Cott, *Adaptive Coloration in Animals,* 1940). According to Cott, adaptive colors and patterns may serve as (1) concealment (to shield an animal from predators or to hide the predator in nature's never-ending game); (2) advertisement, to scare potential predators (as in the prominent false eyespots of so many insects), to maintain territory or social position, or to announce sexual receptivity (as in baboon rump patches); and (3) disguise, as animals mimic unpalatable creatures to gain protection, or resemble an inanimate (and inedible) object (numerous leaf and stick insects, or a bittern, motionless and gazing skyward, lost amidst the reeds). Since disguise lies closer to advertisement than to concealment (a disguised animal does not try to look inconspicuous, but merely like something else), we can immediately appreciate Abbott Thayer's difficulty. He wanted to reduce all three categories to the single purpose of concealment—but fully two-thirds of all color patterns, in conventional accounts, serve the opposite function of increased visibility.

Abbott Thayer, a native of Boston, began his artistic career in the maelstrom of New York City but eventually retreated to a hermitlike existence in rural New Hampshire, where his old interests in natural history revived and deepened. As a committed Darwinian, he believed that all form and pattern must serve some crucial purpose in the unremitting struggle for existence. He also felt that, as a painter, he could interpret the colors of animals in ways and terms unknown to scientists. In 1896, Thayer published his first, landmark article in *The Auk:* "The Law Which Underlies Protective Coloration."

Of course, naturalists had recognized for centuries that many animals blend into their background and become virtually invisible—but scientists had not properly recognized how and why. They tended to think, naively (as I confess I did before my research for this essay), that protection emerged from simple matching between animal and background. But Thayer correctly identified the primary method of concealment as countershading—a device that makes creatures look flat. Animals must indeed share the right color and pattern with their background, but their ghostly disappearance records a loss of dimensionality, not just a matching of color.

In countershading, an animal's colors are precisely graded to counteract the effects of sunlight and shadow. Countershaded animals are darkest on top, where most sunlight falls, and lightest on the bottom (Thayer thereby identified the adaptive significance of light bellies—perhaps the most universal feature of animal coloration). The precise reversal between intensity of coloration and intensity of illumination neatly cancels out all shadow and produces a uniform color from top to bottom. As a result, the animal becomes flat, perfectly two-dimensional, and cannot be seen by observers who have, all their lives, perceived the substantiality of objects by shadow and shading. Artists have struggled for centuries to produce the illusion of depth and roundness on a flat canvas; nature has simply done the opposite—she shades in reverse in order to produce an illusion of flatness in a three-dimensional world.

Contrasting his novel principle of countershading with older ideas about mimicry, Thayer wrote in his original statement of 1896: "Mimicry makes an animal appear to be some other thing, whereas the newly discovered law makes him cease to appear to exist at all."

Thayer, intoxicated with the joy of discovery, attributed his success to his chosen profession and advanced a strong argument about the dangers of specialization and the particular value of "outsiders" to any field of study. He wrote in 1903: "Nature has evolved actual art on the bodies of animals, and only an artist can read it." And later, in his 1909 book, but now with the defensiveness and pugnacity that marked his retreat:

> The entire matter has been in the hands of the wrong custodians. . . . It properly belongs to the realm of pictorial art, and can be interpreted only by painters. For it deals wholly in optical illusion, and this is the very gist of a painter's life. He is born with a sense of it; and, from his cradle to his grave, his eyes, wherever they turn, are unceasingly at work on it—and his pictures live by it. What wonder, then, if it was for him alone to discover that the very art he practices is at full—beyond the most delicate precision of human powers—on almost all animals.

So far, so good. Thayer's first articles and outdoor demonstrations won praise from scientists. He began with relatively modest claims, arguing that he had elucidated the basis for a major principle of concealment but not denying that other patterns of color displayed quite different selective value. Initially, he accepted the other two traditional categories—revealing coloration and mimicry—though he always argued that concealment would gain a far bigger scope than previously admitted. In his most technical paper, published in the *Transactions of the Entomological Society of London* (1903), and introduced favorably by the great English Darwinian E. B. Poulton, Thayer wrote:

> Every possible form of advantageous adaptation must somewhere exist. . . . There must be unpalatability accompanied by warning coloration . . . and equally plain that there must be mimicry.

Indeed, Thayer sought ways to combine ideas of concealment with other categories that he would later deny. He supported, for example, the ingenious speculation of C. Hart Merriam that white rump patches are normally revealing, but that their true value lies in a deer's ability to "erase" the color at moments of danger—a deer "closes down" the patch by lowering its tail over the white blotch and then disappears, invisible, into the forest. In his 1909 book, however, Thayer explicitly repudiated this earlier interpretation and argued for pure concealment—the white patch as "sky mimicking" when seen from below.

Thayer's pathway from insight to ridicule followed a distressingly common route among intellectuals. Countershading for concealment, amidst a host of alternatives, was not enough. Thayer had to have it all. Little by little, plausibly at first, but grading slowly to red wings in the sunset, Thayer laid his battle plans (not an inappropriate metaphor for a father of camouflage). As article succeeded article, Thayer progressively invaded the categories of mimicry and revealing coloration to gain, or so he thought, more cases for concealment. Finally, nothing else remained: *All* patterns of color served to conceal. He wrote in his book: "All patterns and colors whatsoever of all animals that ever prey or are preyed upon are under certain normal circumstances obliterative."

Thayer made his first fateful step in his technical article of 1903. Here, he claimed a second major category of concealing coloration—what he called "ruptive" (we now call them "disruptive") bars, stripes, splotches, and other assorted markings. Disruptive markings make an animal "disappear" by a route different from countershading. They break an animal's coherent outline and produce an insubstantial array of curious and unrelated patches (this principle, more than countershading, became important in military camouflage). A zebra, Thayer argues, does not mimic the reeds in which it hides; rather, the stripes break the animal's outline into bars of light and darkness—and predators see no coherent prey at all.

Again, Thayer had proposed a good idea for some, even many, cases (though not for zebras, who rarely venture into fields of reeds). His 1903 article argues primarily that butterflies carry disruptive pictures of flowers and background scenery upon their wings: "The general aspect of each animal's environment," Thayer wrote, "is found painted upon his coat, in such a way as to minimize his visibility, by making the beholder think he sees *through* him."

But, amidst his good suggestions, Thayer had made his first overextended argument. Countershading could scarcely be mistaken for anything else and offered little scope for claiming too much. But the principle of "ruptive" concealment permitted enormous scope for encompassing other patterns that actually serve to reveal or mimic. Color patches and splotches—the classic domain of warning and revealing patterns (consider the peacock's tail)—could, for an overenthusiast like Thayer, become marks of disruptive concealment. Thus, to cite just one example of overstatement, Thayer argued in a 1909 article, adversarily entitled "An Arraignment of the Theories of Mimicry and Warning Colors," that white patches on a skunk's head mimic the sky when seen by mice from below:

> Such . . . victims as can see would certainly have much more chance to escape were not what would be a dark-looming predator's head converted, by its white sky-counterfeiting, into a deceptive imitation of mere sky.

Still, by 1903, Thayer was not yet ready to claim concealment for all colors. He admitted one category of obvious conspicuousness: "Only unshiny, bright monochrome is intrinsically a revealing coloration."

Now we can finally understand why Thayer was eventually driven to his absurd argument about flamingos and the sunset. (Divorced from the context of Thayer's own personal development, the idea sounds like simple disembodied craziness—as professors always present it for laughs in introductory classes.) Once Thayer decided to go for broke, and to claim that all color works for concealment, flamingos became his crucial test, his do-or-die attempt at exclusivity. As a last shackle before the final plunge, Thayer had admitted that stark monochromes—animals painted throughout with one showy color—were "intrinsically revealing." If he could now show that such monochromes also served for concealment, then his triumph would be complete. Flamingos occupied the center of his daring, not a curious diversion. He had to find a way to fade bright red into ethereal nothingness. Hence the sunset—his as well as the flamingo's.

So Thayer visited the West Indies, got down on his belly in the sulfurous muds, and looked at flamingos—not comfortably down from above (as he always accused lazy and uncritical zoologists of doing), but from the side as might a slithering anaconda or a hungry alligator. And he saw red wings fading into the sunset—the entire feeding flock became a pink cloud, a "sky-matching costume":

> These birds are largely nocturnal, so that the only sky bright enough to show any color upon them is the more or less rosy and golden one that surrounds them from sunset till dark and from dawn until soon after sunrise. They commonly feed in immense, open lagoons, wading in vast phalanxes, while the entire real sky above them and its reflected duplicate below them constitute either one vast hollow sphere of gold, rose, and salmon, or at least glow, on one side or the other, with these tones. Their whole plumage is a most exquisite duplicate of these scenes. . . . This flamingo, having at his feeding time so nearly only sunrise colors to match, wears, as he does, a wonderful imitation of them.

Thayer had finally gone too far and exasperated even his erstwhile supporters. His exaggerations—particularly his flamingos—now brought down a storm of criticism, including Roosevelt's hundred-page barrage. Critics pointed out Thayer's errors in every particular: Flamingos do not concentrate their feeding at dawn and dusk, but are active all day; anacondas and alligators do not inhabit the thin films of saline ponds that flamingos favor; flamingos eat by filtering tiny eyeless animals that cannot enjoy the visual pleasures of sunset.

Most sadly, Thayer's argument even failed in its own terms—and Thayer, who was overenthusiastic to a fault, but neither dishonest nor dishonorable, had to confess. Any object viewed *against* the fading light will appear dark, whatever its actual color. Thayer admitted this explicitly by painting a dark palm tree against the sunset in his infamous and fanciful painting of fading flamingos (reproduced here, for unfortunate practical reasons, in inappropriate black and white). Thus, he could only claim that flamingos looked like the sunset in the *opposite* side of the sky: red clouds of sunset in the west, red masses of flamingos in the east.

Would any animal be so confused by two "sunsets," with flamingos showing dark against the real McCoy? Thayer admitted in his 1909 book:

> Of course a flamingo seen against dawn or evening sky would look dark, like the palm in the lower left-hand figure, no matter what his colors were. The . . . right-hand figures, then, represent the lighted sides of flamingos at morning or evening, and show how closely these tend to reproduce the sky of this time of day; although always, of course, *in the*

White (top) and red (bottom) flamingos fade to invisibility against the sky at sunrise and sunset. From Thayer's 1909 book.

*opposite quarter of the heavens* [Thayer was good enough to underline his admission] from the sunset or dawn itself.

As a final, and feeble, parting salvo, Thayer added: "but the rosy hues very commonly suffuse both sides of the sky, so that . . . the flamingos' illuminated ruddy color very often has a true 'background' of illuminated ruddy sky."

Teddy Roosevelt was particularly perturbed. As an old big-game hunter, he knew that most of Thayer's "ruptive" patterns did not conceal quarries. How could Thayer have it both ways—how could a lion be concealed in the desert, a zebra amongst the reeds when, in fact, they share the same habitat, often to the zebra's fatal disadvantage? Thus, Roosevelt decided to counterattack and wrote his scientific *magnum opus* during some spare time amidst other chores. He saved his best invective for the poor flamingos. Writing on February 2, 1911, to University of California biologist Charles Kofoid, he stated:

> [Thayer's] book shows such a fantastic quality of mind on his part that it is a matter of very real surprise to me that any scientific observer . . . no matter how much credit he may give to Mr. Thayer for certain discoveries and theories, should fail to enter the most emphatic protest against the utter looseness and wildness of his theorizing. Think of being seriously required to consider the theory that flamingos are colored red so that fishes (or oysters for that matter—there is no absurdity of which Mr. Thayer could not be capable) would mistake them for the sunset!

The debate between Roosevelt and Thayer developed into an interesting discussion of scientific methodology, not merely some rhetorical sniping about specifics. To grasp Roosevelt's primarily methodological (and cogent) objections to Thayer's work, consider Thayer's most remarkable painting of all—the frontispiece to his 1909 book, showing a peacock obliterated in the foliage. Here, Thayer argues that every nuance of a peacock's coloration increases his concealment in a particular bit of habitat—the combined effect adding to invisibility. Given the usual interpretation of a peacock's color as revealing, and the gaudy impression that he makes both upon us and, one must assume, the peahen, Thayer's interpretation represents quite a departure from tradition and common sense:

> The peacock's splendor is the effect of a marvelous combination of "obliterative" designs, in forest-colors and patterns. . . . All imaginable forest-tones are to be found in this

A peacock in the woods, showing how, in at least one highly peculiar position, each "showy" feature can help blend the bird into invisibility. From Thayer's 1909 book. NEG. NO. 2A13238. COURTESY DEPARTMENT OF LIBRARY SERVICES, AMERICAN MUSEUM OF NATURAL HISTORY.

bird's costume; and they "melt" him into the scene to a degree past all human analysis.

Thayer then positions his bird so precisely that all features blend with surroundings. He paints the blue neck against a gap in the foliage, so that it may mimic "blue sky seen through the leaves." He matches the golden greens and browns of the back to forest tones. He depicts the white cheek patch as a "ruptive" hole that disaggregates the face. He paints the celebrated ocelli (eyespots) of the tail feathers as leaf mimics. He also notes that ocelli are smallest and dimmest near the body, grading to larger and brighter toward the rear end: "They inevitably lead the eye away from the bird, till it finds itself straying amid the foliage beyond the tail's evanescent border." The spread tail, he claims, may impress the peahen, but it "looks also very much like a shrub bearing some kind of fruit or flower." Finally, he argues that the tail's coppery brown color represents perfectly "the bare ground and tree-trunks seen between the leaves."

What a tour de force, but what can we possibly make of such special pleading? Who would doubt that some conceivable habitat might conceal almost any animal? Note how precisely the peacock must choose his spot to receive the cryptic benefit that Thayer wishes to confer upon him. In particular, he must always place his shimmering blue neck in a gap amidst the foliage where it will vanish against a clear sky (but what does he do on a cloudy day, or in a bush so dense that no holes exist, as seen from all relevant directions at once?). Peacocks, in any case, live primarily in open fields. Their spreading display is a glory to behold—and the very opposite of invisible.

Thayer, of course, knew all this. He didn't claim (as his critics sometimes charged) that a habitat offering protection by concealment must be the usual, or even a common, haunt of its invisible beneficiary. Thayer simply argued that such protection might be important at critical moments occurring only once or twice in an animal's lifetime—at crucial instants of impending death from a stalking predator.

But how could odd and improbable moments shape such complex and intricate patterns as the innumerable details of a peacock's design? With this question, we finally arrive at the key theoretical issue of this debate—the power of natural selection itself. In order to believe that complex designs might be constructed by such rare and momentary benefits as sunsets or particular positions in trees outside an animal's normal habitat, one must have an overarching faith in the power of natural selection. Selection must be so potent that even the rarest of benefits will eventually be engraved into the optimal designs of organisms. Thayer had this faith; Roosevelt, and most biologists then and since, did not. Thayer wrote in 1900: "Of course, to any one who feels the inevitability of natural selection, it is obvious that each organ or structural detail, and likewise each quality of organic forms, owes its existence to the sum of all its uses." Thayer then laid it on the line in stark epitome—patterns of color are built by natural selection, "pure, simple, and omnipotent."

Roosevelt and other acute critics correctly identified the central flaw in Thayer's science—not in his numerous factual errors, but in his methodology. Thayer found a hiding place for all his animals but with a method that made his theory untestable and therefore useless to science. Thayer insisted that he had proved his point simply by finding *any* spot that rendered an animal invisible. He didn't need to show that the creature usually frequented such a place or that the location formed part of a natural habitat at all. For the animal might seek its spot only in the rarest moments of need. But how then could we disprove any of Thayer's claims? We might work for years to show that an animal never entered its domain of invisibility, and Thayer would reply: Wait till tomorrow when urgent need arises. Scientists are trained to avoid such special pleading because it exerts a chilling and stupefying effect upon hypotheses, by rendering them invulnerable to test and potential disproof. Doing is the soul of science and we reject hypotheses that condemn us to impotence.

T. Barbour, former director of Harvard's Museum of Comparative Zoology (where I now sit composing this piece), and J. C. Phillips emphasized this point in reviewing Thayer's book in 1911:

Acquiescence in Mr. Thayer's views throws a pall over the entire subject of animal coloration. Investigation is discouraged; and we find jumbled together a great mass of fascinating and extremely complicated data, all simply ex-

plained by one dogmatic assertion. For we are asked to believe that an animal is protectively clothed whether he is like
his surroundings, or whether he is very unlike them (oblit-
eratively marked) or . . . if he falls between these two classes,
there is still plenty of space to receive him.

Teddy Roosevelt addressed the same issue with more vigor in a
letter to Thayer on March 19, 1912 (just imagine any presidential
candidate taking time out to pursue natural history more than a
month after the New Hampshire primary—oh, I know, campaigns
were shorter then):

> There is in Africa a blue rump baboon. It is also true that
> the Mediterranean Sea bounds one side of Africa. If you
> should make a series of experiments tending to show that if
> the blue rump baboon stood on its head by the Mediterra-
> nean you would mix up his rump and the Mediterranean,
> you might be illustrating something in optics, but you
> would not be illustrating anything that had any bearing
> whatsoever on the part played by the coloration of the ani-
> mal in actual life. . . . My dear Mr. Thayer, if you would face
> facts, you might really help in elucidating some of the prob-
> lems before me, but you can do nothing but mischief, and
> not very much of that, when conducting such experiments.
> . . . Your experiments are of no more real value than the
> experiment of putting a raven in a coal scuttle, and then
> claiming that he is concealed.

Contemporary (and later) accounts of Thayer's debacle rest
largely upon a red herring, concealed in more than the sunset,
that will not explain his failure and only reinforces a common and
harmful stereotype about the intrinsic differences among intellectual styles. In short, we are told that Abbott Thayer ultimately
failed because he possessed an artist's temperament—good for
an initial insight perhaps, but with no staying power for the hard
(and often dull) work of real science.

Such charges were often lodged against Thayer, and with undoubted rhetorical effect, but they represent a dangerous use of
ad hominem argument with anti-intellectual overtones. Thayer

may have laid himself open to such ridicule with a passionate temperament that he made no effort to control in a more formal age. John Jay Chapman, the acerbic essayist, wrote of Thayer (admittedly in a fit of pique when his wife, at his great displeasure, decided to study art in Thayer's studio):

> Thayer by the way, is a hipped egoist who paints three hours, has a headache, walks four hours—holds his own pulse, wants to save his sacred light for the world, cares for nobody, and has fits of dejection during which forty women hold his hand and tell him not to despair—for humanity's sake.

But is such passion the exclusive birthright of artists? I have known many scientists equally insufferable.

Thayer's scientific critics also raised the charge of artistic temperament. Roosevelt wrote, in a statement that might have attracted more attention in our litigious age: Thayer's misstatements "are due to the enthusiasm of a certain type of artistic temperament, an enthusiasm also known to certain types of scientific and business temperaments, and which when it manifests itself in business is sure to bring the owner into trouble as if he were guilty of deliberate misconduct." Barbour and Phillips argued that "Mr. Thayer, in his enthusiasm, has ignored or glossed over with an artistic haze. . . . This method of persuasion, while it does appeal to the public, is—there is no other word—simple charlatanry however unwitting." Barbour and Phillips then defended the cold light of dispassionate science in a bit of self-serving puffery:

> [Our statements] are simply the impressions made upon open-minded observers who have no axe to grind, and who have no reason to take sides on the question, one way or another. They have been written in a friendly spirit, and we hope they will be received in the same way.

Do friendly spirits ever accuse their opponents of "simple charlatanry"?

The charge of artistic temperament may be convenient and

effective, especially since it appeals to a common stereotype—but it won't wash. The facile interpretation that scientists wouldn't give Thayer a hearing because he was an "outsider" won't work either—for contemporary accounts belie such charges of territoriality and narrowness. Even though Thayer made such strong claims—quoted above—for scientists' incompetence in a domain accessible only to artists, naturalists welcomed his insights about countershading and enjoyed both his initial articles and his outdoor demonstrations. E. B. Poulton, one of England's greatest evolutionists, warmly supported Thayer and wrote introductions to his publications. Frank M. Chapman, great ornithologist and editor of *The Auk,* wrote in his *Autobiography of a Bird Lover:*

> As an editor, doubtless my most notable contributions to the Auk's pages were Abbott Thayer's classical papers on protective coloration. . . . I knew little of Thayer's eminence as an artist. It was the man himself who impressed me by the overwhelming force of his personality. He made direct and inescapable demands on one's attention. He was intensely vital and lived normally at heights which I reached only occasionally and then only for short periods.

Thayer's ultimate failure reflects a more universal tendency, distributed without reference to profession among all kinds of people. Nothing but habit and tradition separate the "two cultures" of humanities and science. The processes of thought and modes of reason are similar—so are the people. Only subject matter differs. Science may usually treat the world's empirical information; art may thrive on aesthetic judgment. But scientists also traffic in ideas and opinions, and artists surely respect fact.

The *idée fixe* is a common intellectual fault of all professions, not a characteristic failure of artists. I have often written about scientists as single-mindedly committed to absurd unities and false simplifications as Thayer was devoted to the exclusivity of concealing coloration in nature. Some are charming and a bit dotty—such as old Randolph Kirkpatrick, who thought that all rocks were made of single-celled nummulospheres (see Essay 22 in *The Panda's Thumb,* 1980). Others are devious and more than a bit dangerous—such as Cyril Burt, who fabricated data to prove that all intelligence resided in heredity (see my book *The Mismeasure of Man*).

Abbott Thayer had an *idée fixe;* he burned with desire to reduce a messy and complex world to one beautiful, simple principle of explanation. Such monistic schemes never work. History has built irreducible complexity and variety into the bounteous world of organisms. Diversity reigns at the superficial level of overt phenomena—animal colors serve many different functions. The unifying principles are deeper and more abstract—may I suggest evolution itself for starters.

---

## Postscript

Abbott H. Thayer extended his flamingo theory ever further than I had realized. Historian of science Sharon Kingsland, who wrote an excellent technical article on "Abbott Thayer and the Protective Coloration Debate" in 1978 (and who therefore would have made my own work ever so much easier had I known of her prior efforts), sent me a 1911 note by Thayer triumphantly announcing a new genre of paintings with backgrounds made from the actual skins of animals supposedly concealed by their colors and markings. Thayer wrote, including flamingos of course:

> The public will soon be astonished when I show them a dawn picture made out of the entire skin of one of these birds [flamingos] simply "mosaicked" into the sky of a painting of one of their lagoons. I am now making such a picture. I have already nearly finished a picture of a Himalayan gorge made wholly of the skins of Monaul pheasants; and another one of a New Hampshire snow scene similarly done with magpies. Artists are positively amazed by both of them.

On a more practical and positive note, I learned from my correspondents that Thayer's views on concealment were far more important in the history of naval camouflage than I had realized or that the biological literature had recorded. I received two fascinating letters from Lewis R. Melson, USNR. He wrote:

> Many years ago, I was summarily ordered to assume the responsibility for directing the efforts of the U.S. Navy's Ship Concealment and Camouflage Division, relieving the genius who had guided this effort throughout World War II, Commander Dayton Reginald Evans Brown. Dayton had perfected the camouflage patterns employed on all naval ships and aircraft throughout the war. In his briefing of what I could expect in directing the continuation of his work, I found his theories and designs were based upon Abbott H. Thayer's earlier work in the field of concealment and camouflage. . . . Despite whatever everyone thought and thinks about Thayer's theories, both his "protective coloration" and "ruptive" designs were vital for concealing ships and aircraft.

Melson continued:

> All naval concealment and camouflage is designed for protection against the horizon in the case of shipping and either for concealment against a sea or sky background, again at long ranges, for aircraft. [Note from my essay how much of Thayer's work involved "disappearance" of an image seen against the sky or horizon.] Thayer's "protective coloration" designs were outstanding for aircraft, light undersides and dark above [as fish, seen from below by their predators, tend to display]. Ship concealment for temperate and tropical oceans employed the "protective coloration" designs, while "ruptive" or "disruptive" designs worked best against polar backgrounds.

Melson also taught me some history of camouflage during the two world wars. Despite our later and fruitful use in World War II, the U.S. Navy had originally rejected Thayer's proposal during World War I. However, Thayer had greater success in Britain, where his designs proved highly valuable during the First World War. Melson wrote:

Thayer's suggestions . . . called for very light colored ships using broken patterns of white and pale blue. The intent of this pattern was to blend the ship against the background at night and in overcast weather. In the high northern latitudes surrounding the British Isles with its frequent storms, fogs, and long periods of darkness, these patterns proved very successful. HMS *Broke* was the first ship so painted and it was rammed twice by sister ships of the Royal Navy, whose captains protested that they had been unable to see *Broke*.

Melson ended his letter with a fine affirmation of potential interaction between pure and applied science:

Thanks again for the article on Thayer. It will join my mementos of those heady days when we were able to contribute bits and pieces to the world of science and engineering.

# 15 | Petrus Camper's Angle

I REMEMBER WATCHING Toscanini, a little old man made even smaller on the tiny screen of our first television set. I understood nothing of classical music when I was nine, but Toscanini's intensity nearly moved me to tears—a man older than my grandfather and scarcely bigger than me, drawing such concerted sound from his players. I remember how he stepped off the podium after each piece and mopped his brow with a handkerchief.

Classical music had little currency in those days just before the long-playing record. In television's only other foray into this arcane world, we could watch the annual Christmas presentation of Menotti's opera *Amahl and the Night Visitors*. Amahl, the young cripple with a passion for embellishment, tells his mother that two kings are outside, requesting entry to their humble cottage. She chides him, laments his disinclination to speak truly, and sends him to the door again. Amahl returns to admit that, indeed, two kings do not stand outside. His mother rejoices, but Amahl proclaims: "There are three kings . . . and one of them is black."

I remembered this line when I started to visit art museums much later and soon realized that Menotti had been following an old tradition, not making a modern plea for racial harmony. One of the Magi is always depicted as a black man. This traditional iconography is not biblical, but a later interpretation. The gospel writers do not even specify the number of wise men who saw the star in the east and came to worship. Some early sources cite up to twelve, but the number soon stabilized at three. Later, this trio received names—Balthazar, Melchior, and Gaspar, first specified in a sixth-century mosaic in Ravenna—and then, symbolic inter-

229

pretations. As these allegories moved from the specific to the general, the portrayal of one Magus as black stabilized. The three were first seen as kings of Arabia, Persia, and India, then (by the Venerable Bede, for example) as symbols for the three great continents of Africa, Europe, and Asia, and finally, as representatives of the three major human races: white, yellow, and black.

I was reminded of this iconographical tidbit recently as I read one of the classic works of physical anthropology—the historical beginning of scientific measurement of the human skull. Petrus Camper, Dutch anatomist and painter, was born in Leiden in 1722. He studied both art and science, then trained as a midwife before receiving his degree as a physician. (Men may be midwives. The name refers to a person, male or female, who stays with [*mit,* as in modern German] a woman [*wif*] during birth—so the female end of the etymology refers to the mother, not the attendant.) In 1755, he became a professor of anatomy in Amsterdam and spent the rest of his comfortable life alternating between his country home and his professional duties in Amsterdam and Groningen. Camper, who discovered the air spaces in bird bones and studied the hearing of fishes and the croaking of frogs, was revered as one of the great intellects of Europe during his own lifetime. The busy life that such attention brings, made even more hectic by the political career that he forged during his later years, left Camper little time to write and publish his scientific studies. At his death in 1789, he left his major work on the measurement of human anatomy in manuscript. His son published this posthumous document in 1791, both in the original Dutch and in French translation. (I read the French, an edition printed in Utrecht, presumably by typesetters who didn't know the language, and so full of errors that I almost decided it might be easier to learn Dutch and work from the other version.)

This work bears an extended title, both characteristic of the age and expressive of the contents: *Physical dissertation on the real differences that men of different countries and ages display in their facial traits; on the beauty that characterizes the statues and engraved stones of antiquity; followed by the proposition of a new method for drawing human heads with the greatest accuracy.* Camper's treatise is remembered today for one primary achievement—the definition of the so-called facial angle, the first widely accepted measurement for comparing the skulls of different races and nationalities.

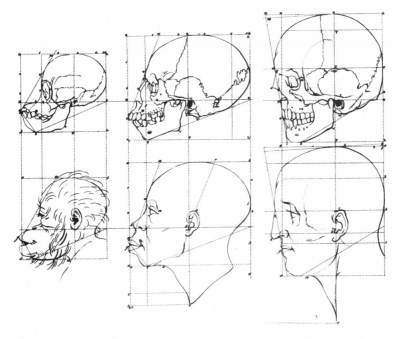

Illustration from Camper's original work showing increasing value for the facial angle of (left to right) an ape, an African, and a Grecian head. NEG. NO. 337249. COURTESY DEPARTMENT OF LIBRARY SERVICES, AMERICAN MUSEUM OF NATURAL HISTORY.

Camper's facial angle is the traditional beginning of craniometry, or the science of measuring human skulls, a major subdiscipline of physical anthropology.

The human skull may be divided into two basic components: the vault of the cranium itself and the face in front. Camper's facial angle sought to specify the relationship between these two parts. Camper first drew a line connecting the ear opening with the base of the nose (the so-called horizontal, or "h–k" on his illustration of an African head). He then constructed another line joining the most forward projection of the upper jaw (the bottom of the upper lip in living heads, usually the edge of the incisor teeth on skulls) with the most protruding point of the brow above the eyes ("h–n" on his African head), and called ever since the facial line. Camper then defined the facial angle as the intersection of the horizontal (his basis of reference) with the facial line (roughly, the forward slant of the face).

In a general way, the facial angle measures the relative flatness versus forward extension of the face. A low value means that the jaws extend far in front of the cranial vault, giving the entire skull a sloping appearance. A high value indicates a flat face with jaws projecting no farther forward than the brow itself. When a facial angle exceeds 90 degrees, the vault of the cranium projects farther forward than the underlying jaws.

The facial angle soon became the first widely accepted tool for quantitative comparison of human skulls. It spawned an immense literature, a host of proposals for slight improvements based on different criteria, and a bevy of instruments designed to measure this fundamental aspect of human life. Camper's angle became the first quantitative device for establishing invidious comparisons, based on inherent distinctions, among human races. The early craniometricians reported that African blacks possessed the lowest facial angles (farthest forward projection of the jaws), with Orientals in the middle, and Europeans on top, with facial angles sometimes approaching 90 degrees. Since apes had facial angles even lower than blacks, and since the facial angles for ancient statues of Greek deities exceeded those of all living Europeans, the smooth ascent from monkey to majesty seemed assured. Historian John S. Haller writes:

> The facial angle was the most extensively elaborated and artlessly abused criterion for racial somatology. . . . By 1860 the facial angle had become the most frequent means of explaining the graduation of species. Like the Chain of Being, the races of man consisted of an ordered hierarchy in which the Hottentot, the Kaffir, the Chinaman, and the Indian held a specific position in the order of life.

I had never read Camper's original definition of the facial angle or his own recommendations for its use and meaning. Neither had most of the nineteenth-century craniometricians who established the facial angle as a primary instrument of scientific racism (Camper's posthumous work has always been rare and difficult to obtain). I am no longer surprised when the study of a neglected original shows, as in this case, that later interpretations departed from an author's own intentions. Such stories cannot rank as news; they fall into the category of "dog bites man." And besides, maybe the later readings were correct and the initial proposal wrong—good insights for bad reasons are legion in the world of intellect.

The story of Camper's own interpretation of his facial angle is interesting for another reason. The archaeology of knowledge assumes greatest importance when it seeks new insights from our past. By cultural heritage and proven reliability, we approach problems in a set of stereotyped ways and often assume (following the cardinal sin of pride) that our modern conventions exhaust the domain of possible inquiry. We should study the past for the simplest of reasons—to increase our "sample size" in modes of thought, for we need all the help we can get. Camper's own rationale is instructive because it departs fundamentally from the sociology and a conceptual basis of modern scientific inquiry into human variability. We should recover and understand Camper's reasoning both to pay proper respect to a fine thinker and to expand our own sense of possibility.

Camper did not define the facial angle as a device for ranking races or nations by innate worth or intellect. He did not even approach the problem of human variability with motives that we would now recognize as scientific. In Camper's day, anthropology did not exist as a discipline; science had not been defined, either as a word or as a separate domain of knowledge. Scholars often worked simultaneously in areas now walled off into separate faculties of universities. Such "cross-disciplinary" work seemed neither odd nor prodigious to eighteenth-century thinkers.

Camper was a professor of anatomy and medicine, but he was also an accomplished artist, good enough to win admission to the Painter's Academy during a long visit to England (1748 to 1750). Camper defined his facial angle with the requisite precision of geometry and the quantitative preferences of science, but his motive lay in the domain of art. (He saw no contradiction, and neither should we.) We may now return to the issue of the black Magus and to Camper's own statement about his intentions.

Camper tells us right at the outset of his treatise that his desire to quantify human variation first arose in response to a minor annoyance with Western painting. He had studied the black

Magus in many classical paintings and noted that, while his color matched the hues of Africa, his face almost always displayed the features of European whites—a kind of Renaissance minstrel show with whites in blackface. (Since few Africans then lived in Europe, Camper reasoned that most artists had used white models, faithfully copying the facial features of a European and then painting the figure black.)

Camper wanted to prevent such errors by establishing a set of simple guidelines (lengths and angles) to define the chief characters of each human group. His treatise devotes more space to differences between old and young than to disparity among races or nations—for Camper was also distressed that the infant Jesus had often been drawn from an older model (no photographic surrogates to keep a baby's image still in Camper's day).

Yet Camper did not locate his immediate motive for the facial angle in descriptive anthropology of actual humans, but in a much loftier problem—no less than the definition of beauty itself. Like so many of his contemporaries, Camper believed that the cultures of ancient Greece and Rome had reached a height of refinement never since repeated and perhaps not even subject to recapture. (This is a difficult concept to grasp in the light of our later cultural preference for progress as a feature of technological history—old must mean inferior. But our forebears were not so encumbered, and ideas of a previous golden age, surpassing anything achieved since, had great power and attraction. The Renaissance [literally, rebirth] received its name from this conviction, and its heroes were trying, or so they thought, to recapture the knowledge and glory of antiquity, not to create novel improvements in art or architecture.)

Camper was obsessed with a particular issue arising from this reverence for antiquity. We can all agree, he states, that the great sculptors of ancient Greece achieved a beauty and nobility that we have not been able to match in our new art or often even to duplicate in simple attempts to copy ancient statues. One might take the easy route out of this dilemma and argue that Phidias and his brethren were just good copyists and that the people of ancient Greece surpassed all modern folk in beauty and proportion. But Camper had evidence against this proposition, for he noted that the few Greek attempts at actual portraiture (on coins, for example) showed people much like ourselves, warts and all.

Moreover, the Greeks had made no secret of their preference for idealization. Camper quotes Lysippus's desire "not to represent men as they are, but as they present themselves to our imagination." "The ideal of Antique Beauty," Camper writes, "does not exist in nature; it is purely a concept of the imagination."

But how shall we define this ideal of beauty? Camper traces the sorry attempts by poets, artists, and philosophers throughout the centuries. He notes that, in the absence of a firm criterion, each field has tried to fob off the definition by analogy—poets exemplify beauty by reference to art; artists by reference to poetry. Explicit attempts often foundered in nonsense, as in this example from 1584: "Beauty is only beautiful by its own beauty," a motto that inspired Camper's appropriate riposte, "Can there be a greater absurdity?"

And yet, Camper argues, we all agree about the beauty of certain objects, so some common criterion must exist. He writes:

> A beautiful starry sky pleases everyone. A sunrise, a calm sea, excites a sensation of pleasure in all people, and we all agree that these phenomena convey an impression of beauty.

Camper therefore decided to abandon the overarching attempts that had always devolved into nonsense and to concentrate instead on something specific that might be defined precisely—the human head.

Again he argues (incorrectly, I think, but I am explicating, not judging) that common standards exist and that, in particular, we all agree about the maximal beauty of Grecian statuary:

> We will not find a single person who does not regard the head of Apollo or Venus as possessing a superior Beauty, and who does not view these heads as infinitely superior to those of the most beautiful men and women [of our day].

Since the Greek achievement involved abstraction, not portraiture, some secret knowledge must have allowed them to improve the actual human form. Camper longed to recover their rule book. He did not doubt that the great sculptures of antiquity had proceeded by mathematical formulas, not simple intuition—for proportion and harmony, geometrically expressed, were hall-marks of Greek thought. Camper would, therefore, try to infer their physical rules of ratios and angles: "It is difficult to imitate the truly sublime beauty that characterizes Antiquity until we have discovered the true physical reasons on which it was founded."

Camper therefore devised an ingenious method of inference (also a good illustration of the primary counterintuitive principle that marks true excellence in science). When faced with a grand (but intractable) issue—like *the* definition of beauty—don't seek the ultimate, general solution; find a corner that can be defined precisely and, as our new cliché proclaims, go for it. He decided to draw, in profile and with great precision, a range of human heads spanning nations and ages. He would then characterize these heads by various angles and ratios, trying to establish simple gradations from what we regard as least to most pleasing. He would then extrapolate this gradient in the "more pleasing" direction to construct idealized heads that exaggerate those features regarded as most beautiful in actual people. Perhaps the Greeks had sculpted their deities in the same manner.

With this background, we can grasp Camper's own interpretation of the facial angle. Camper held that modern humans range from 70 degrees to somewhere between 80 and 90 degrees in this measure. He also made two other observations: first, that monkeys and other "brutes" maintained lower angles in proportion to their rank in the scale of nature (monkeys lower than apes, dogs lower than monkeys, and birds lower than dogs); second, that higher angles characterize smaller faces tucked below a more bulging cranium—a sign of mental nobility on the ancient theme of more is better.

Having established this range of improvement for living creatures, Camper extrapolated his facial angle in the favorable direction toward higher values. *Voilà.* He had found the secret. The beautiful skulls of antiquity had achieved their pleasing proportions by exaggerating the facial angle beyond values attained by real people. Camper could even define the distinctions that had eluded experts and made for such difficulty in attempts to copy and define. Romans, he found, preferred an angle of 95 degrees, but the ancient Greek sculptors all used 100 degrees as their ideal—and this difference explains both our ease in distinguishing Greek originals from Roman copies and our aesthetic preference for Greek statuary. (Proportion, he also argued, is always a balance between too little and too much. We cannot extrapolate the facial angle forever. At values of more than 100 degrees, a human skull begins to look displeasing and eventually monstrous—as in individuals afflicted with hydrocephalus. The peculiar genius of the Greeks, Camper argued, lay in their precise understanding of the facial angle. The great Athenian sculptors could push its value right to the edge, where maximal beauty switches to deformity. The Romans had not been so brave, and they paid the aesthetic price.)

Thus, Camper felt that he had broken the code of antiquity and offered a precise definition of beauty (at least for the human head): "What constitutes a beautiful face? I answer, a disposition of traits such that the facial line makes an angle of 100 degrees with the horizontal." Camper had defined an abstraction, but he had worked by extrapolation from nature. He ended his treatise with pride in this achievement: "I have tried to establish on the foundation of Nature herself, the true character of Beauty in faces and heads."

This context explains why the later use of facial angles for racist rankings represents such a departure from Camper's convictions and concerns. To be sure, two aspects of Camper's work could be invoked to support these later interpretations, particularly in quotes taken out of context. First, he did, and without any explicit justification, make aesthetic judgments about the relative beauty of races—never doubting that Nordic Europeans must top the scale objectively and never considering that other folks might advocate different standards. "A Lapplander," he writes, "has always been regarded, and without exception throughout the world, as more ugly than a Persian or a Georgian." (One wonders if anyone had ever sent a packet of questionnaires to the Scandinavian tundra; Camper, in any case at least, does not confine his accusations of ugliness to non-Caucasians.)

Second, Camper did provide an ordering of human races by facial angle—and in the usual direction of later racist rankings, with Africans at the bottom, Orientals in the middle, and Europeans on top. He also did not fail to note that this ordering placed Africans closest to apes and Europeans nearest to Greek gods. In discussing the observed range of facial angles (70 to 100 degrees in statues and actual heads), Camper notes that "It [this range] constitutes the entire gradation from the head of the Negro to the sublime beauty of the Greek of Antiquity." Extrapolating further, Camper writes:

> As the facial line moves back [for a small face tucked under bulging skull] I produce a head of Antiquity; as I bring it forward [for a larger, projecting face] I produce the head of a Negro. If I bring it still further forward, the head of a monkey results, more forward still, and I get a dog, and finally a woodcock; this, now, is the primary basis of my edifice.

(Our deprecations never cease. The French word for woodcock— *bécasse*—also refers to a stupid woman in modern French slang.)

I will not defend Camper's view of human variation any more than I would pillory Lincoln for racism or Darwin for sexism (though both are guilty by modern standards). Camper lived in a different world, and we cannot single him out for judgment when he idly repeats the commonplaces of his age (nor, in general, may we evaluate the past by the present, if we hope to understand our forebears).

Camper's comments on racial rankings are fleeting and stated *en passant.* He makes no major point of African distinctions except to suggest that artists might now render the black Magus correctly in painting the Epiphany. He does not harp upon differences among human groups and entirely avoids the favorite theme of all later writings in craniometric racism—finer scale distinctions between "inferior" and "superior" Europeans. His text contains not a whiff or hint of any suggestion that low facial angles imply anything about moral worth or intellect. He charges Africans with nothing but maximal departure from ideal beauty. Moreover, and most important, Camper's clearly stated views on the nature of human variability preclude, necessarily and a priori, any equation of difference with innate inferiority. This is the key point that later commentators have missed because we have lost Camper's world view and cannot interpret his text without recovering the larger structure of his ideas.

We now live in a Darwinian world of variation, shadings, and continuity. For us, variation among human groups is fundamental, both as an intrinsic property of nature and as a potential substratum for more substantial change. We see no difference in principle between variation within a species and established differences between species—for one can become the other via natural selection. Given this potential continuity, both kinds of variation may record an underlying and basically similar genetic inheritance. To us, therefore, linear rankings (like Camper's for the facial angle) quite properly smack of racism.

But Camper dwelt in the pre-Darwinian world of typology. Species were fixed and created entities. Differences among species recorded their fundamental natures. But variation within a species could only be viewed as a series of reversible "accidents" (departures from a species' essence) imposed by a variety of factors, including climate, food, habits, or direct manipulation. If all humans represented but a single species, then our variation could only be superficial and accidental in this Platonic sense. Physical differences could not be tokens of innate inferiority. (By "accidental," Camper and his contemporaries did not mean capricious or devoid of immediate import in heredity. They knew that black parents had black children. Rather, they argued that these traits, impressed into heredity by climate or food, had no fixed status and could be easily modified by new conditions of life. They were often wrong, of course, but that's not the point.)

Therefore, to understand Camper's views about human variability, we must first learn whether he regarded all humans as members of one species or as products of several separate creations (a popular position known at the time as polygeny). Camper recognized these terms of the argument and came down strongly and incisively for human unity as a single species (monogeny). In designating races by the technical term "variety," Camper used the jargon of his day to underscore his conviction that our differences are accidental and imposed departures from an essence shared by all; our races are not separated by differences fixed in heredity. "Blacks, mulattos, and whites are not diverse species of men, but only varieties of the human species. Our skin is constituted exactly like that of the colored nations; we are therefore only less black than they." We cannot even know, Camper adds, whether Adam and Eve were created white or black since transitions between superficial varieties can occur so easily (an attack on those who viewed blacks as degenerate and Adam and Eve as necessarily created in Caucasian perfection):

> Whether Adam and Eve were created white or black is an entirely indifferent issue without consequences, since the passage from white to black, considerable though it be, operates as easily as that from black to white.

Misinterpretation may be more common than accuracy, but a misreading precisely opposite to an author's true intent may still excite our interest for its sheer perversity. When, in order to grasp this inversion, we must stretch our minds and learn to understand some fossil systems of thinking, then we may convert a simple correction to a generality worthy of note. Poor Petrus Camper. He became the semiofficial grandpappy of the quantitative approach to scientific racism, yet his own concept of human variability precluded judgments about innate worth a priori. He developed a measure later used to make invidious distinctions among actual groups of people, but he pressed his own invention to the service of abstract beauty. He became a villain of science when he tried to establish criteria for art. Camper got a bad posthumous shake on earth; I only hope that he met the right deity on high (facial angle of 100 degrees, naturally), the God of Isaiah, who also equated beauty with number and proportion—he "who hath measured the waters in the hollow of his hand, and meted out heaven with the span."

# 16 | Literary Bias on the Slippery Slope

EVERY PROFESSION has its version: Some speak of "Sod's law"; others of "Murphy's law." The formulations vary, but all make the same point—if anything bad can happen, it will. Such universality of attribution can only arise for one reason— the principle is true (even though we know that it isn't).

The fieldworker's version is simply stated: You always find the most interesting specimens at the very last moment, just when you absolutely must leave. The effect of this phenomenon can easily be quantified. It operates weakly for localities near home and easily revisited and ever more strongly for distant and exotic regions requiring great effort and expense for future expeditions. Everyone has experienced this law of nature. I once spent two weeks on Great Abaco, visiting every nook and cranny of the island and assiduously proving that two supposed species of *Cerion* (my favorite land snail) really belonged to one variable group. On the last morning, as the plane began to load, we drove to the only unexamined place, an isolated corner of the island with the improbable name Hole-in-the-Wall. There we found hundreds of large white snails, members of the second species.

Each profession treasures a classic, or canonical, version of the basic story. The paleontological "standard," known to all my colleagues as a favorite campfire tale and anecdote for introductory classes, achieves its top billing by joining the most famous geologist of his era with the most important fossils of any time. The story, I have just discovered, is also entirely false (more than a bit embarrassing since I cited the usual version to begin an earlier essay in this series).

241

Charles Doolittle Walcott (1850–1927) was both the world's leading expert on Cambrian rocks and fossils (the crucial time for the initial flowering of multicellular life) and the most powerful scientific administrator in America. Walcott, who knew every president from Teddy Roosevelt to Calvin Coolidge, and who persuaded Andrew Carnegie to establish the Carnegie Institute of Washington, had little formal education and began his career as a fieldworker for the United States Geological Survey. He rose to chief, and resigned in 1907 to become secretary (their name for boss) of the Smithsonian Institution. Walcott had his finger, more accurately his fist, in every important scientific pot in Washington.

Walcott loved the Canadian Rockies and, continuing well into his seventies, spent nearly every summer in tents and on horseback, collecting fossils and indulging his favorite hobby of panoramic photography. In 1909, Walcott made his greatest discovery in Middle Cambrian rocks exposed on the western flank of the ridge connecting Mount Field and Mount Wapta in eastern British Columbia.

The fossil record is, almost exclusively, a tale told by the hard parts of organisms. Soft anatomy quickly disaggregates and decays, leaving bones and shells behind. For two basic reasons, we cannot gain an adequate appreciation for the full range of ancient life from these usual remains. First, most organisms contain no hard parts at all, and we miss them entirely. Second, hard parts, especially superficial coverings, often tell us very little about the animal within or underneath. What could you learn about the anatomy of a snail from the shell alone?

Paleontologists therefore treasure the exceedingly rare soft-bodied faunas occasionally preserved when a series of unusual circumstances coincide—rapid burial, oxygen-free environments devoid of bacteria or scavengers, and little subsequent disturbance of sediments.

Walcott's 1909 discovery—called the Burgess Shale—surpasses all others in significance because he found an exquisite fauna of soft-bodied organisms from the most crucial of all times. About 570 million years ago, virtually all modern phyla of animals made their first appearance in an episode called "the Cambrian explosion" to honor its geological rapidity. The Burgess

Shale dates from a time just afterward and offers our only insight into the true range of diversity generated by this most prolific of all evolutionary events.

Walcott, committed to a conventional view of slow and steady progress in increasing complexity and diversity, completely misinterpreted the Burgess animals. He shoehorned them all into modern groups, interpreting the entire fauna as a set of simpler precursors for later forms. A comprehensive restudy during the past twenty years has inverted Walcott's view and taught us the most surprising thing we know about the history of life: The fossils from this one small quarry in British Columbia exceed, in anatomical diversity, all modern organisms in the world's oceans today. Some fifteen to twenty Burgess creatures cannot be placed into any modern phylum and represent unique forms of life, failed experiments in metazoan design. Within known groups, the Burgess range far exceeds what prevails today. Taxonomists have described almost a million living species of arthropods, but all can be placed into three great groups—insects and their relatives, spiders and their kin, and crustaceans. In Walcott's single Canadian quarry, vastly fewer species include about twenty more basic anatomical designs! The history of life is a tale of decimation and later stabilization of few surviving anatomies, not a story of steady expansion and progress.

But this is another story for another time (see my book *Wonderful Life,* 1989). I provide this epitome only to emphasize the context for paleontology's classic instance of Sod's law. These are no ordinary fossils, and their discoverer was no ordinary man.

I can provide no better narration for the usual version than the basic source itself—the obituary notice for Walcott published by his longtime friend and former research assistant Charles Schuchert, professor of paleontology at Yale. (Schuchert was, by then, the most powerful paleontologist in America, and Yale became the leading center of training for academic paleontology. The same story is told far and wide in basically similar versions, but I suspect that Schuchert was the primary source for canonization and spread. I first learned the tale from my thesis adviser, Norman D. Newell. He heard it from his adviser, Carl Dunbar, also at Yale, who got it directly from Schuchert.) Schuchert wrote in 1928:

One of the most striking of Walcott's faunal discoveries came at the end of the field season of 1909, when Mrs. Walcott's horse slid in going down the trail and turned up a slab that at once attracted her husband's attention. Here was a great treasure—wholly strange Crustacea of Middle Cambrian time—but where in the mountain was the mother rock from which the slab had come? Snow was even then falling, and the solving of the riddle had to be left to another season, but next year the Walcotts were back again on Mount Wapta, and eventually the slab was traced to a layer of shale—later called the Burgess shale—3,000 feet above the town of Field, British Columbia, and 8,000 feet above the sea.

Stories are subject to a kind of natural selection. As they propagate in the retelling and mutate by embellishment, most eventually fall by the wayside to extinction from public consciousness. The few survivors hang tough because they speak to deeper themes that stir our souls or tickle our funnybones. The Burgess legend is a particularly good story because it moves from tension to resolution, and enfolds within its basically simple structure two of the greatest themes in conventional narration—serendipity and industry leading to its just reward. We would never have known about the Burgess if Mrs. Walcott's horse hadn't slipped going downslope on the very last day of the field season (as night descended and snow fell, to provide a dramatic backdrop of last-minute chanciness). So Walcott bides his time for a year in considerable anxiety. But he is a good geologist and knows how to find his quarry (literally in this case). He returns the next summer and finally locates the Burgess Shale by hard work and geological skill. He starts with the dislodged block and traces it patiently upslope until he finds the mother lode. Schuchert doesn't mention a time, but most versions state that Walcott spent a week or more trying to locate the source. Walcott's son Sidney, reminiscing sixty years later, wrote in 1971: "We worked our way up, trying to find the bed of rock from which our original find had been dislodged. A week later and some 750 feet higher we decided that we had found the site."

I can imagine two basic reasons for the survival and propagation of this canonical story. First, it is simply too good a tale to pass into oblivion. When both good luck and honest labor combine to produce victory, we all feel grateful to discover that fortune occasionally smiles, and uplifted to learn that effort brings reward. Second, the story might be true. And if dramatic and factual value actually coincide, then we have a real winner.

I had always grasped the drama and never doubted the veracity (the story is plausible, after all). But in 1988, while spending several days in the Walcott archives at the Smithsonian Institution, I discovered that all key points of the story are false. I found that some of my colleagues had also tracked down the smoking gun before me, for the relevant pages of Walcott's diary had been earmarked and photographed before.

Walcott, the great conservative administrator, left a precious gift to future historians by his assiduous recordkeeping. He never missed a day of writing in his diary. Even at the very worst moment of his life, July 11, 1911, he made the following, crisply factual entry about his wife: "Helena killed at Bridgeport Conn. by train being smashed up at 2:30 A.M. Did not hear of it until 3 P.M. Left for Bridgeport 5:35 P.M." (Walcott was meticulous, but please do not think him callous. Overcome with grief the next day, he wrote on July 12: "My love—my wife—my comrade for 24 years. I thank God I had her for that time. Her untimely fate I cannot now understand.")

Walcott's diary for the close of the 1909 field season neatly dismisses part one of the canonical tale. Walcott found the first soft-bodied fossils on Burgess ridge either on August 30 or 31. His entry for August 30 reads:

> Out collecting on the Stephen formation [the unit that includes what Walcott later called the Burgess Shale] all day. Found many interesting fossils on the west slope of the ridge between Mounts Field and Wapta [the right locality for the Burgess Shale]. Helena, Helen, Arthur, and Stuart [his wife, daughter, assistant, and son] came up with remainder of outfit at 4 P.M.

On the next day, they had clearly discovered a rich assemblage of soft-bodied fossils. Walcott's quick sketches (see figure) are so clear that I can identify the three genera he depicts—*Marrella* (upper left), the most common Burgess fossil and one of the

The smoking gun for exploding a Burgess Shale legend. Walcott's diary for the end of August and the beginning of September, 1909. He collected for an entire week in good weather. SMITHSONIAN INSTITUTION.

unique arthropods beyond the range of modern designs; *Waptia*, a bivalved arthropod (upper right); and the peculiar trilobite *Naraoia* (lower left). Walcott wrote: "Out with Helena and Stuart collecting fossils from the Stephen formation. We found a remarkable group of Phyllopod crustaceans. Took a large number of fine specimens to camp."

What about the horse slipping and the snow falling? If this incident occurred at all, we must mark the date as August 30, when Walcott's family came up the slope to meet him in the late afternoon. They might have turned up the slab as they descended for the night, returning the next morning to find the specimens that Walcott drew on August 31. This reconstruction gains some support from a letter that Walcott wrote to Marr (for whom he later named the "lace crab" *Marrella*) in October 1909:

When we were collecting from the Middle Cambrian, a stray slab of shale brought down by a snow slide showed a fine Phyllopod crustacean on a broken edge. Mrs. W. and I worked on that slab from 8 in the morning until 6 in the evening and took back with us the finest collection of Phyllopod crustaceans that I have ever seen.

(Phyllopod, or "leaf-footed," is an old name for marine arthropods with rows of lacy gills, often used for swimming, on one branch of their legs.)

Transformation can be subtle. A snow slide becomes a snowstorm, and the night before a happy day in the field becomes a forced and hurried end to an entire season. But far more important, Walcott's field season did not finish with the discoveries of August 30 and 31. The party remained on Burgess ridge until September 7! Walcott was thrilled by his discovery and collected with avidity every day thereafter. The diaries breathe not a single word about snow, and Walcott assiduously reported the weather in every entry. His happy week brought nothing but praise for Mother Nature. On September 1 he wrote: "Beautiful warm days."

Finally, I strongly suspect that Walcott located the source for his stray block during the last week of his 1909 field season—at least the basic area of outcrop, if not the very richest layers. On September 1, the day after he drew the three arthropods, Walcott wrote: "We continued collecting. Found a fine group of sponges on slope (in situ) [meaning undisturbed and in their original position]." Sponges, containing some hard parts, extend beyond the richest layers of soft-bodied preservation, but the best specimens come from the strata of the Burgess mother lode. On each subsequent day, Walcott found abundant soft-bodied specimens, and his descriptions do not read like the work of a man encountering a lucky stray block here and there. On September 2, he discovers that the supposed shell of an ostracode really houses the body of a Phyllopod: "Working high up on the slope while Helena collected near the trail. Found that the large so-called Leperditia-like test is the shield of a Phyllopod." The Burgess quarry is "high up on the slope," while stray blocks would slide down toward the trail.

On September 3, Walcott was even more successful: "Found a

fine lot of Phyllopod crustaceans and brought in several slabs of rock to break up at camp." In any event, he continued to collect, and put in a full day for his last hurrah on September 7: "With Stuart and Mr. Rutter went up on fossil beds. Out from 7 A.M. to 6:30 P.M. Our last day in camp for 1909."

If I am right about his discovery of the main beds in 1909, then the second part of the canonical tale—the week-long patient tracing of errant block to source in 1910—should be equally false. Walcott's diary for 1910 supports my interpretation. On July 10, champing at the bit, he hiked up to the Burgess Pass campground, but found the area too deep in snow for any excavations. Finally, on July 29, Walcott reports that his party set up "at Burgess Pass campground of 1909." On July 30, they climbed neighboring Mount Field and collected fossils. Walcott indicates that they made their first attempt to locate the Burgess beds on August 1:

> All out collecting the Burgess formation until 4 P.M. when a cold wind and rain drove us into camp. Measured section of the Burgess formation—420 feet thick. Sidney with me. Stuart with his mother and Helen puttering about camp.

("Measuring a section" is geological jargon for tracing the vertical sequence of strata and noting the rock types and fossils. If you wished to find the source of an errant block dislodged and tumbled below, you would measure the section above, trying to match your block to its most likely layer.)

I think that Charles and Sidney Walcott located the Burgess beds on this very first day, because Walcott writes for his next entry of August 2: "Out collecting with Helena, Stuart, and Sidney. We found a fine lot of 'lace crabs' and various odds and ends of things." "Lace crab" was Walcott's informal field term for *Marrella,* and *Marrella* is the marker of the mother lode—the most common animal in the Burgess Shale. If we wish to give the canonical tale all benefit of doubt, and argue that these lace crabs of August 2 came from dislodged blocks, we still cannot grant a week of strenuous effort for locating the mother lode, for Walcott writes just two days later on August 4: "Helena worked out a lot of Phyllopod crustaceans from 'Lace Crab layer.'" From then on, until the end of summer, they quarried the lace crab layer, now known as the Burgess Shale.

The canonical tale is more romantic and inspiring, but the plain factuality of the diary makes more sense. I have been to the Burgess ridge. The trail lies just a few hundred feet below the main Burgess beds. The slope is simple and steep, with strata well exposed. Tracing an errant block to its source should not have presented a major problem—for Walcott was more than a good geologist; he was a great geologist. He should have located the main beds right away, in 1909, since he had a week to work after first discovering soft-bodied fossils. He was not able to quarry in 1909—the only constraint imposed by limits of time. But he found many fine fossils and probably the main beds themselves. He knew just where to go in 1910 and set up shop in the right place as soon as the snows melted.

Memory is a fascinating trickster. Words and images have enormous power and can easily displace actual experience over the years. As an intriguing testimony to the power of legend, consider the late memories of Walcott's son Sidney. In 1971, more than sixty years after the events, Sidney wrote a short article for *Smithsonian*, "How I Found My Own Fossil." (The largest Burgess arthropod bears the name *Sidneyia inexpectans* in honor of his discovery.) Sidney must have heard the canonical tale over and over again across the many years (think of him enduring mounds of rubber chicken and endless repetitions of the anecdote in after-dinner speeches)—and his actual experience faded as the conventional myth took root.

Sidney's version includes the two main ingredients—serendipity in the chance discovery of a dislodged slab blocking the pathway of packhorses, and assiduous effort in the patient, week-long tracing of block to source. But Sidney places the packhorse incident on his watch in 1910, not on his mother's the previous year:

> Father suddenly told me to halt the packtrain. I signaled, and the horses started to browse at the side of the trail. Often on our summer camping trips I had seen Father throw stones and logs out of the trail to make the going a bit easier for the horses. So it was no surprise to see him upend a slab, worn white by the shoes of horses slipping on it for

years. He hit it a few times along its edge with his geological hammer and it split open. "Look Sidney," he called. I saw several extraordinary fossils on the rock surface. "Let's look further tomorrow. . . . We won't go to Field tonight." To our family, back in 1910, it seemed a miracle that Father's simple act of thoughtfulness for the comfort and safety of a few packhorses led to this discovery.

A lovely story, but absolutely nothing about it can be true. Sidney knew the canonical yarn about slabs and packhorses, but moved the tale a year forward. We cannot believe that slabs could have blocked paths for two years running, with fossils always on their upturned edges, especially since an unanticipated discovery in 1909 precludes a similar surprise the next year. Moreover, Sidney could not have remembered an actual incident of the first season, and then mixed up the years, because he wasn't there in 1909!

Sidney's second ingredient, his tale of a week-long search for the mother lode (cited previously in this essay), is equally false from the evidence of Walcott's diary, and similarly read into memory from the repetition of legend, not the recall of actual events.

Why am I bothering with all this detail? To be sure, truth has a certain moral edge over falsehood, but few people care much about corrections to stories they never heard about people they never knew. If the only lesson in this little reversal of Burgess orthodoxy exhorts us to be careful lest a tendency to embellish or romanticize stifle the weakly flickering flame of truth, then this essay is as banal as the sentence I just wrote. But I would defend my effort on two grounds. First, the Burgess animals happen to be the world's most important fossils, and the purely factual issues surrounding their discovery therefore demand more than the usual care and attention to accuracy. We might not challenge a family legend about Uncle Joe in the interests of domestic peace and benevolence, but we really would like to know how Jesus lived and died because different views have had such palpable effects upon billions of lives. Second, I believe that our tendencies to construct legends raise an issue far more interesting than watchdog warnings about eternal verity.

I would begin by asking why almost every canonical tale is false in the same way—a less interesting reality converted to a simple story with a message. Do we need these stories so badly because life isn't heroic or thrilling most of the time? Sean O'Casey said that the stage must be larger than life, and few poets or playwrights can succeed by fidelity to the commonplace. It takes the artistry of James Joyce to make a masterpiece from one day in the life of an ordinary man. Most of our existence is eating, sleeping, walking, and breathing. Even the life of a soldier, if expressed in real time, would be almost uninterrupted tedium—for an old motto identifies this profession as long periods of boredom interspersed with short moments of terror.

Astute scientists understand that political and cultural bias must impact their ideas, and they strive to recognize these inevitable influences. But we usually fail to acknowledge another source of error that might be called literary bias. So much of science proceeds by telling stories—and we are especially vulnerable to constraints of this medium because we so rarely recognize what we are doing. We think that we are reading nature by applying rules of logic and laws of matter to our observations. But we are often telling stories—in the good sense, but stories nonetheless. Consider the traditional scenarios of human evolution—tales of the hunt, of campfires, dark caves, rituals, and toolmaking, coming of age, struggle and death. How much is based on bones and artifacts and how much on the norms of literature?

If these reconstructions are stories, then they are bound by the rules of canonical legendmaking. And if we construct our stories to be *unlike* life—the main point of this essay—then our literary propensities are probably derailing our hope to understand the quotidian reality of our evolution. Stories only go in certain ways—and these paths do not conform to patterns of actual life.

This constraint does not apply only to something so clearly ripe for narration and close to home as "the rise of man from the apes" (to choose a storylike description that enfolds biases of gender and progress into its conventionality). Even the most distant and abstract subjects, like the formation of the universe or the principles of evolution, fall within the bounds of necessary narrative. Our images of evolution are caught in the web of tale telling. They involve progress, pageant; above all, ceaseless motion somewhere. Even revisionist stories that question ideas of gradual progress—the sort that I have been spinning for years in these essays—are tales of another kind about good fortune, unpredictability, and contingency (the kingdom lost for want of a horseshoe nail). But focus on almost any evolutionary moment, and nothing much is happening. Evolution, like soldiering and life itself, is daily repetition almost all the time. Evolutionary days may be generations, but as the Preacher said, one passeth away and another cometh, but the earth abideth forever. The fullness of time, of course, does provide a sufficient range for picking out rare moments of activity and linking them together into a story. But we must understand that nothing happens most of the time— and we don't because our stories don't admit this theme—if we hope to grasp the dynamics of evolutionary change. (This sentence may sound contradictory, but it isn't. To know the reasons for infrequent change, one must understand the ordinary rules of stability.) The Burgess Shale teaches us that, for the history of basic anatomical designs, almost everything happened in the geological moment just before, and almost nothing in more than 500 million years since.

Included in this "almost nothing," as a kind of geological afterthought of the last few million years, is the first development of self-conscious intelligence on this planet—an odd and unpredictable invention of a little twig on the mammalian evolutionary bush. Any definition of this uniqueness, embedded as it is in our possession of language, must involve our ability to frame the world as stories and to transmit these tales to others. If our propensity to grasp nature as story has distorted our perceptions, I shall accept this limit of mentality upon knowledge, for we receive in trade both the joys of literature and the core of our being.

# 6 | Down Under

# 17 | Glow, Big Glowworm

SMALL MISUNDERSTANDINGS are often a prod to insight or victory. For such a minor error with major consequences, Laurel and Hardy got into terminal trouble with the toymaster in *March of the Wooden Soldiers*—they got fired for building 100 soldiers six feet high, when Santa had ordered 600 at one foot. But the six-footers later saved Toyland from the invasion of Barnaby and his bogeymen.

In insects that undergo a complete metamorphosis, cells that will form adult tissues are already present in the bodies of larvae as isolated patches called imaginal disks. For many years, I regarded this term as one of the oddest in all biology—for I always read "imaginal" as "imaginary" and thought I was being told that this substrate of maturity really didn't exist at all.

When I learned the true origin of this term, I realized that I had not only misunderstood but had made an absolutely backward interpretation. I also discovered that my resolution had taught me something interesting—about ways of looking at the world, not about any facts of nature per se—and I therefore judged my former error as fruitful.

Linnaeus himself, father of taxonomy, named the stages of insect development. He designated the feeding stage that hatched from the egg as a larva (the caterpillar of a moth or the maggot of a housefly), and he called the sexually mature adult an imago, hence imaginal disk for precursors of adult tissues within the larva.

The etymologies of these terms provided my insight—a larva is a mask; an imago, the image or essential form of a species. Linnaeus, in other words, viewed the development of insects as progress toward fulfillment. The first stage is only preparatory; it hides the true and complete representation of a species. The final form embodies the essence of louseness, thripsness, or flyness. Imaginal disks, by both etymology and concept, are bits of higher reality lurking within initial imperfection—no sign of "let's pretend" here.

Most impediments to scientific understanding are conceptual locks, not factual lacks. Most difficult to dislodge are those biases that escape our scrutiny because they seem so obviously, even ineluctably, just. We know ourselves best and tend to view other creatures as mirrors of our own constitution and social arrangements. (Aristotle, and nearly two millennia of successors, designated the large bee that leads the swarm as a king.)

Few aspects of human existence are more basic than our life cycle of growth and development. For all the glories of childhood, we in the West have generally viewed our youngsters as undeveloped and imperfect adults—smaller, weaker, and more ignorant. Adulthood is a termination; childhood, an upward path. How natural, then, that we should also interpret the life cycles of other organisms as a linear path from imperfect potential to final realization—from the small, ill-formed creature that first develops from an egg to the large and complex fruition that produces the egg of the next generation.

How obvious, in particular, that insect larvae are imperfect juveniles and imagoes realized adults. Linnaeus's etymology embodies this traditional interpretation imposed from human life upon the development of insects. When we combine this dubious comparison of human and insect life cycles with our more general preference for viewing developmental sequences as ladders of progress (a prejudice that has hampered our understanding of evolution even more than our resolution of embryology), insect larvae seem doomed to easy dismissal by an aggregation of biases—etymological, conceptual, and parochial.

If we turn to two leading works of popular science, published five years after Darwin's *Origin of Species*—one on life cycles in general, the other on insects—we obtain a good sense of these traditional biases. A. de Quatrefages, great French student of that economic leader among insect larvae, the silkworm, wrote in his *Metamorphosis of Man and the Lower Animals* (1864) that "larvae . . . are always incomplete beings; they are true first sketches, which are rendered more and more perfect at each developmental phase."

*An Introduction to Entomology,* by William Kirby, rector of Barham, and William Spence, wins first prize among British works of popular science for celebrity, for longevity (its first edition appeared in 1815), and for prose in the most preciously purple tradition of "nature writing," as satirized by example in James Joyce's *Ulysses:* "Note the meanderings of some purling rill as it babbles on its way, fanned by the gentlest zephyrs tho' quarrelling with the stony obstacles, to the tumbling waters of Neptune's blue domain. . . ." To which, Mr. Dedalus replies: "Agonizing Christ, wouldn't it give you a heartburn on your arse." And for which (among other things) *Ulysses* was once banned from the United States as obscene—although I would sooner exclude that purling rill than a heartburn on any part of the anatomy.

In their first post-Darwinian edition (1863), Kirby and Spence make no bones about their preference for well-formed imagoes and their distaste for grubby larvae (a redundancy for emphasis of my point—grubs are larvae, and we owe this adjective to the same prejudice):

> That active little fly, now an unbidden guest at your table, whose delicate palate selects your choicest viands, while extending his proboscis to the margin of a drop of wine, and then gaily flying to take a more solid repast from a pear or peach; now gamboling with his comrades in the air, now gracefully currying his furled wings with his taper feet, was but the other day a disgusting grub, without wings, without legs, without eyes, wallowing, well pleased, in the midst of a mass of excrement.

The adult, they write, is called an imago "because, having laid aside its mask [larva], and cast off its swaddling bands [the pupal cocoon, or chrysalis], being no longer disguised [larva] or confined [pupa], or in any other respect imperfect, it is now become a true representative or image of its species."

The burden of metaphor becomes immeasurably heavier for

larvae when Kirby and Spence then drag out that oldest of all insect analogies from an age of more pervasive Christianity—the life cycle of a butterfly to the passage of a soul from first life in the imperfect prison of a human body (larval caterpillar), to death and entombment (pupal chrysalis), to the winged freedom of resurrection (imago, or butterfly). This simile dates to the great Dutch biologist Jan Swammerdam, child of Cartesian rationalism but also, at heart, a religious mystic, who first discovered the rudimentary wings of butterflies, enfurled in late stages of larval caterpillars. Swammerdam wrote near the end of the seventeenth century: "This process is formed in so remarkable a manner in butterflies, that we see therein the resurrection painted before our eyes, and exemplified so as to be examined by our hands." Kirby and Spence then elaborated just a bit:

> To see a caterpillar crawling upon the earth sustained by the most ordinary kinds of food, which when . . . its appointed work being finished, passes into an intermediate state of seeming death, when it is wound up in a kind of shroud and encased in a coffin, and is most commonly buried under the earth . . . then, when called by the warmth of the solar beam, they burst from their sepulchres, cast off their raiments . . . come forth as a bride out of her chamber—to survey them, I say, arrayed in their nuptial glory, prepared to enjoy a new and more exalted condition of life, in which all their powers are developed, and they are arrived at the perfection of their nature . . . who that witnesses this interesting scene can help seeing in it a lively representation of man in his threefold state of existence. . . .The butterfly, the representative of the soul, is prepared in the larva for its future state of glory; . . . it will come to its state of repose in the pupa, which is its Hades; and at length, when it assumes the imago, break forth with new powers and beauty to its final glory and the reign of love.

But must we follow this tradition and view larvae as harbingers of better things? Must all life cycles be conceptualized as paths of progress leading to an adult form? Human adults control the world's media—and the restriction of this power to one stage of our life cycle imposes a myopic view. I would be happy to counter this prejudice (as many have) by emphasizing the creativity and specialness of human childhood, but this essay speaks for insects.

I will admit that our standard prejudice applies, in one sense, to creatures like ourselves. Our bodies do grow and transform in continuity. A human adult is an enlarged version of its own childhood; we grown-ups retain the same organs, reshaped a bit and often increased a great deal. (Many insects with simple life cycles, or so-called incomplete metamorphoses, also grow in continuity. This essay treats those insects that cycle through the classic stages of complete metamorphosis: egg, larva, pupa, and imago.)

But how can we apply this bias of the upward path to complex life cycles of other creatures? In what sense is the polyp of a cnidarian (the phylum of corals and their allies) more—or less—complete than the medusa that buds from its body? One stage feeds and grows; the other mates and lays eggs. They perform different and equally necessary functions. What else can one say? Insect larvae and imagoes perform the same division—larvae eat and imagoes reproduce. Moreover, larvae do not grow into imagoes by increase and complication of parts. Instead, larval tissues are sloughed off and destroyed during the pupal stage, while the imago largely develops from small aggregations of cells—the imaginal disks of this essay's beginning—that resided, but did not differentiate, within the larva. Degenerating larval tissues are often used as a culture medium for growth of the imago within the pupa. Larva and imago are different and discrete, not before and shadowy versus later and complete.

Even Kirby and Spence sensed this true distinction between objects equally well suited for feeding and reproduction, though they soon buried their insight in cascading metaphors about progress and resurrection:

> Were you . . . to compare the internal conformation of the caterpillar with that of the butterfly, you would witness changes even more extraordinary. In the former you would find some thousands of muscles, which in the latter are replaced by others of a form and structure entirely different. Nearly the whole body of the caterpillar is occupied by a capacious stomach. In the butterfly it has become converted into an almost imperceptible thread-like viscus; and the abdomen is now filled by two large packets of eggs.

If we break through the tyranny of our usual bias, to a different view of larvae and imagoes as separate and potentially equal devices for feeding and reproduction, many puzzles are immediately resolved. Each stage adapts in its own way, and depending upon ecology and environment, one might be emphasized, the other degraded to insignificance in our limited eyes. The "degraded" stage might be the imago as well as the larva—more likely, in fact, since feeding and growth can be rushed only so much, but mating, as poets proclaim, can be one enchanted evening. Thus, I used to feel sorry for the mayfly and its legendary one day of existence, but such brevity only haunts the imago, and longer-lived larvae also count in the total cycle of life. And what about the seventeen-year "locust" (actually a cicada)? Larvae don't lie around doing nothing during this dog's age, waiting patiently for their few days of visible glory. They have an active life underground, including long stretches of dormancy to be sure, but also active growth through numerous molts.

Thus, we find our best examples of an alternative and expansive view of life cycles among species that emphasize the size, length, and complexity of larval life at the apparent expense of imaginal domination—where, to borrow Butler's famous line with only minor change in context, a hen really does seem to be the egg's way of manufacturing another egg. I recently encountered a fine case during a visit to New Zealand—made all the more dramatic because human perceptions focus entirely upon the larva and ignore the imago.

After you leave the smoking and steaming, the boiling and puffing, the sulfurous stench of geysers, fumaroles, and mud pots around Rotorua, you arrive at the second best site on the standard tourist itinerary of the North Island—the glowworm grotto of Waitomo Cave. Here, in utter silence, you glide by boat into a spectacular underground planetarium, an amphitheater lit with thousands of green dots—each the illuminated rear end of a fly larva (not a worm at all). (I was dazzled by the effect because I found it so unlike the heavens. Stars are arrayed in the sky at random with respect to the earth's position. Hence, we view them as clumped into constellations. This may sound paradoxical, but my statement reflects a proper and unappreciated aspect of random distributions. Evenly spaced dots are well ordered for cause. Random arrays always include some clumping, just as we will flip several heads in a row quite often so long as we can make enough tosses—and our sky is not wanting for stars. The glowworms, on the other hand, are spaced more evenly because larvae compete with, and even eat, each other—and each constructs an exclusive territory. The glowworm grotto is an ordered heaven.)

These larval glowworms are profoundly modified members of the family Mycetophilidae, or fungus gnats. Imagoes of this species are unremarkable, but the larvae rank among the earth's most curious creatures. Two larval traits (and nothing imaginal) inspired the name for this peculiar species—*Arachnocampa luminosa,* honoring both the light and the silken nest that both houses the glowworm and traps its prey (for Arachne the weaver, namesake of spiders, or arachnids, as well). The imagoes of *Arachnocampa luminosa* are small and short-lived mating machines. The much larger and longer-lived larvae have evolved three complex and coordinated adaptations—carnivory, light, and webbing—that distinguish them from the simpler larval habits of ancestral fungus gnats: burrowing into mushrooms, munching all the way.

In a total life cycle (egg to egg) often lasting eleven months, *Arachnocampa luminosa* spends eight to nine months as a larval glowworm. Larvae molt four times and grow from 3- to 5-millimeter hatchlings to a final length of some 30 to 40 millimeters. (By contrast, imagoes are 12 to 16 millimeters in length, males slightly smaller than females, and live but one to four days, males usually longer than females.)

Carnivory is the focus of larval existence, the coordinating theme behind a life-style so different from the normal course of larval herbivory in fungus gnats. Consider the three principal ingredients:

*Luminescence:* The light organ of *A. luminosa* forms at the rear end of the larva from enlarged tips of four excretory tubes. These tubes carry a waste product that glows in the presence of luciferase, an enzyme also produced by the larva. This reaction requires a good supply of oxygen, and the four excretory tubes lie embedded in a dense network of respiratory tubules that both supply oxygen to fuel the reaction and then reflect and direct the light downward. This complex and specially evolved system functions to attract insects (mostly small midges) to the nest. Pupae and imagoes retain the ability to luminesce. The light of female pupae and adults attracts males, but the glow of adult males has no known function.

*The Nest and Feeding Threads:* From glands in its mouth, the glowworm exudes silk and mucus to construct a marvel of organic architecture. The young larva first builds the so-called nest—really more of a hollow tube or runway—some two to three times the length of its body. A network of fine silk threads suspends this nest from the cave's ceiling. The larva drops a curtain of closely spaced feeding threads from its nest. These "fishing lines" may number up to seventy per nest and may extend almost a foot in length (or ten times the span of the larva itself). Each line is studded along its entire length with evenly spaced, sticky droplets that catch intruding insects; the entire structure resembles, in miniature, a delicate curtain of glass beads. Since the slightest current of air can cause these lines to tangle, caves, culverts, ditches, and calm spaces amidst vegetation provide the limited habitats for *A. luminosa* in New Zealand.

*Carnivory:* Using its lighted rear end as a beacon, *A. luminosa* attracts prey to its feeding threads. Two posterior papillae contain sense organs that detect vibrations of ensnared prey. The larva then crawls partway down the proper line, leaving half to two-thirds of its rear in the nest, and hauls up both line and meal at a rate of some 2 millimeters per second.

The rest of the life cycle pales by comparison with this complexity of larval anatomy and behavior. The pupal stage lasts a bit less than two weeks and already records a marked reduction in size (15 to 18 millimeters for females, 12 to 14 for males). I have already noted the imago's decrease in body size and duration of life. Imaginal behavior also presents little in the way of diversity or complexity. Adult flies have no mouth and do not feed at all. We commit no great exaggeration by stating that they behave as unipurpose mating and egg-laying machines during their brief existence. Up to three males may congregate at a female pupa, awaiting her emergence. They jockey for positon and fight as the female fly begins to break through her encasement. As soon as the tip of her abdomen emerges, males (if present) begin to mate. Thus, females can be fertilized even before they break fully from the pupal case. Females may then live for less than a day (and no more than three), doing little more before they expire than finding an appropriate place for some 100 to 300 eggs, laid one at a time in clumps of 40 to 50. Males may live an additional day (up to four); with luck, they may find another female and do it again for posterity.

As a final and grisly irony, emphasizing larval dominance over the life cycle of *A. luminosa,* a rapacious glowworm will eat anything that touches its feeding threads. The much smaller imagoes often fly into the lines and end up as just another meal for their own children.*

Please do not draw from this essay the conclusion that larvae

---

*To throw in a tidbit for readers interested in the history of evolutionary theory, this tightly coordinated complex of larval adaptations so intrigued Richard Goldschmidt that he once wrote an entire article to argue that light, carnivory, and nest building could not have arisen by gradual piecemeal, since each makes no sense without the others—and that all, therefore, must have appeared at once as a fortuitous consequence of a large mutational change, a "hopeful monster," in his colorful terminology.

This proposal (published in English in *Revue Scientifique,* 1948) inspired a stern reaction from orthodox Darwinians. Although I have great sympathy for Goldschmidt's iconoclasm, he was, I think, clearly wrong in this case. As J. F. Jackson pointed out (1974), Goldschmidt made an error in the taxonomic assignment of *A. luminosa* among the Mycetophilidae. He ranked this species in the subfamily Bolitophilinae. All larvae of this group burrow into soft mushrooms, and none shows even incipient development of any among the three linked features that mark the unique form and behavior of *A. luminosa.* Hence, Goldschmidt argued for all or nothing.

But *A. luminosa* probably belongs in another subfamily, the Keroplatinae—and, unknown to Goldschmidt, several species within this group do display a series of plausible transitions. *Leptomorphus* catches and eats fungal spores trapped on a sheetlike nest slung below a mushroom. Some species of *Macrocera* and *Keroplatus* also build trap nets for fungal spores but will eat small arthropods that also become ensnared. Species of *Orfelia, Apemon,* and Platyura build webs of similar form but not associated with mushrooms—and they live exclusively on a diet of trapped insects. Finally, *Orfelia aeropiscator* (literally, air fisher) both builds a nest and hangs vertical feeding threads but does not possess a light.

These various "intermediates" are, of course, not ancestral to *A. luminosa.* Each represents a well-adapted species in its own right, not a transitional stage to the threefold association of New Zealand glowworms. But this array does show that each step in a plausible sequence of structurally intermediate stages can work as a successful organism. This style of argument follows Darwin's famous resolution for a potential evolutionary origin for the extraordinary complexity of the vertebrate eye. Darwin identified a series of structural intermediates, from simple light-sensitive dots to cameralike lens systems—not actual ancestors (for these are lost among nonpreservable eyes in a fossil record of hard parts) but plausible sequences disproving the "commonsense" notion that nothing in between is possible in principle.

are really more important than imagoes, either in *A. luminosa* or in general. I have tried to show that larvae must not be dismissed—as preparatory, undeveloped, or incomplete—by false analogy to a dubious (but socially favored) interpretation of human development. If any "higher reality" exists, we can only specify the life cycle itself. Larva and imago are but two stages of a totality—and you really can't have one without the other. Eggs need hens as much as hens need eggs.

I do try to show that child-adult is the wrong metaphor for understanding larva-imago. I have proceeded by discussing a case where larvae attract all our attention—literally as a source of beauty; structurally in greater size, length of life, and complexity of anatomy and behavior; and evolutionarily as focus of a major transformation from a simpler and very different ancestral style—while imagoes have scarcely modified their inherited form and behavior at all. But our proper emphasis on the larva of *A. luminosa* does not mark any superiority.

We need another metaphor to break the common interpretation that degrades larvae to a penumbra of insignificance. (How many of you include maggot in your concept of fly? And how many have ever considered the mayfly's longer larval life?) The facts of nature are what they are, but we can only view them through spectacles of our mind. Our mind works largely by metaphor and comparison, not always (or often) by relentless logic. When we are caught in conceptual traps, the best exit is often a change in metaphor—not because the new guideline will be truer to nature (for neither the old nor the new metaphor lies "out there" in the woods), but because we need a shift to more fruitful perspectives, and metaphor is often the best agent of conceptual transition.

If we wish to understand larvae as working items in their own right, we should replace the developmental metaphor of child-adult with an economic simile that recognizes the basic distinction in function between larvae and imagoes—larvae as machines built for feeding and imagoes as devices for reproduction. Fortunately, an obvious candidate presents itself on the very first page of the founding document itself—Adam Smith's *Wealth of Nations*. We find our superior metaphor in the title of Chapter 1, "On the Division of Labor," and in Smith's opening sentence:

The greatest improvement in the productive powers of labor, and the greater part of the skill, dexterity, and judgment with which it is anywhere directed, or applied, seem to have been the effects of the division of labor.

By allocating the different, sometimes contradictory, functions of feeding and reproduction to sequential phases of the life cycle, insects with complete metamorphosis have achieved a division of labor that permits a finer adaptive honing of each separate activity.

If you can dredge up old memories of your first college course in economics, you will remember that Adam Smith purposely chose a humble example to illustrate the division of labor—pin making. He identifies eighteen separate actions in drawing the wire, cutting, pointing, manufacture of the head, fastening head to shaft, and mounting the finished products in paper for sale. One man, he argues, could make fewer than twenty pins a day if he performed all these operations himself. But ten men, sharing the work by rigid division of labor, can manufacture about 48,000 pins a day. A human existence spent pointing pins or fashioning their heads or pushing them into paper may strike us as the height of tedium, but larvae of *A. luminosa* encounter no obvious psychic stress in a life fully devoted to gastronomy.

Hobbyists and professional entomologists will, no doubt, have recognized an unintended irony in Smith's selection of pin making to illustrate the division of labor. Pins are the primary stock-in-trade of any insect collector. They are used to fasten the dry and chitinous imagoes—but not the fat and juicy larvae—to collecting boards and boxes. Thus, the imagoes of *A. luminosa* may end their natural life caught in a larval web, but if they happen to fall into the clutches of a human collector, they will, instead, be transfixed by the very object that symbolizes their fall from conceptual dominance to proper partnership.

---

## Postscript

Nothing brings greater pleasure to a scholar than utility in extension—the fruitfulness of a personal thought or idea when devel-

Stars

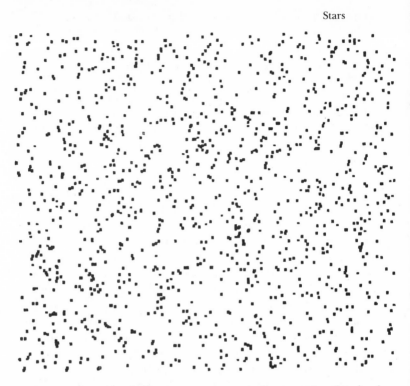

Output from Ed Purcell's computer program for arranging dots by the "stars," ABOVE (random), and the "worms," FACING PAGE (ordered by fields of inhibition around each dot), options. Note the curious psychological effect. Most of us would see order in the strings and clumps of the figure just above, and would interpret the figure on the opposite page, with its lack of apparent pattern, as random. In fact, the opposite is true, and our ordinary conceptions are faulty.

oped by colleagues beyond the point of one's own grasp. I make a tangential reference in this essay to a common paradox—the apparent pattern of random arrays versus the perceived absence of sensible order in truly rule-bound systems. This paradox arises because random systems are highly clumped, and we perceive clumps as determined order. I gave the example of the heavens—

Worms

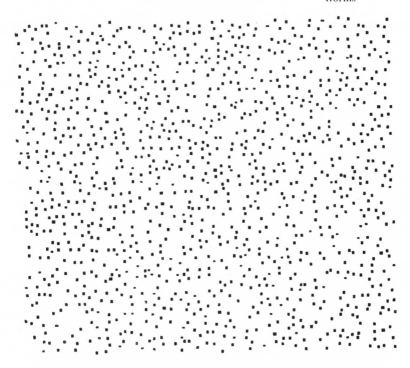

where we "see" constellations because stars are distributed at random relative to the earth's position. I contrasted our perception of heavenly order with the artificial "sky" of Waitomo Cave—where "stars" are the self-illuminated rear ends of fly larvae. Since these carnivorous larvae space themselves out in an ordered array (because they eat anything in their vicinity and therefore set up "zones of inhibition" around their own bodies), the Waitomo "sky" looks strange to us for its absence of clumping.

My favorite colleague, Ed Purcell (Nobel laureate in physics and sometime collaborator on baseball statistics), read this tangential comment and wrote a quick computer program to illustrate the effect. Into an array of square cells (144 units on the X-axis and 96 on the Y-axis for a total of 13,824 positions), Purcell placed either "stars" or "worms" by the following rules of randomness and order (following the heavens versus the fly larvae of Waitomo). In the stars option, squares are simply occupied at random (a random number generator spits out a figure between 1 and 13,824 and the appropriate square is inked in). In the worms option, the same generator spits out a number, but the appropriate square is inked in only if it and all surrounding squares are unoccupied (just as a worm sets up a zone of inhibition about itself). Thus, worm squares are spaced out by a principle of order; star squares are just filled in as the random numbers come up.

Now examine the patterns produced with 1,500 stars and worms (still less than 50 percent capacity for worms, since one in four squares could be occupied, and 3,456 potential worm holes therefore exist). By ordinary vernacular perception, we could swear that the "stars" program must be generating causal order, while the "worms" program, for apparent lack of pattern, seems to be placing the squares haphazardly. Of course, exactly the opposite is true. In his letter to me, Ed wrote:

> What interests me more in the random field of "stars" is the overpowering impression of "features" of one sort or another. It is hard to accept the fact that any perceived feature—be it string, clump, constellation, corridor, curved chain, lacuna—is a totally meaningless accident, having as its cause the avidity for pattern of my eye and brain! Yet that is perfectly true in this case.

I don't know why our brains (by design or culture) equip us so poorly as probability calculators—but this nearly ubiquitous failure constitutes one of the chief, and often dangerous, dilemmas of both intellectual and everyday life (the essays of Section 9, particularly number 31 on Joe DiMaggio's hitting streak, discuss this subject at greater length). Ed Purcell adds, emphasizing the pervasiveness of misperception, even among people trained in probability:

> If you ask a physics student to take pen in hand and sketch a random pattern of 1,500 dots, I suspect the result will look more like the "worms" option than the "stars."

# 18 | To Be a Platypus

LONG AGO, garrulous old Polonius exalted brevity as the soul of wit, but later technology, rather than sweet reason, won his day and established verbal condensation as a form of art in itself. The telegram, sent for cash on the line and by the word, made brevity both elegant and economical—and the word *telegraphic* entered our language for a style that conveys bare essentials and nothing else.

The prize for transmitting most meaning with least verbosity must surely go to Sir Charles Napier, who subdued the Indian province of Sind and announced his triumph, via telegram to his superiors in London, with the minimal but fully adequate *"Peccavi."* This tale, in its own telegraphic way, speaks volumes about the social order and education of imperial Britain. In an age when all gentlemen studied Latin, and could scarcely rise in government service without a boost from the old boys of similar background in appropriate public schools, Napier never doubted that his superiors would remember the first-person past tense of the verb *peccare*—and would properly translate his message and pun: I have sinned.

The most famous telegram from my profession did not quite reach this admirable minimum, but it must receive honorable mention for conveying a great deal in few words. In 1884, W. H. Caldwell, a young Cambridge biologist, sent his celebrated telegram from Australia to a triumphant reading at the Annual Meeting of the British Association in Montreal. Caldwell wired: "Monotremes oviparous, ovum meroblastic."

This message may lack the ring of *peccavi* and might be viewed by the uninitiated as pure mumbo jumbo. But all professional biologists could make the translation and recognize that Caldwell had solved a particularly stubborn and vexatious problem of natural history. In essence, his telegram said: The duckbilled platypus lays eggs.

(Each word of Caldwell's telegram needs some explication. Oviparous animals lay eggs, while viviparous creatures give birth to live young; ovoviviparous organisms form eggs within their bodies, and young hatch inside their mothers. Sorry for the jargon so early in the essay, but these distinctions become important later on. Monotremes are that most enigmatic group of mammals from the Australian region—including the spiny echidna, actually two separate genera of anteaters, and the duckbilled platypus, an inhabitant of streams and creeks. An ovum is an egg cell, and meroblastic refers to a mode of cleavage, or initial division into embryonic cells, after fertilization. Yolk, the egg's food supply, accumulates at one end of the ovum, called the vegetal pole. Cleavage begins at the other end, called the animal pole. If the egg is very yolky, the cleavage plane cannot penetrate and divide the vegetal end. Such an egg shows incomplete, or meroblastic, cleavage—division into discrete cells at the animal pole but little or no separation at the yolky end. Egg-laying land vertebrates, reptiles and birds, tend to produce yolky egg cells with meroblastic cleavage, while most mammals show complete, or holoblastic, cleavage. Therefore, in adding "ovum meroblastic" to "monotremes oviparous," Caldwell emphasized the reptilian character of these paradoxical mammals—not only do they lay eggs but the eggs are typically reptilian in their yolkiness.)

The platypus surely wins first prize in anybody's contest to identify the most curious mammal. Harry Burrell, author of the classic volume on this anomaly (*The Platypus: Its Discovery, Position, Form and Characteristics, Habits and Life History,* 1927), wrote: "Every writer upon the platypus begins with an expression of wonder. Never was there such a disconcerting animal!" (I guess I just broke tradition by starting with the sublime *Hamlet.*)

The platypus sports an unbeatable combination for strangeness: first, an odd habitat with curiously adapted form to match; second, the real reason for its special place in zoological history—its engimatic mélange of reptilian (or birdlike), with obvious mammalian, characters. Ironically, the feature that first suggested premammalian affinity—the "duckbill" itself—supports no such meaning. The platypus's muzzle (the main theme of this column) is a purely mammalian adaptation to feeding in fresh waters, not a throwback to ancestral form—although the duckbill's formal name embodies this false interpretation: *Ornithorhynchus anatinus* (or the ducklike bird snout).

Chinese taxidermists had long fooled (and defrauded) European mariners with heads and trunks of monkeys stitched to the hind parts of fish—one prominent source for the persistence of mermaid legends. In this context, one can scarcely blame George Shaw for his caution in first describing the platypus (1799):

> Of all the Mammalia yet known it seems the most extraordinary in its conformation, exhibiting the perfect resemblance of the beak of a Duck engrafted on the head of a quadruped. So accurate is the similitude, that, at first view, it naturally excites the idea of some deceptive preparation.

But Shaw could find no stitches, and the skeleton was surely discrete and of one functional piece (the premaxillary bones of the upper jaw extend into the bill and provide its major support). Shaw concluded:

> On a subject so extraordinary as the present, a degree of scepticism is not only pardonable but laudable; and I ought perhaps to acknowledge that I almost doubt the testimony of my own eyes with respect to the structure of this animal's beak; yet must confess that I can perceive no appearance of any deceptive preparation . . . nor can the most accurate examination of expert anatomists discover any deception.

The frontal bill may have provoked most astonishment, but the rear end also provided numerous reasons for amazement. The platypus sported only one opening, the cloaca, for all excretory and reproductive business (as in reptiles, but not most mammals, with their multiplicity of orifices for birth and various forms of excretion; Monotremata, or "one-holed," the technical name for the platypus and allied echidna, honors this unmammalian feature).

Internally, the puzzle only increased. The oviducts did not

unite into a uterus, but extended separately into the cloacal tube. Moreover, as in birds, the right ovary had become rudimentary, and all egg cells formed in the left ovary. This configuration inevitably led to a most troubling hypothesis for biologists committed, as most were in these pre-Darwinian days, to the division of nature into unambiguous, static categories: no uterus, no internal space to form a placenta, a reproductive tract reptilian in form. All this suggested the unthinkable for a mammal—birth from eggs. The neighboring marsupials, with their pouches and tiny joeys, had already compromised the noble name of mammal. Would Australia also yield the ultimate embarrassment of fur from eggs?

As anatomists studied this creature early in the nineteenth century, the mystery only deepened. The platypus looked like a perfectly good mammal in all "standard" nonreproductive traits. It sported a full coat of hair and the defining anatomical signature of mammals—one bone, the dentary, in its lower jaw and three, the hammer, anvil, and stirrup, in its middle ear. (Reptiles have several jawbones and only one ear bone. Two reptilian jawbones became the hammer and anvil of the mammalian ear.) But premammalian characters also extended beyond the reproductive system. In particular, the platypus grew an interclavicle bone in its shoulder girdle—a feature of reptiles shared by no placental mammal.

What could this curious mélange be, beyond a divine test of faith and patience? Debate centered on modes of reproduction, for eggs had not yet been found and Caldwell's telegram lay half a century in the future. All three possibilities boasted their vociferous and celebrated defenders—for no great biologist could avoid such a fascinating creature, and all leaders of natural history entered the fray. Meckel, the great German anatomist, and his French colleague Blainville predicted viviparity, argued that eggs would never be found, and accommodated the monotremes among ordinary mammals. E. Home, who first described the platypus in detail (1802), and the renowned English anatomist Richard Owen chose the middle pathway of ovoviviparity and argued that failure to find eggs indicated their dissolution within the female's body. But the early French evolutionists, Lamarck and Etienne Geoffroy Saint-Hilaire, insisted that anatomy could not lie and that the platypus must be oviparous. Eggs, they argued, would eventually be found.

Geoffroy, by the way, coined the name *monotreme* in an interesting publication that reveals as much about French social history as *peccavi* indicated for imperial Britain. This issue of the *Bulletin des sciences* is labeled *Thermidor, an 11 de la République*. With revolutionary fervor at its height, France broke all ties with the old order and started counting again from year one (1793). They also redivided the year into twelve equal months, and renamed the months to honor the seasons rather than old gods and emperors. Thus, Geoffroy christened the monotremes in a summer month (Thermidor) during the eleventh year (1803) of the Republic (see Essay 24 for more on the French revolutionary calendar).

Just one incident in the pre-Caldwell wars will indicate the intensity of nineteenth-century debate about platypuses and the relief at Caldwell's resolution. When the great naturalists delineated their positions and defined the battleground, mammary glands had not been found in the female platypus—an apparent argument for those, like Geoffroy, who tried to distance monotremes as far as possible from mammals. Then, in 1824, Meckel discovered mammary glands. But since platypuses never do anything by the book, these glands were peculiar enough to spur more debate rather than conciliation. The glands were enormous, extending nearly from the forelegs to the hind limbs—and they led to no common opening, for no nipples could be found. (We now know that the female excretes milk through numerous pores onto a portion of her ventral surface, where the baby platypus laps it up.) Geoffroy, committed to oviparity and unwilling to admit anything like a mammalian upbringing, counterattacked. Meckel's glands, he argued, were not mammary organs, but homologues of the odiferous flank glands of shrews, secreting substances for attraction of mates. When Meckel then extracted a milky substance from the mammary gland, Geoffroy admitted that the secretion must be food of some sort, but not milk. The glands, he now argued, are not mammary but a special feature of monotremes, used to secrete thin strands of mucus that thicken in water to provide food for young hatched from the undiscovered eggs.

Owen then counterattacked to support Meckel for three reasons: The glands are largest shortly after the inferred time of birth (though Geoffroy expected the same for mucus used in feeding). The female echidna, living in sand and unable to thicken mucus in water, possesses glands of the same form. Finally, Owen suspended the secretion in alcohol and obtained globules, like milk, not angular fragments, like mucus (an interesting commentary upon the rudimentary state of chemical analysis during the 1830s).

Geoffroy held firm—both to oviparity (correctly) and to the special status of feeding glands (incorrectly, for they are indeed mammary). In 1822, Geoffroy formally established the Monotremata as a fifth class of vertebrates, ranking equally with fishes, reptiles (then including amphibians), birds, and mammals. We may view Geoffroy as stubborn, and we certainly now regard the monotremes as mammals, however peculiar—but he presents a cogent and perceptive argument well worth our attention. Don't shoehorn monotremes into the class Mammalia to make everything neat and foreclose discussion, he pleads. Taxonomies are guides to action, not passive devices for ordering. Leave monotremes separate and in uncomfortable limbo—"which suggests the necessity of further examination [and] is far better than an assimilation to normality, founded on strained and mistaken relations, which invites indolence to believe and slumber" (letter to the Zoological Society of London, 1833).

Geoffroy also kept the flame of oviparity alive, arguing that the cloaca and reproductive tract bore no other interpretation: "Such as the organ is, such must be its function; the sexual apparatus of an oviparous animal can produce nothing but an egg." So Caldwell arrived in Australia in September 1883—and finally resolved the great debate, eighty years after its inception.

Caldwell, though barely a graduate, proceeded in the grand imperial style (he soon disappeared from biological view and became a successful businessman in Scotland). He employed 150 aboriginals and collected nearly 1,400 echidnas—quite a hecatomb for monotreme biology. On the subject of social insights, this time quite uncomfortable, Caldwell described his colonial style of collecting:

> The blacks were paid half-a-crown for every female, but the price of flour, tea, and sugar, which I sold to them, rose with

the supply of Echidna. The half-crowns were, therefore, always just sufficient to buy food enough to keep the lazy blacks hungry.

It was, of course, often done—but rarely said so boldly and without apology. In any case, Caldwell eventually found the eggs of the platypus (usually laid two at a time and easily overlooked at their small size of less than an inch in length).

Caldwell solved a specific mystery that had plagued zoology for nearly a century, but he only intensified the general problem. He had proved irrevocably that the platypus is a mélange, not available for unambiguous placement into any major group of vertebrates. Geoffroy had been right about the eggs; Meckel about the mammary glands.

The platypus has always suffered from false expectations based on human foibles. (This essay discusses the two stages of this false hoping, and then tries to rescue the poor platypus in its own terms.) During the half-century between its discovery and Darwin's *Origin of Species,* the platypus endured endless attempts to deny or mitigate its true mélange of characters associated with different groups of vertebrates. Nature needed clean categories established by divine wisdom. An animal could not both lay eggs and feed its young with milk from mammary glands. So Geoffroy insisted upon eggs and no milk; Meckel upon milk and live birth.

Caldwell's discovery coincided with the twenty-fifth anniversary of Darwin's *Origin.* By this time, evolution had made the idea of intermediacy (and mélanges of characters) acceptable, if not positively intriguing. Yet, freed of one burden, the platypus assumed another—this time imposed by evolution, the very idea that had just liberated this poor creature from uncongenial shoving into rigid categories. The platypus, in short, shouldered (with its interclavicle bone) the burden of primitiveness. It would be a mammal, to be sure—but an amoeba among the gods; a tawdry, pitiable little fellow weighted down with the reptilian mark of Cain.

Caldwell dispatched his epitome a century ago, but the platypus has never escaped. I have spent the last week as a nearly full-time reader of platypusology. With a few welcome exceptions (mostly among Australian biologists who know the creature intimately), nearly every article identifies something central about the platypus as undeveloped or inefficient relative to placental mammals—as if the undoubted presence of premammalian characters condemns each feature of the platypus to an unfinished, blundering state.

Before I refute the myth of primitiveness for the platypus in particular, I should discuss the general fallacy that equates early with inefficient and still underlies so much of our failure to understand evolution properly. The theme has circulated through these essays for years—ladders and bushes. But I try to provide a new twist here—the basic distinction between *early branching* and *undeveloped,* or *inefficient, structure.*

If evolution were a ladder toward progress, with reptiles on a rung below mammals, then I suppose that eggs and an interclavicle would identify platypuses as intrinsically wanting. But the Old Testament author of Proverbs, though speaking of wisdom rather than evolution, provided the proper metaphor, *etz chayim:* She is a *tree of life* to them who take hold upon her. Evolution proceeds by branching, and not (usually) by wholesale transformation and replacement. Although a lineage of reptiles did evolve into mammals, reptiles remain with us in all their glorious abundance of snakes, lizards, turtles, and crocodiles. Reptiles are doing just fine in their own way.

The presence of premammalian characters in platypuses does not brand them as inferior or inefficient. But these characters do convey a different and interesting message. They do signify an early branching of monotreme ancestors from the lineage leading to placental mammals. This lineage did not lose its reptilian characters all at once, but in the halting and piecemeal fashion so characteristic of evolutionary trends. A branch that split from this central lineage after the defining features of mammals had evolved (hair and an earful of previous jawbones, for example) might retain other premammalian characters (birth from eggs and an interclavicle) as a sign of early derivation, not a mark of backwardness.

The premammalian characters of the platypuses only identify the antiquity of their lineage as a separate branch of the mammalian tree. If anything, this very antiquity might give the platypus more scope (that is, more time) to become what it really is, in opposition to the myth of primitivity: a superbly engineered creature for a particular, and unusual, mode of life. The platypus is an elegant solution for mammalian life in streams—not a primitive relic of a bygone world. Old does not mean hidebound in a Darwinian world.

Once we shuck the false expectation of primitiveness, we can view the platypus more fruitfully as a bundle of adaptations. Within this appropriate theme of *good* design, we must make one further distinction between shared adaptations of all mammals and particular inventions of platypuses. The first category includes a coat of fur well adapted for protecting platypuses in the (often) cold water of their streams (the waterproof hair even traps a layer of air next to the skin, thus providing additional insulation). As further protection in cold water and on the same theme of inherited features, platypuses can regulate their body temperatures as well as most "higher" mammals, although the assumption of primitivity stalled the discovery of this capacity until 1973—before that, most biologists had argued that platypus temperatures plummeted in cold waters, requiring frequent returns to the burrow for warming up. (My information on the ecology of modern platypuses comes primarily from Tom Grant's excellent book, *The Platypus,* New South Wales University Press, 1984, and from conversations with Frank Carrick in Brisbane. Grant and Carrick are Australia's leading professional students of platypuses, and I thank them for their time and care.)

These features, shared by passive inheritance with other mammals, certainly benefit the platypus, but they provide no argument for my theme of direct adaptation—the replacement of restraining primitivity by a view of the platypus as actively evolving in its own interest. Many other features, however, including nearly everything that makes the platypus so distinctive, fall within the second category of special invention.

Platypuses are relatively small mammals (the largest known weighed just over five pounds and barely exceeded two feet from tip to tail). They construct burrows in the banks of creeks and rivers: long (up to sixty feet) for nesting; shorter for daily use. They spend most of their life in the water, searching for food (primarily insect larvae and other small invertebrates) by probing into bottom sediments with their bills.

The special adaptations of platypuses have fitted them in a subtle and intricate way for aquatic life. The streamlined body moves easily through water. The large, webbed forefeet propel the animal forward by alternate kicks, while the tail and partially webbed rear feet act as rudders and steering devices (in digging a burrow, the platypus anchors with its rear feet and excavates with its forelimbs). The bill works as a feeding structure par excellence, as I shall describe in a moment. Other features undoubtedly serve in the great Darwinian game of courtship, reproduction, and rearing—but we know rather little about this vital aspect of platypus life. As an example, males bear a sharp, hollow spur on their ankles, attached by a duct to a poison gland in their thighs. These spurs, presumably used in combat with competing males, grow large during the breeding season. In captivity, males have killed others with poison from their spurs, and many platypuses, both male and female, sport distinctive punctures when captured in the wild.

Yet even this long and impressive list of special devices has been commonly misrepresented as yet another aspect (or spinoff) of pervasive primitiveness. Burrell, in his classic volume (1927), actually argued that platypuses develop such complex adaptations because simple creatures can't rely upon the flexibility of intelligence and must develop special structures for each required action. Burrell wrote:

> Man . . . has escaped the need for specialization because his evolution has been projected outside himself into an evolution of tools and weapons. Other animals in need of tools and weapons must evolve them from their own bodily parts; we therefore frequently find a specialized adaptation to environmental needs grafted on to primitive simplicity of structure.

You can't win in such a world. You are either primitive prima facie or specialized as a result of lurking and implicit simplicity! From such a Catch-22, platypuses can only be rescued by new concepts, not additional observations.

As a supreme irony, and ultimate defense of adaptation versus ineptitude, the structure that built the myth of primitivity—the misnamed duckbill itself—represents the platypus's finest special invention. The platypus bill is not a homologue of any feature in birds. It is a novel structure, uniquely evolved by monotremes (the echidna carries a different version as its long and pointed snout). The bill is not simply a hard, inert horny structure. Soft skin covers the firm substrate, and this skin houses a remarkable array of sensory organs. In fact, and strange to tell, the platypus, when under water, shuts down all its other sensory systems and relies entirely upon its bill to locate obstacles and food. Flaps of skin cover tiny eyes and nonpinnate ears when a platypus dives, while a pair of valves closes off the nostrils under water.

E. Home, in the first monograph of platypus anatomy (1802), made an astute observation that correctly identified the bill as a complex and vital sensory organ. He dissected the cranial nerves and found almost rudimentary olfactory and optic members but a remarkably developed trigeminal, carrying information from the face to the brain. With great insight, Home compared the platypus bill to a human hand in function and subtlety. (Home never saw a live platypus and worked only by inference from anatomy.) He wrote:

> The olfactory nerves are small and so are the optic nerves; but the fifth pair which supplies the muscles of the face are uncommonly large. We should be led from this circumstance to believe, that the sensibility of the different parts of the bill is very great, and therefore it answers to the purpose of a hand, and is capable of nice discrimination in its feeling.

Then, in the same year that Caldwell discovered eggs, the English biologist E. B. Poulton found the primary sensory organs of the bill. He located numerous columns of epithelial cells, each underlain by a complex of neural transmitters. He called them "push rods," arguing by analogy with electrical bells that a sensory stimulus (a current of water or an object in bottom sediments) would depress the column and ignite the neural spark.

A set of elegant experiments in modern neurophysiology by R. C. Bohringer and M. J. Rowe (1977 and 1981) can only increase our appreciation for the fine-tuned adaptation of the platypus bill. They found Poulton's rods over the bill's entire surface, but four to six times more densely packed at the anterior border of the upper bill, where platypuses must first encounter obstacles and food items. They noted different kinds of nerve receptors under the rods, suggesting that platypuses can distinguish varying kinds of signals (perhaps static versus moving components or live versus dead food). Although individual rods may not provide sufficient information for tracing the direction of a stimulus, each rod maps to a definite location on the brain, strongly implying that the sequence of activation among an array of rods permits the platypus to identify the size and location of objects.

Neurophysiologists can locate areas of the brain responsible for activating definite parts of the body and draw a "map" of the body upon the brain itself. (These experiments proceed from either direction. Either one stimulates a body part and records the pattern of activity in a set of electrodes implanted into the brain, or one pulses a spot on the brain and determines the resulting motion of body parts.) We have no finer demonstrations of evolutionary adaptation than numerous brain maps that record the importance of specially developed organs by their unusually enlarged areas of representation upon the cortex. Thus, a raccoon's brain map displays an enormous domain for its forepaws, a pig's for its snout, a spider monkey's for its tail. Bohringer and Rowe have added the platypus to this informative array. A map of the platypus's cortex is mostly bill.

We have come a long way from the first prominent evolutionary interpretation ever presented for the platypus bill. In 1844, in the major pre-Darwinian defense of evolution written in English, Robert Chambers tried to derive a mammal from a bird in two great leaps, via the intermediate link of a duckbilled platypus. One step, Chambers wrote,

> would suffice in a goose to give its progeny the body of a rat, and produce the ornithorhynchus, or might give the progeny of an ornithorhynchus the mouth and feet of a true rodent, and thus complete at two stages the passage from the aves to the mammalia.

The platypus, having suffered such slings and arrows of outrageous fortune in imposed degradation by human hands, has cast its arms (and its bill) against a sea of troubles and vindicated itself. The whips and scorns of time shall heal. The oppressor's wrong, the proud man's contumely have been reversed by modern studies—enterprises of great pith and moment. The platypus is one honey of an adaptation.

# 19 | Bligh's Bounty

IN 1789, a British naval officer discovered some islands near Australia and lamented his inability to provide a good description:

> Being constantly wet, it was with the utmost difficulty I could open a book to write, and I am sensible that what I have done can only serve to point out where these lands are to be found again, and give the idea of their extent.

As he wrote these lines, Captain William Bligh was steering a longboat with eighteen loyal crew members into the annals of human heroism at sea—via his 4,000-mile journey to Timor, accomplished without loss of a single man, and following the seizure of his ship, *The Bounty*, in history's most famous mutiny.

Bligh may have been overbearing; he surely wins no awards for insight into human psychology. But history and Charles Laughton have not treated him fairly either. Bligh was committed, meticulous, and orderly to a fault—how else, in such peril, could he have bothered to describe some scattered pieces of new Pacific real estate.

Bligh's habit of close recording yielded other benefits, including one forgotten item to science. Obsessed by the failure of his *Bounty* mission to bring Tahitian breadfruit as food for West Indian slaves, Bligh returned to Tahiti aboard the *Providence* and successfully unloaded 1,200 trees at Port Royal, Jamaica, in 1793 (his ship was described as a floating forest). En route, he stopped in Australia and had an interesting meal.

281

George Tobin, one of Bligh's officers, described their quarry as

> a kind of sloth about the size of a roasting pig with a proboscis 2 or 3 inches in length. . . . On the back were short quills like those of the Porcupine. . . . The animal was roasted and found of a delicate flavor.

Bligh himself made a drawing of his creature before the banquet. The officers of the *Providence* had eaten an echidna, one of Australia's most unusual mammals—an egg-laying anteater closely related to the duckbilled platypus.

Bligh brought his drawing back to England. In 1802, it appeared as a figure (reproduced here) accompanying the first technical description of the echidna's anatomy by Everard Home in the *Philosophical Transactions of the Royal Society* (G. Shaw had published a preliminary and superficial description in 1792).

Home discovered the strange mix of reptilian and mammalian features that has inspired interest and puzzlement among biologists ever since. He also imposed upon the echidna, for the first time, the distinctive burden of primitivity that has continually hampered proper zoological understanding of all monotremes, the egg-laying mammals of Australia. Home described the echidna as not quite all there in mammalian terms, a lesser form stamped with features of lower groups:

> These characters distinguish [the echidna] in a very remarkable manner, from all other quadrupeds, giving this new tribe a resemblance in some respects to birds, in others to the Amphibia; so that it may be considered as an intermediate link between the classes of Mammalia, Aves, and Amphibia.

Unfortunately, Home could not study the organ that most clearly belies the myth of primitivity. "The brain," he wrote, "was not in a state to admit of particular examination." Home did have an opportunity to infer the echidna's anomalously large brain from the internal form of its skull, well drawn on the plate just preceding Bligh's figure (and also reproduced here). But Home said nothing about this potential challenge to his general interpretation.

Original drawing of an echidna by none other than Captain Bligh of *Bounty* fame. NEG. NO. 337535. COURTESY DEPARTMENT OF LIBRARY SERVICES, AMERICAN MUSEUM OF NATURAL HISTORY.

And so the burden of primitivity stuck tenaciously to echidnas, and continues to hold fast in our supposedly more sophisticated age. Some great zoologists have struggled against this convenient fallacy, most notably the early French evolutionist Etienne Geoffroy Saint-Hilaire, who coined the name Monotremata (see

Essay 18) and labored unsuccessfully to establish the echidna and platypus as a new class of vertebrates, separate from both mammals and reptiles and not merely inferior to placentals. By his own manifesto, he chose his strategy explicitly to avoid the conceptual lock that assumptions of primitivity would clamp upon our understanding of monotremes. He wrote in 1827:

> What is defective, I repeat, is our manner of perception, our way of conceiving the organization of monotremes; that is, our determination, made a priori, to join them violently to mammals [by *violemment*, Geoffroy means, of course, "without any conceptual justification"], to place them in the same class and, after our disappointments and false judgments, then to make our unjust grievances heard, as when we speak of them as mammals essentially and necessarily outside the rules.

But Geoffroy's legitimate complaint, so eloquently expressed, did not prevail, and the myth of primitivity continues, despite its blatant flaw. As I argue in the preceding essay on platypuses, the myth of primitivity rests upon a logical confusion between early branching from the ancestors of placental mammals (the true meaning of reptilian characters retained by monotremes) and structural inferiority. Unless geological age of branching is a sure guide to level of anatomical organization—as it is not—egg laying and interclavicle bones do not brand platypuses and echidnas as inferior mammals.

Everard Home's 1802 figure of an echidna's skull. The large size of the brain was apparent even then. NEG. NO. 337429. COURTESY DEPARTMENT OF LIBRARY SERVICES, AMERICAN MUSEUM OF NATURAL HISTORY.

Beyond this general defense, echidnas can provide ample specific evidence of their adequacy. They are, first of all, a clear success in ecological terms. Echidnas live all over the Australian continent (and extend into Papua-New Guinea), the only native mammal with such a wide range. Moreover, *the* echidna, as a single struggling relict, ranks with *the* rat and *the* monkey (those meaningless synecdoches of the psychological literature) as an absurd abstraction of nature's richness. Echidnas come as two species in two separate genera and with quite different habits. *Tachyglossus aculeatus* (the Australian form with Papuan extensions) rips apart ant and termite nests with its stout forelimbs and collects the inhabitants on its sticky tongue. The larger and longer-snouted *Zaglossus bruijni* of Papua-New Guinea lives on a nearly exclusive diet of earthworms. Moreover, three other species, including the "giant" echidna, *Zaglossus hacketti,* have been found as fossils in Australia. Echidnas are a successful and at least modestly varied group.

But echidnas hold a far more important ace in the hole as their ultimate defense against charges of primitivity. The same cultural biases that lead us to classify creatures as primitive or advanced have established the form and function of brains as our primary criterion of ranking. Echidnas have big and richly convoluted brains. Scientists have recognized this anomaly in the tale of primitivity for more than a century—and they have developed an array of arguments, indeed a set of traditions, for working around such an evident and disconcerting fact. Large brains undoubtedly serve echidnas well; but they also help to instruct us about an important issue in the practice of science—how do scientists treat factual anomalies? What do we do with evidence that challenges a comfortable view of nature's order?

The echidna's brain refutes the myth of primitivity with a double whammy—size and conformation. (I discuss only the Australian species, *Tachyglossus aculeatus;* its larger Papua-New Guinea relative, *Zaglossus,* remains virtually unknown to science—for basic information about echidnas, see the two books by M. Griffiths listed in the bibliography.) Since mammalian brains increase more slowly than body weight along the so-called mouse-to-elephant curve, we can use neither absolute nor relative brain weight as a criterion. (Big mammals have absolutely large brains as an uninteresting consequence of body size, while small mammals have relatively large brains because brains increase more slowly than bodies.) Biologists have therefore developed a standard criterion: measured brain weight relative to expected brain weight for an average mammal of *the same* body size. This ratio, dubbed EQ (or encephalization quotient) in amusing analogy with you know what, measures 1.0 for mammals right on the mouse-to-elephant curve, above 1.0 for brainier than average mammals, and less than 1.0 for brain weights below the norm.

To provide some feel for the range of EQ's, so-called basal insectivores—a selected stem group among the order traditionally ranked lowest among placental mammals—record a mean of 0.311. Adding advanced insectivores, the average rises to 0.443. Rodents, a perfectly respectable group (and dominant among mammals by sheer number), weigh in with a mean EQ of 0.652. (Primates and carnivores rank consistently above 1.0.) Monotremes are not, by this criterion, mental giants—their EQ's range from 0.50 to 0.75—but they rank way above the traditional primitives among placentals and right up there with rodents and other "respected" groups. Monotremes continue to shine by other standards of size as well. Some neurologists regard the ratio of brain to spinal cord as a promising measure of mental advance. Fish generally dip below 1:1 (spinal cord heavier than brain). We top-heavy humans tip the scale at 50:1; cats score 4:1. The "lowly" echidna waddles in front of tabby at approximately 6:1.

By conformation, rather than simple size alone, echidnas are even more impressive. The neocortex, the putative site of higher mental functions, occupies a larger percentage of total brain weight in supposedly advanced creatures. The neocortex of basal insectivores averages 13 percent of brain weight; the North American marsupial opossum records 22 percent. Echidnas score 43 percent (platypuses 48 percent), right up there with the prosimians (54 percent), basal group of the lordly primates. (All my figures for brain sizes come from H. J. Jerison, 1973, and P. Pirlot and J. Nelson, 1978.)

The neocortex of echidnas is not only expanded and nearly spherical as in primates; its surface is also richly convoluted in a series of deep folds and bumps (sulci and gyri), a traditional criterion of mental advance in mammals. (Curiously, by comparison, the platypus neocortex, while equally expanded and spherical, is almost completely smooth.)

Many famous nineteenth-century neuroanatomists studied monotreme brains, hoping to understand the basis of human mental triumph by examining its lowly origins. Echidnas provided an endless source of puzzlement and frustration. William Henry Flower dissected an echidna in 1865 and wrote of "this most remarkable brain, with its largely developed and richly convoluted hemispheres." He admitted: "It is difficult to see in many of the peculiarities of their brain even an approach in the direction of that of the bird." And Grafton Elliot Smith, the great Australian anatomist who later fell for Piltdown Man in such a big way, wrote with evident befuddlement in 1902:

> The most obtrusive feature of this brain is the relatively enormous development of the cerebral hemispheres. . . . In addition, the extent of the cortex is very considerably increased by numerous deep sulci. The meaning of this large neopallium is quite incomprehensible. The factors which the study of other mammalian brains has shown to be the determinants of the extent of the cortex fail completely to explain how it is that a small animal of the lowliest status in the mammalian series comes to possess this large cortical apparatus.

One might have anticipated that scientists, so enlightened by monotreme mentality, would simply abandon the myth of primitivity. But prompt submission to items of contrary evidence is not, despite another prominent myth (this time about scientific procedure), the usual response of scientists to nature's assaults upon traditional beliefs. Instead, most students of monotreme brains have recorded their surprise and then sought different criteria, again to affirm the myth of primitivity.

A favorite argument cites the absence in monotremes (and marsupials as well) of a corpus callosum—the bundle of fibers connecting the right and left hemispheres of "higher" mental processing in placental mammals. In a wonderful example of blatantly circular logic, A. A. Abbie, one of Australia's finest natural historians, wrote in a famous article of 1941 (commissures, to a neuroanatomist, are connecting bands of neural tissue, like the corpus callosum):

> Since in mammals cerebral evolution and with it any progressive total evolution is reflected so closely in the state of the cerebral commissures it is clear that the taxonomic significance of these commissures far transcends that of any other physical character.

In other words, since we know (a priori) that monotremes are primitive, search for the character that affirms a lowly status (lack of a corpus callosum) and proclaim this character, ipso facto, more important than any other (size of brain, convolutions, or any other indication of monotreme adequacy). (I shall have more to say about commissures later on, but let me just mention for now that lack of a corpus callosum does not preclude communication across the cerebral hemispheres. Monotremes possess at least two other commissures—the hippocampal and the anterior—capable of making connections, though by a route more circuitous than the pathway of the corpus callosum.)

This tradition of switching to another criterion continues in modern studies. In their 1978 article on monotreme brain sizes, for example, Pirlot and Nelson admit, after recording volumes and convolutions for echidnas: "It is very difficult to isolate criteria that clearly establish the 'primitiveness' of monotreme brains." But they seek and putatively find, though they honorably temper their good cheer with yet another admission of the puzzling size of the monotreme neocortex:

> This cortex could be considered to be among the most primitive mammalian cortices on the basis of the low number and low density of large, especially pyramidal neurons. It is surprising to find that a very high proportion of cortex is neocortex. This does not necessarily mean an advanced degree of progressiveness, although the two are usually related.

The basic data on size and external conformation of echidna brains have been recorded (and viewed as troubling) for more than a century. More sophisticated information on neural fine structure and actual use of the cortical apparatus in learning has been gathered during the past twenty years—all affirming, again and again, the respectability of echidna intelligence.

In 1964, R. A. Lende published the first extensive map of localized sensory and motor areas on the echidna's cerebral cortex. (I discuss the general procedures of such study in the preceding essay on platypus brains. P. S. Ulinski, 1984, has recently confirmed and greatly extended Lende's work in a series of elegant experiments.) Lende discovered a surprising pattern of localization, basically mammalian in character but different from placental mappings. He identified separate areas for visual, auditory, and sensory control (the motor area overlapped the sensory region and extended forward to an additional section of the cortex), all demarcated one from the other by constant sulci (fissures of the cortex) and located together at the rear of the cortex.

Most surprisingly, these areas abut one another without any so-called association cortex in between. (Association cortex includes areas of the cerebral surface that do not control any specific sensory or motor function and may play a role in coordinating and integrating the basic inputs. For this reason, amount and position of association cortex have sometimes been advanced as criteria of "higher" mental function. But such negative definitions are troubling and should not be pushed too hard or far.) In any case, Lende identified a relatively enormous area of unspecified (perhaps association) cortex in front of his mapped sensory and motor areas. Lende concluded, in a statement oft-quoted against those who maintain the myth of primitivity:

> Ahead of the posteriorly situated sensory and motor areas established in this study there is relatively more "frontal cortex" than in any other mammal, including man, the function of which remains unexplained.

Other studies have tried to push the echidna brain to its practical limits by imposing upon these anteaters all the modern apparatus of mazes, levers, and food rewards so favored by the science of comparative psychology. Echidnas have performed remarkably well in all these studies, again confuting the persistent impression of stupidity still conveyed by textbooks, and even by the most "official" of all sources—the Australian Museum's *Complete*

*Book of Australian Mammals,* edited by R. Strahan (1983), which insists without evidence:

> In this last respect [brainpower], monotremes are inferior to typical placental mammals and, probably, to typical marsupials. The paucity of living monotremes may therefore be due to their being less bright, less adaptable in their behavior, than other mammals.

To cite just three studies among several of similar intent and conclusion:

1. Saunders, Chen, and Pridmore (1971) ran echidnas through a simple two-choice T-maze (down a central channel, then either right or left into a bin of food or a blank wall). They trained echidnas to move in one direction (location of the chow, of course), then switched the food box to the other arm of the T. In such studies of so-called habit-reversal learning, most fish never switch, birds learn very slowly, mammals rapidly. Echidnas showed quick improvement with a steady reduction in errors—and at typically mammalian rates. Half the experiments (seven of fifteen) on well-trained echidnas yielded the optimal performance of "one-trial reversal" (you switch the food box and the animal goes the wrong way—where the food used to be—the first time, then immediately cottons on and heads in the other direction, toward the chow, each time). Rats often show one-trial reversal learning, birds never.

2. Buchmann and Rhodes (1978) tested echidnas for their ability to learn positional (right or left) and visual-tactile (black and rough versus white and smooth) cues—with echidnas pushing the appropriate lever to gain their food reward. As an obvious testimony to mental adequacy, they report that "unsuccessful (unrewarded) responses were often associated with vigorous kicking at the operanda." Echidnas learned at a characteristic rate for placental mammals and also remembered well. One animal, retested a month later, performed immediate one-trial reversals.

Buchmann and Rhodes compared their echidnas with other animals tested in similar procedures. Crabs and goldfish did not show improvement (did not learn) over time. Echidnas displayed great variation in their speed of learning—one improved faster (and one slower) than rats; all echidnas performed better than cats. Take these results (and the reward for success as well) with a grain of salt because numbers are limited and procedures varied widely among studies—but still, the single best performer on the entire chart was an echidna.

Buchmann and Rhodes conclude: "There is no evidence that the performance of echidnas is inferior to eutherian [placental] or metatherian [marsupial] mammals." They end by ridiculing the "quaint, explicitly or tacitly-held views that echidnas are little more than animated pin-cushions, or, at the best, glorified reptiles."

3. Gates (1978) studied learning in visual discrimination (black versus white, and various complex patterns of vertical and horizontal striping). His results parallel the other studies—echidnas learned quickly, at typical mammalian rates. But he added an interesting twist that confutes the only serious, direct argument ever offered from brain anatomy for monotreme inferiority—the claim that absence of a corpus callosum precludes transfer between the cerebral hemispheres, thereby compromising "higher" mental functions.

Gates occluded one eye and taught echidnas to distinguish black from white panels with the other eye. They reached "criterion performance" in an average of 100 trials. He then uncovered the occluded eye, bandaged the one that had overseen the initial learning, and did the experiment again. If no information passes from one cerebral hemisphere to the other, then previous learning on one side of the brain should offer no help to the other, and the 100-trial average should persist. But echidnas only needed 40 trials to reach criterion with the second eye.

Gates conjectures that information is either passing across the other two commissures in the absence of a corpus callosum, or via the few optic fibers that do not cross to the other side of the brain. (In vertebrate visual systems, inputs from the right eye go to the left hemisphere of the brain, left eye to the right hemisphere; thus, each eye "informs" the opposite hemisphere. But about 1 percent of optic fibers do not cross over, and therefore map to their own hemisphere. These few fibers may sneak a little learning to the hemisphere dependent upon the occluded eye.) In addition, direct evidence of electrical stimulation has shown

that inputs to one hemisphere can elicit responses in corresponding parts of the other hemisphere—information clearly gets
across in the absence of a corpus callosum.

The solution to the paradox of such adequate intelligence in
such a primitive mammal is stunningly simple. The premise—the
myth of primitivity itself—is dead wrong. To say it one more, and
one last, time: The reptilian features of monotremes only record
their early branching from the ancestry of placental mammals—
and time of branching is no measure of anatomical complexity or
mental status.

Monotremes have evolved separately from placentals for a
long time—more than enough for both groups to reach, by parallel evolution in independent lineages, advanced levels of mental
functioning permitted by their basic, shared mammalian design.
The primary evidence for parallel evolution has been staring us
in the face for a century, forming part of the standard literature
on echidnas, well featured even in primary documents that uphold the myth of primitivity. We know that the echidna's brain
attained its large size by an independent route. The platypus has
a smooth (if bulbous) brain. The echidna evolved complex ridges
and folds on its cerebral surface as a special feature of its own
lineage. These sulci and gyri cannot be identified (homologized)
with the well-known convolutions of placental brains. The
echidna brain is so different, by virtue of a separate evolution to
large size, that its convolutions have been named by Greek letters
to avoid any misplaced comparison with the different ridges
and folds of placental brains. And Grafton Elliot Smith, the man
most puzzled by echidna brains, did the naming—apparently
without realizing that the very need for such separate designations provided the direct evidence that could refute the myth of
primitivity.

In his eloquent plea for monotremes (1827), Geoffroy Saint-
Hilaire wrote brilliantly about the subtle interplay of fact and
theory in science. He recognized the power of theory to guide the
discovery of fact and to set a context for fruitful interpretation.
("To limit our efforts to the simple practicalities of an ocular
examination would be to condemn the activities of the mind.")
But he also acknowledged the flip side of useful guidance, the
extraordinary power of theory to restrict our vision, in particular
to render "obvious" facts nearly invisible, by denying them a

sensible context. ("At first useless, these facts had to remain un-perceived until the moment when the needs and progress of science provoked us to discover them.") Or as Warner Oland, the Swedish pseudo-Oriental Charlie Chan, once said in one of his most delightfully anachronistic pseudo-Confucian sayings (*Charlie Chan in Egypt,* 1935): "Theory like mist on eyeglasses. Obscure facts."

# 20 | Here Goes Nothing

GOLIATH PAID THE HIGHEST of prices to learn the most elementary of lessons—thou shalt not judge intrinsic quality by external appearance. When the giant first saw David, "he disdained him: for he was but a youth, and ruddy, and of a fair countenance" (1 Sam. 17:42). Saul had been similarly unimpressed when David presented himself as an opponent for Goliath and savior of Israel. Saul doubted out loud: "for thou art but a youth, and he a man of war from his youth" (1 Sam. 17:33). But David persuaded Saul by telling him that actions speak louder than appearances—for David, as a young shepherd, had rescued a lamb from a predatory lion: "I went out after him, and smote him, and delivered it out of his mouth" (1 Sam. 17:35).

This old tale presents a double entendre to introduce this essay—first as a preface to my opening story about a famous insight deceptively clothed in drab appearance; and second as a quirky lead to the body of this essay, a tale of animals that really do deliver from their mouths: *Rheobatrachus silus,* an Australian frog that swallows its fertilized eggs, broods tadpoles in its stomach, and gives birth to young frogs through its mouth.

Henry Walter Bates landed at Pará (now Belém), Brazil, near the mouth of the Amazon, in 1848. He arrived with Alfred Russel Wallace, who had suggested the trip to tropical jungles, arguing that a direct study of nature at her richest might elucidate the origin of species and also provide many fine specimens for sale. Wallace returned to England in 1852, but Bates remained for eleven years, collecting nearly 8,000 new species (mostly insects) and exploring the entire Amazon valley.

294

In 1863, Bates published his two-volume classic, perhaps the greatest work of nineteenth-century natural history and travel, *The Naturalist on the River Amazons*. But two years earlier, Bates had hidden his most exciting discovery in a technical paper with a disarmingly pedestrian title: "Contributions to an Insect Fauna of the Amazon Valley," published in the *Transactions of the Linnaean Society*. The reviewer of Bates's paper (*Natural History Reviews*, 1863, pp. 219–224) lauded Bates's insight but lamented the ill-chosen label: "From its unpretending and somewhat indefinite title," he wrote, "we fear [that Bates's work] may be overlooked in the ever-flowing rush of scientific literature." The reviewer therefore sought to rescue Bates from his own modesty by providing a bit of publicity for the discovery. Fortunately, he had sufficient oomph to give Bates a good send-off. The reviewer was Charles Darwin, and he added a section on Bates's insight to the last edition of the *Origin of Species*.

Bates had discovered and correctly explained the major style of protective mimicry in animals. In Batesian mimicry (for the phenomenon now bears his name), uncommon and tasty animals (the mimics) gain protection by evolving uncanny resemblance to abundant and foul-tasting creatures (the models) that predators learn to avoid. The viceroy butterfly is a dead ringer for the monarch, which, as a caterpillar, consumes enough noxious poisons from its favored plant foods to sicken any untutored bird. (Vomiting birds have become a cliché of natural history films. Once afflicted, twice shy, as the old saying goes. The tale may be more than twice told, but many cognoscenti do not realize that the viceroy's name memorializes its mimicry—for this butterfly is the surrogate, or vice-king, to the ruler, or monarch, itself.)

Darwin delighted in Bates's discovery because he viewed mimicry as such a fine demonstration of evolution in action. Creationism, Darwin consistently argued, cannot be disproved directly because it claims to explain everything. Creationism becomes impervious to test and, therefore, useless to science. Evolutionists must proceed by showing that any creationist explanation becomes a *reductio ad absurdum* by twists of illogic and special pleading required to preserve the idea of God's unalterable will in the face of evidence for historical change.

In his review of Bates's paper, Darwin emphasizes that creationists must explain the precision of duplicity by mimics as a simple act of divine construction—"they were thus clothed from the hour of their creation," he writes. Such a claim, Darwin then argues, is even worse than wrong because it stymies science by providing no possible test for truth or falsity—it is an argument "made at the expense of putting an effectual bar to all further inquiry." Darwin then presents his *reductio ad absurdum,* showing that any fair-minded person must view mimicry as a product of historical change.

Creationists had made a central distinction between true species, or entities created by God, and mere varieties, or products of small changes permitted within a created type (breeds of dogs or strains of wheat, for example). But Bates had shown that some mimics are true species and others only varieties of species that lack mimetic features in regions not inhabited by the model. Would God have created some mimics from the dust of the earth but allowed others to reach their precision by limited natural selection within the confines of a created type? Is it not more reasonable to propose that mimicking species began as varieties and then evolved further to become separate entities? And much worse for creationists: Bates had shown that some mimicking species resemble models that are only varieties. Would God have created a mimic from scratch to resemble another form that evolved (in strictly limited fashion) to its current state? God may work in strange ways, his wonders to perform—but would he really so tax our credulity? The historical explanation makes so much more sense.

But if mimicry became a source of delight for Darwin, it also presented a serious problem. We may easily grasp the necessity for a historical account. We may understand *how* the system works once all its elements develop, but *why* does this process of mimicry ever begin? What starts it off, and what propels it forward? Why, in Darwin's words, "to the perplexity of naturalists, has nature condescended to the tricks of the stage?" More specifically: Any butterfly mimic, in the rich faunas of the Amazon valley, shares its space with many potential models. Why does a mimic converge upon one particular model? We can understand how natural selection might perfect a resemblance already well established, but what begins the process along one of many potential pathways—especially since we can scarcely imagine that a 1 or 2 percent resemblance to a model provides much, if any, advantage for a mimic. This old dilemma in evolutionary theory even has a name in the jargon of my profession—the problem of the "incipient stages of useful structures." Darwin had a good answer for mimicry, and I will return to it after a long story about frogs—the central subject of this essay and another illustration of the same principle that Darwin established to resolve the dilemma of incipient stages.

We remember Darwin's *Beagle* voyage primarily for the big and spectacular animals that he discovered or studied: the fossil *Toxodon* and the giant Galápagos tortoises. But many small creatures, though less celebrated, brought enormous scientific reward—among them a Chilean frog appropriately named *Rhinoderma darwini*. Most frogs lay their eggs in water and then allow the tadpoles to make their own way, but many species have evolved various styles of parental care, and the range of these adaptations extols nature's unity in diversity.

In *R. darwini,* males ingest the fertilized eggs and brood them in the large throat pouches usually reserved for an earlier act of courtship—the incessant croaking that defines territory and attracts females. Up to fifteen young may fill the pouch, puffing out all along the father's ventral (lower) surface and compressing the vital organs above. G. B. Howes ended his classic account of this curious life-style (*Proceedings of the Zoological Society of London,* 1888) with a charming anthropomorphism. Previous students of *Rhinoderma,* he noted, had supposed that the male does not feed while carrying his young. But Howes dissected a brooding male and found its stomach full of beetles and flies and its large intestine clogged with "excreta like that of a normal individual." He concluded, with an almost palpable sigh of relief, "that this extraordinary paternal instinct does not lead up to that self-abnegation" postulated by previous authors.

But nature consistently frustrates our attempts to read intrinsic solicitude into her ways. In November 1973, two Australian scientists discovered a form of parental care that must preclude feeding, for these frogs brood their young in their stomachs and then give birth through their mouths. And we can scarcely imagine that a single organ acts as a nurturing uterus and a site of acid digestion at the same time.

*Rheobatrachus silus,* a small aquatic frog living under stones or in rock pools of shallow streams and rills in a small area of southeast

Queensland, was first discovered and described in 1973. Later that year, C. J. Corben and G. J. Ingram of Brisbane attempted to transfer a specimen from one aquarium to another. To their astonishment, it "rose to the surface of the water and, after compression of the lateral body muscles, propulsively ejected from the mouth six living tadpoles" (from the original description published by Corben, Ingram, and M. J. Tyler in 1974). They initially assumed, from their knowledge of *Rhinoderma,* that their brooder was a male rearing young in its throat pouch. Eighteen days later, they found a young frog swimming beside its parent; two days later, a further pair emerged unobserved in the night. At that point, they decided (as the euphemism goes) to "sacrifice" their golden goose. But the parent, when grasped, "ejected by propulsive vomiting eight juveniles in the space of no more than two seconds. Over the next few minutes a further five juveniles were ejected." They then dissected the parent and received their biggest surprise. The frog had no vocal sac. It was a female with "a very large, thin-walled, dilated stomach"—the obvious home of the next generation.

Natural birth had not yet been observed in *Rheobatrachus.* All young had either emerged unobserved or been vomited forth as a violent reaction after handling. The first young had greeted the outside world prematurely as tadpoles (since development clearly proceeds all the way to froghood in the mother's stomach, as later births demonstrated).

Art then frustrated nature, and a second observation also failed to resolve the mode of natural birth. In January 1978, a pregnant female was shipped express airfreight from Brisbane to Adelaide for observation. But the poor frog was—yes, you guessed it—"delayed" by an industrial dispute. The mother, still hanging on, eventually arrived surrounded by twenty-one dead young; a twenty-second frog remained in her stomach upon dissection. Finally, in 1979, K. R. McDonald and D. B. Carter successfully transported two pregnant females to Adelaide—and the great event was finally recorded. The first female, carefully set up for photography, frustrated all hopes by vomiting six juveniles "at great speed, flying upwards . . . for approximately one meter . . . a substantial distance relative to the body size of the female." But the second mother obliged. Of her twenty-six offspring, two appeared gently and, apparently, voluntarily. The mother "partially emerged from the water, shook her head, opened her mouth, and two babies actively struggled out." The photo of a fully formed baby frog, resting on its parent's tongue before birth, has already become a classic of natural history. This second female, about two inches long, weighed 11.62 grams after birth. Her twenty-six children weighed 7.66 grams, or 66 percent of her weight without them. An admirable effort indeed!

*Rheobatrachus* inspired great excitement among Australian scientists, and research groups in Adelaide and Brisbane have been studying this frog intensively, with all work admirably summarized and discussed in a volume edited by M. J. Tyler (1983). Rarely has such extensive and coordinated information been presented on a natural oddity, and we are grateful to these Australian scientists for bringing together their work in such a useful way.

This volume also presents enough detail (usually lacking in technical publications) to give nonscientists a feel for the actual procedures of research, warts and all (an appropriate metaphor for the subject). Glen Ingram's article on natural history, for example, enumerates all the day-to-day dilemmas that technical papers rarely mention: slippery bodies that elude capture; simple difficulties in seeing a small, shy frog that lives in inaccessible places (Ingram learned to identify *Rheobatrachus* by characteristic ripples made by its jump into water); rain, fog, and dampness; and regeneration that frustrates identification (ecologists must recognize individual animals in order to monitor size and movement of populations by mark-recapture techniques; amphibians and reptiles are traditionally marked by distinctive patterns of toe clipping, a painless and unobtrusive procedure, but *Rheobatrachus* frustrates tradition by regenerating its clipped toes, and Ingram could not reidentify his original captures). To this, we must add the usually unacknowledged bane of all natural history: boredom. You don't see your animals most of the time; so you wait and wait and wait (not always pleasant on the boggy banks of a stream in rainy season). Somehow, though, such plagues seem appropriate enough, given the subject. Frogs, after all, stand among the ten Mosaic originals: "I will smite all thy borders with frogs: And the river shall bring forth frogs abundantly, which shall go up and come into thine house, and into thy bedchambers, and upon thy bed . . . and into thine ovens, and into thy kneading troughs" (Exod. 8:2–3).

The biblical author of Exodus was, unfortunately, not describing *Rheobatrachus,* a rare animal indeed. Not a single *Rheobatrachus silus* has been seen in its natural habitat since 1981. A series of dry summers and late rains has restricted the range of this aquatic frog—and five years of no sitings must raise fears about extinction. Fortunately, a second species, named *R. vitellinus,* was discovered in January 1984, living in shallow sections of fast-flowing streams, about 500 miles north of the range of *R. silus.* This slightly larger version (up to three inches in length) also broods in its stomach; twenty-two baby frogs inhabited a pregnant female.

When discussed as a disembodied oddity (the problem with traditional writing in natural history), *Rheobatrachus* may pique our interest but not our intellect. Placed into a proper context among other objects of nature's diversity—the "comparative approach" so characteristic of evolutionary biology—gastric brooding in *Rheobatrachus* embodies a message of great theoretical interest. *Rheobatrachus,* in one sense, stands alone. No other vertebrate swallows its own fertilized eggs, converts its stomach into a brood pouch, and gives birth through its mouth. But in another sense, *Rheobatrachus* represents just one solution to a common problem among frogs.

In his review of parental care, R. W. McDiarmid argues that frogs display "the greatest array of reproductive modes found in any vertebrates" (see his article in G. M. Burghardt and M. Bekoff, 1978). Much inconclusive speculation has been devoted to reasons for the frequent and independent evolution of brooding (and other forms of parental care) in frogs—a profound departure, after all, from the usual amphibian habit of laying eggs in water and permitting the young to pass their early lives as unattended aquatic tadpoles. Several authors have suggested the following common denominator: In many habitats, and for a variety of reasons, life as a free-swimming tadpole may become sufficiently uninviting to impose strong evolutionary pressure for bypassing this stage and undergoing "direct development" from egg to completed frog. Brooding is an excellent strategy for direct development—since tadpole life may be spent in a brood pouch, and the bad old world need not be faced directly before froghood.

In any case, brooding has evolved often in frogs, and in an astonishing variety of modes. As a minimal encumbrance and modification, some frogs simply attach eggs to their exteriors. Males of the midwife toad *Alytes obstetricans* wrap strings of eggs about their legs and carry them in tow.

At the other extreme of modification, some frogs have evolved special brood pouches in unconventional places. The female *Gastrotheca riobambae,* an Ecuadorean frog from Andean valleys, develops a pouch on her back, with an opening near the rear and an internal extension nearly to her head. The male places fertilized eggs in her pouch, where they develop under the skin of her back for five to six weeks before emerging as late-stage tadpoles.

In another Australian frog, *Assa darlingtoni,* males develop pouches on their undersides, opening near their hind legs but extending forward to the front legs (see article by G. J. Ingram, M. Anstis, and C. J. Corben, 1975). Females lay their eggs among leaves. When they hatch, the male places himself in the middle of the mass and either coats himself with jelly from the spawn or, perhaps, secretes a slippery substance himself. The emerging tadpoles then perform a unique act of acrobatics among amphibians: they move in an ungainly fashion by bending their bodies, head toward tail, and then springing sideways and forward. In this inefficient manner, they migrate over the slippery body of their father and enter the brood pouch under their own steam. (I am almost tempted to say, given the Australian venue, that these creatures have been emboldened to perform in such unfroglike ways by watching too many surrounding marsupials, for the kangaroo's undeveloped, almost larval joey also must endure a slow and tortuous crawl to the parental pouch!)

In a kind of intermediate mode, some frogs brood their young internally but use structures already available for other purposes. I have already discussed *Rhinoderma,* the vocal-pouch brooder of Chile. Evolution seizes its opportunities. The male vocal pouch is roomy and available; in a context of strong pressure for brooding, some lineage will eventually overcome the behavioral obstacles and grasp this ready possibility. The eggs of *R. darwini* develop for twenty-three days before the tadpoles hatch. For the first twenty days, tadpoles grow within eggs exposed to the external environment. But tadpoles then begin to move, and this behavior apparently triggers a response from the male parent. He then takes the advanced eggs into his vocal pouch. They hatch there three days later and remain for fifty-two days until the end of metamorphosis, when the young emerge through their father's mouth as perfectly formed little froglets. In the related species *R. rufum,* muscular activity begins after eight days within the egg, and males keep the tadpoles in their vocal sacs for much shorter periods, finally expelling them, still in the tadpole stage, into water (see article by K. Busse, 1970).

In this context, *Rheobatrachus* is less an oddity than a fulfillment. Stomachs provide the only other large internal pouch with an egress of sufficient size. Some lineage of frogs was bound to exploit this possibility. But stomachs present a special problem not faced by vocal sacs or novel pouches of special construction—and we now encounter the key dilemma that will bring us back to mimicry in butterflies and the evolutionary problem of incipient stages. Stomachs are already doing something else—and that something is profoundly inimical to the care and protection of fragile young. Stomachs secrete acid and digest food—and eggs and tadpoles are, as they say down under, mighty good tucker.

In short, to turn a stomach into a brood pouch, something must turn off the secretion of hydrochloric acid and suppress the passage of eggs into the intestine. At a minimum, the brooding mother cannot eat during the weeks that she carries young in her stomach. This inhibition may arise automatically and present no special problem. Stomachs contain "stretch receptors" that tell an organism when to stop eating by imposing a feeling of satiety as the mechanical consequence of a full stomach. A batch of swallowed eggs will surely set off this reaction and suppress further eating.

But this fact scarcely solves our problem—for why doesn't the mother simply secrete her usual acid, digest the eggs, and relieve her feeling of satiety? What turns off the secretion of hydrochloric acid and the passage of eggs into the intestine?

Tyler and his colleagues immediately realized, when they discovered gastric brooding in *Rheobatrachus,* that suppression of stomach function formed the crux of their problem. "Clearly," they wrote, "the intact amphibian stomach is likely to be an alien environment for brooding." They began by studying the changes induced by brooding in the architecture of the stomach. They found that the secretory mucosa (the lining that produces acid) regresses while the musculature strengthens, thus converting the stomach into a strong and chemically inert pouch. Moreover, these changes are not "preparatory"—that is, they do not occur before a female swallows her eggs. Probably, then, something in the eggs or tadpoles themselves acts to suppress their own destruction and make a congenial place of their new home. The Australian researchers then set out to find the substance that suppresses acid secretion in the stomach—and they have apparently succeeded.

P. O'Brien and D. Shearman, in a series of ingenious experiments, concentrated water that had been in contact with developing *Rheobatrachus* embryos to test for a chemical substance that might suppress stomach function in the mothers. They dissected out the gastric mucosa (secreting surface) of the toad *Bufo marinus* (*Rheobatrachus* itself is too rare to sacrifice so many adult females for such an experiment) and kept it alive in vitro. They showed that this isolated mucosa can function normally to secrete stomach acids and that well-known chemical inhibitors will suppress the secretion. They then demonstrated that water in contact with *Rheobatrachus* tadpoles suppresses the mucosa, while water in contact with tadpoles of other species has no effect. Finally, they succeeded in isolating a chemical suppressor from the water—prostaglandin $E_2$. (The prostaglandins are hormonelike substances, named for their first discovery as secretions of the human prostate gland—though they form throughout the body and serve many functions.)

Thus, we may finally return to mimicry and the problem of incipient stages. I trust that some readers have been bothered by an apparent dilemma of illogic and reversed causality. The eggs of *Rheobatrachus* must contain the prostaglandin that suppresses secretion of gastric acid and allows the stomach to serve as an inert brood pouch. It's nice to know that eggs contain a substance for their own protection in a hostile environment. But in a world of history—not of created perfection—how can such a system arise? The ancestors of *Rheobatrachus* must have been conventional frogs, laying eggs for external development. At some point, a female *Rheobatrachus* must have swallowed its fertilized eggs (presumably taking them for food, not with the foresight of evolutionary innovation)—and the fortuitous presence of prostaglandin suppressed digestion and permitted the eggs to develop in their mother's stomach.

The key word is *fortuitous*. One cannot seriously believe that ancestral eggs actively evolved prostaglandin because they knew that, millions of years in the future, a mother would swallow them and they would then need some inhibitor of gastric secretion. The eggs must have contained prostaglandin for another reason or for no particular reason at all (perhaps just as a metabolic by-product of development). Prostaglandin provided a lucky break with respect to the later evolution of gastric brooding—a historical precondition fortuitously available at the right moment, a sine qua non evolved for other reasons and pressed into service to initiate a new evolutionary direction.

Darwin proposed the same explanation for the initiation of mimicry—as a general solution to the old problem of incipient stages. Mimicry works splendidly as a completed system, but what gets the process started along one potential pathway among many? Darwin argued that a mimicking butterfly must begin with a slight *and fortuitous* resemblance to its model. Without this leg up for initiation, the process of improvement to mimetic perfection cannot begin. But once an accidental, initial resemblance provides some slight edge, natural selection can improve the fit from imperfect beginnings.

Thus, Darwin noted with pleasure Bates's demonstration that mimicry always arose among butterflies more prone to vary than others that never evolve mimetic forms. This tendency to vary must be the precondition that establishes fortuitous initial resemblance to models in some cases. "It is necessary to suppose," Darwin wrote, that ancestral mimics "accidently resembled a member of another and protected group in a sufficient degree to afford some slight protection, this having given the basis for the subsequent acquisition of the most perfect resemblance." Ancestral mimics happened to resemble a model in some slight manner—and the evolutionary process could begin. The eggs of *Rheobatrachus* happened to contain a prostaglandin that inhibited gastric secretion—and their mother's stomach became a temporary home, not an engine of destruction.

New evolutionary directions must have such quirky beginnings based on the fortuitous presence of structures and possibilities evolved for other reasons. After all, in nature, as in human invention, one cannot prepare actively for the utterly unexpected. Gastric brooding must be an either-or, a quantum jump in evolutionary potential. As Tyler argues, what intermediary stage can one imagine? Many fishes (but no frogs) brood young in their mouths—while only males possess throat pouches, but only female *Rheobatrachus* broods in its stomach. Eggs can't develop halfway down the esophagus.

We glimpse in the story of *Rheobatrachus* a model for the introduction of creativity and new directions in evolution (not just a tale of growing bigger or smaller, fiercer or milder, by the everyday action of natural selection). Such new directions, as Darwin argued in resolving the problem of incipient stages, must be initiated by fortuitous prerequisites, thus imparting a quirky and unpredictable character to the history of life. These new directions may involve minimal changes at first—since the fortuitous prerequisites are already present, though not so utilized, in ancestors. A female *Rheobatrachus* swallowed its fertilized eggs, and a striking new behavior and mode of brooding arose at once by virtue of a chemical fortuitously present in eggs, and by the automatic action of stretch receptors in the stomach. Such minimal changes are pregnant with possibilities. Most probably lead nowhere beyond a few oddballs—as with *Rheobatrachus,* probably already well on its way to extinction.

But a few quirky new directions may become seeds of major innovations and floods of diversity in life's history. The first protoamphibian that crawled out of its pond has long been a favorite source of evolutionary cartoon humor. The captions are endless—from "see ya later as alligator" to "because the weather's better out here." But my favorite reads "here goes nothing." It doesn't happen often, but when nothing becomes something, the inherent power of evolution, normally an exquisitely conservative force, can break forth. Or, as Reginald Bunthorne proclaims in Gilbert and Sullivan's *Patience* (which evolution must have above all else): "Nature for restraint too mighty far, has burst the bonds of art—and here we are."

———————

## *Postscript*

It is my sad duty to report a change of state, between writing and republishing this essay, that has made its title eerily prophetic. *Rheobatrachus silus,* the stomach-brooding frog and star of this essay, has apparently become extinct. This species was discovered in 1973, living in fair abundance in a restricted region of southeast Queensland, Australia. In early 1990, the National Research Council (of the United States) convened a conference to discuss "unexplained losses of amphibian populations around the world" (as reported in *Science News,* March 3, 1990). Michael J. Tyler, member of the team that discovered stomach brooding in *Rheobatrachus,* reported that 100 specimens could easily be observed per night when the population maintained fair abundance during the mid-1970s. Naturalists have not found a single individual since 1981, and must now conclude that the species is extinct (for several years they hoped that they were merely observing a sharp and perhaps cyclical reduction in numbers). Even more sadly, this loss forms part of a disturbing and unexplained pattern in amphibian populations throughout the world. In Australia alone, 20 of 194 frog species have suffered serious local drops in population size during the past decade, and at least one other species has become extinct.

# 7 | Intellectual Biography

# 21 | In a Jumbled Drawer

AS MY SON GROWS, I have monitored the changing fashions in kiddie culture for words expressing deep admiration—what I called "cool" in my day, and my father designated "swell." The half-life seems to be about six months, as "excellent" (with curious lingering emphasis on the first syllable) gave way to "bad" (extended, like a sheep bleat, long enough to turn into its opposite), to "wicked," to "rad" (short for radical). The latest incumbent—"awesome"—possesses more staying power, and has been reigning for at least two years. My only objection, from the fuddy-duddy's corner, lies in kiddie criteria for discernment. Ethan's buddies require such a tiny extension beyond the ordinary to proclaim something "awesome"—just a little bit bigger, brighter, and, especially, louder will do. This or that is proclaimed awesome every second sentence—and we have a lost a wonderful English word.

Now let me tell you about awesome—the real thing, when adults still held possession of the concept. I collected fossils all my youthful life, or at least on those rare occasions of departure from the asphalt of New York City. I had amassed, by the end of college, five cartonsful, all ordered and labeled—and I was pretty proud of both quantity and quality. Then I got my present job as curator of fossil invertebrates at Harvard's Museum of Comparative Zoology. I came to Cambridge with my five cartons and discovered that my new stewardship extended to 15,000 drawers of fossils, including some of the world's finest and oldest specimens, brought from Europe by Louis Agassiz more than a century ago. I put the cartons in a back corner of my office twenty years ago this month. I have never opened them. Me with my five cartons facing those 15,000 drawers—that is awe.

But when awe subsides, ecstasy creeps in. For I had 15,000 drawers to open, each harboring a potential discovery or insight. Raise to the $n$th power any simile you ever heard for "as happy as"—a boy in a candy store, a pig in . . . well, you know what. I spent two weeks pulling out every last drawer, and I found a cornucopia of disparate objects that have fueled my aesthetic and intellectual pleasure ever since.

The fossils were sublime, but I found as much fascination in the odd paraphernalia of culture that, for various reasons, end up in museum drawers. Late eighteenth century apothecary boxes, thread cases from the mills of Lawrence, Victorian cigar boxes of gaudy Cuban design—all the better to house fossils. Tickets to Lowell Institute lecture series by Gray, Agassiz, and Lyell, invitations to a ball honoring Napoleon III, merchants' calling cards from Victorian Cincinnati—all the better (on their blank obverses) to label fossils. Pages from the Sears catalogue for 1903, snippets of nineteenth-century newspapers—all the better to wrap fossils. The most interesting news item, a headline from a Cincinnati paper for July 11, 1881, read "Garfield's Grit" and announced that the president, though severely wounded in the recent assassination attempt, "is now on the sunny side of life again," and would almost surely recover—the flip side to a happy Harry Truman holding that 1948 *Chicago Tribune* headline announcing Dewey's victory.

For my most interesting discovery, I opened a drawer late one night and found only a jumble of specimens inside. Someone had obviously overturned the drawer and dumped the contents. But the thick layer of dust identified the disordered pile as a very old jumble. Inside, I found the following note:

This incident was the result of the carelessness of the Janitor Eli Grant who managed to overturn about half a dozen drawers of specimens by undertaking to move certain trays which he was not authorized to touch. The accident happened during my absence but I judge that it arose from an excess of zeal rather than from any recklessness. I have deemed it best to leave the specimens exactly as I found them awaiting an opportunity to have them arranged by Mr. Hartt.

I developed an immediate dislike for this pusillanimous assistant—fingering the janitor, distancing himself even further from responsibility by assuring the boss that he hadn't been there at

The guilty note from N.S. Shaler, left in a drawer to forestall Agassiz's wrath.

the time, then feeling a bit guilty for placing Mr. Grant's job in jeopardy and praising him for zeal through the back door. I then looked at the date and signature—Cambridge, April 26, 1869, N. S. Shaler.

David lamented over Saul: "How are the mighty fallen." But one might look the other way in ontogeny and observe, "How meek are the mighty when young and subservient." Nathaniel Southgate Shaler became one of the greatest and most popular teachers in the history of Harvard University. He was a giant among late nineteenth century American naturalists. But in 1869, Shaler was just a junior professor without tenure, and his superior was the most powerful and imperious biologist in America—none other than Louis Agassiz himself. Obviously, Shaler had written that note in mortal fear of Agassiz's celebrated wrath. Equally obviously, Agassiz had never found out—for Shaler became professor of paleontology later that year, while a century of undisturbed dust still lies atop the jumbled specimens.

N. S. Shaler reaped the rewards of his unflinching loyalty to Agassiz. The path of devotion was not smooth. Agassiz was a transplanted European with an Old World sense of professorial authority. He told students what they would study, awarded degrees by oral examination and direct assessment of competence, and insisted upon personal approval for any publication based on material at his museum. He never failed in encouragement, warmth, and enthusiasm—and he was a beloved teacher. But he never relinquished one iota of authority. These attitudes might only have yielded a tightly run ship in times of intellectual quiescence, but Agassiz was captain on the most troubled waters of biological history. Agassiz opened his museum in 1859, the same year that Darwin published *The Origin of Species.* He gathered around himself the most promising, and therefore most independently minded, group of young zoologists in America, Shaler included. Inevitably, evolution became the chief subject of discussion. With equal inevitability, students flocked eagerly to this beacon of intellectual excitement and became enthusiastic converts. But Agassiz had built both a career and a coherent philosophy upon the creationist premise that species are ideas in God's mind, made incarnate by his hand in a world of material objects. Sooner shall a camel pass through the eye of a needle than the old lion and young wolf cubs shall dwell in harmony amidst such disagreement.

And so, inevitably once more, Agassiz's students revolted—against both his overweening authority and his old-fashioned ideas. In 1863, they formed what they called, in half-jest, a committee for the protection of American students from foreign-born professors. Agassiz, however, held all the cards in a hierarchical world, and he booted the rebels out, much to the benefit of American science, as they formed departments and centers at other great universities. Agassiz then staffed his museum with older and uncontroversial professionals, bringing peace and mediocrity once again to Harvard.

Of his truly excellent students, only Shaler remained loyal. And Shaler reaped his earthly reward. He received his bachelor of science in geology, *summa cum laude,* in July 1862. After a spell of service in the Civil War, fighting for the Union from his native Kentucky, Shaler returned to Harvard in 1864. Agassiz, describing Shaler as "the one of my American students whom I love the best," appointed him assistant in paleontology at the Museum of Comparative Zoology. In 1869, soon after he penned the guilty note that would lie unread for exactly 100 years (I found it in 1969), Shaler received his lifetime appointment as professor of geology, succeeding Agassiz (who continued to lecture in zoology until he died in 1873). There Shaler remained until his death in 1906, writing numerous treatises on everything from the geology of Martha's Vineyard to the nature of morality and immortality. He also became, by far, Harvard's most popular professor. His classes overflowed, and his students poured forth praise for his enthusiasm, his articulateness, and the comfort, optimism, and basic conventionality of his words, spoken to the elite at the height of America's gilded age. On the day of his funeral, flags in city buildings and student fraternities flew at half-mast, and many shops closed. Thirty years later, at the Harvard tercentenary of 1936, Shaler was named twelfth among the fifty people most important to the history of Harvard. To this day, his bust rests, with only fourteen others, including Franklin's, Longfellow's, and, of course, Agassiz's, in the faculty room of Bullfinch's University Hall (and you can take my word for it; I made a special field trip over there and counted).

Shaler's loyalty to Agassiz, and to comfortable convention in general, held as strongly in ideology as in practice. Shaler wrote these words of condolence to Agassiz's widow, Elizabeth Cary, founder of Radcliffe College, when Louis died in 1873: "He never was a greater teacher than now. He never was more truly at his chosen work. . . . While he lived I always felt myself a boy beside him." (See David N. Livingstone, *Nathaniel Southgate Shaler and the Culture of American Science,* University of Alabama Press, 1987, for the source of this quotation and an excellent account of Shaler's intellectual life.)

I don't think that Shaler, in his eulogy to Elizabeth, either erred or exaggerated in his chosen metaphor of subservience to Agassiz's vision. While Shaler remained subordinate, he followed Agassiz's intellectual lead, often with the epigone's habit of exaggerating his master's voice. Shaler's very first publication provides an interesting example ("Lateral Symmetry in Brachiopoda," 1861). Here Shaler supports both Agassiz's creationism and his zoological classification. Brachiopods, once a dominant group in the fossil record of marine invertebrates, are now a minor component of oceanic faunas. With their bivalved shells, they look superficially like clams, but their soft anatomy is entirely distinct, and they are now classified as a separate phylum. But Georges Cuvier, Agassiz's great mentor, had placed brachiopods with clams and snails in his phylum Mollusca—and Agassiz, whose loyalty to Cuvier matched any devotion of Shaler's, wished both to uphold Cuvier's classification and to use his concept of Mollusca as an argument against Darwin.

Shaler obliged in his first public performance. He affirmed Cuvier and Agassiz's inclusion of brachiopods in the Mollusca by claiming a bilateral symmetry of soft parts similar enough to the symmetry of such "standard" forms as clams and squids to justify a conclusion of common plan in design. But he then took a swipe at Darwin's reason for including separate groups in a single phylum by arguing that no evolutionary transition could possibly link adult brachiopod and clam. (Shaler was quite right about this, but not for his stated reason. You cannot transform a brachiopod into a clam, but then nature never did because brachiopods aren't mollusks and the two groups are entirely separate—contrary to Shaler's first conclusion.) The planes of bilateral symmetry are different for the two groups, Shaler argued correctly, and no transition could occur because any smooth intermediate would have to pass through a nonbilateral stage entirely inconsistent with molluscan design. Shaler wrote:

> Such a transition would require a series of forms, each of which must present a negation of that very principle of bilateral symmetry which we have found of so much importance. And must we not, therefore, conclude that the series which united these two orders is a series of thought, which is itself connected, though manifested by two structures which have no genetic relations.

Now if you're holding a nineteenth-century scorecard, and therefore know the players, only one man could be lurking behind this statement. Only one real Platonist of this ilk operated in America, only one leading biologist still willing to designate species as thoughts of a Creator, and taxonomic relationships as the interconnections within His mind—Louis Agassiz. Shaler, with the true zeal of the acolyte, even out-Agassized Agassiz in referring to the central character of bilateral symmetry as "the fundamental thought of the type" and then designating animal taxonomy as "a study of personified thought." Even Agassiz was not so explicit in specifying the attributes of his God.

When the winds of inevitability blew strongly enough, and when Shaler's own position became secure in the late 1860s, he finally embraced evolution, but ever so gently, and in a manner that would cause minimal offense to Agassiz and to any Brahmin member of the old Boston order. After Agassiz's death, Shaler continued to espouse a version of evolution with maximal loyalty to Agassiz's larger vision of natural harmony, and marked aversion to all Darwinian ideas of chanciness, contingency, unpredictability, opportunism, and quirkiness. He led the American Neo-Lamarckian school—a powerful group of anti-Darwinian evolutionists who held out for order, purpose, and progress in nature through the principle of inheritance for features acquired by the effort of organisms. Progress in mentality might be predictably ordained if some organisms strove for improvement during their lives and passed their achievements to their offspring. No waiting for the Darwinian chanciness of favorable environments and fortuitous variations.

Shaler's loyalty to Agassiz persisted right through this fundamental change from creationism to evolution. For example, though he could scarcely deny the common origin of all humans in the light of evolutionary theory, Shaler still advocated Agassiz's distinctive view (representing the "polygenist" school of pre-Darwinian anthropology) that human races are separate species, properly and necessarily kept apart both on public conveyances and in bedrooms. Shaler argued for an evolutionary separation of races so long ago that accumulated differences had become, for all practical purposes, permanent.

Practical purposes, in the genteel racism of patrician Boston, abetted by a slaveholding Kentucky ancestry, meant "using biology as an accomplice" (in Condorcet's words) to advocate a "nativist" social policy (where "natives" are not the truly indigenous American Indians, but the earliest immigrants from Protestant western and northern Europe). Shaler reserved his lowest opinion for black Americans, but invested his social energies in the Immigration Restriction League and its attempts to prevent dilution of American whites (read WASPs) by the great Catholic and Jewish unwashed of southern and eastern Europe.

One can hardly fathom the psychological and sociological complexities of racism, but the forced intellectual rationales are always intriguing and more accessible. Shaler's own defense merged his two chief interests in geography and zoology. He argued that we live in a world of sensible and optimal pattern, devoid of quirk or caprice. People differ because they have adapted by Lamarckian means to their local environments; our capacities are a map of our original homes—and we really shouldn't live elsewhere (hence the biological rectitude of restricting immigration). The languid tropics cannot inspire genius, and you cannot contemplate the Pythagorean absolute while trying to keep body and soul together in an igloo. Hence the tough, but tractable, lands of northern Europe yielded the best of humanity. Shaler wrote:

> Our continents and seas, cannot be considered as physical accidents in which, and on which, organic beings have found an ever-perilous resting place, but as great engines operating in a determined way to secure the advance of life.

Shaler then applied this cardinal belief in overarching order (against the Darwinian specter of unpredictable contingency) to the largest question of all—the meaning of human life as a proof of God's existence and benevolence. In so doing, he completed the evolutionary version of Agassiz's dearest principle—the infusion of sensible, progressive, divine order into the cosmos, with the elevation of "man" (and I think he really meant only half of us) to the pinnacle of God's intent. Shaler could not deny his generation's proof of evolution, and had departed from his master in this conviction, but he had been faithful in constructing a vision of evolution so mild that it left all cosmic comfort intact, thereby affirming the deepest principle of Agassiz's natural theology.

Shaler rooted his argument in a simple claim about probability. (Shaler often repeated this line of reasoning. My quotations come from his last and most widely read book—*The Individual: A Study of Life and Death,* 1901.) Human life is the end result of an evolutionary sequence stretching back into the immensity of time and including thousands of steps, each necessary as a link in the rising sequence:

> The possibility of man's development has rested on the successive institution of species in linked order. . . . If, in this succession of tens of thousands of species, living through a series of millions of years, any of these links of the human chain had been broken; if any one of the species had failed to give birth to its successor, the chance of the development of man would have been lost.

Human evolution, Shaler holds, would have been "unattainable without the guidance of a controlling power intent on the end." If one sequence alone could have engendered us, and if the world be ruled by Darwinian caprice and contingency, our appearance would have been "essentially impossible." For surely, one link would have failed, one step in ten thousand been aborted, thus ending forever the ascent toward consciousness. Only divine watchfulness and intent could have produced the human mind (not a direct finger in the pot, perhaps, but at least an intelligent construction of nature's laws with a desired end in view):

The facts connected with the organic approach to man afford what is perhaps the strongest argument, or at least the most condensed, in favor of the opinion that there is an intelligent principle in control of the universe.

Nathaniel Southgate Shaler was one of the most influential American intellectuals of his time. Today, he is unknown. I doubt that one in a hundred readers of this essay (geologists and Harvardians excepted) has ever heard of him. His biography rates thirteen lines in the *Encyclopaedia Britannica,* more than half devoted to a listing of book titles. Why has he faded, and what does his eclipse teach us about the power and permanence of human thought? We can, perhaps, best approach this question by considering one of Shaler's best friends, a man also influenced by Agassiz, but in a different way—William James. In their day, Shaler and James were peas in a pod of Harvard fame. Now Shaler is a memory for a few professionals, and James is one of America's great gifts to the history of human thought. Why the difference?

William James also came under Agassiz's spell during his student years. Agassiz decided to take six undergraduates along on his famous Thayer Expedition to Brazil (1866). They would help the trained scientists in collecting specimens and, in return, hear lectures from Agassiz on all aspects of natural history. William James, among the lucky six, certainly appreciated the value of Agassiz's formidable intellect and pedagogical skill. He wrote to his father: "I am getting a pretty valuable training from the Prof. who pitches into me right and left and makes me [own] up to a great many of my imperfections. This morning he said I was 'totally uneducated.' "

But James maintained his critical perspective, while Shaler became an acolyte and then an epigone. James wrote:

> I have profited a great deal by hearing Agassiz talk, not so much by what he says, for never did a man utter a greater amount of humbug, but by learning the way of feeling of such a vast practical engine as he is. . . . I delight to be with him. I only saw his defects at first, but now his wonderful qualities throw them quite in the background. . . . I never saw a man work so hard.

Was James "smarter" than Shaler? Does their difference in renown today reflect some basic disparity in amount of intellectual power? This is a senseless question for many reasons. Intelligence is too complex and multifaceted a thing to reduce to any single dimension. What can we say? Both men had certain brilliance, but they used their skills differently. Shaler was content to follow Agassiz throughout his career, happy to employ his formidable intellect in constructing an elaborate rationale for contemporary preferences, never challenging the conservative assumptions of his class and culture. James questioned Agassiz from day one. James probed and wondered, reached and struggled every day of his life. Shaler built pretty buildings to house comfortable furniture. Intelligence or temperament; brains or guts? I don't know. But I do know that oblivion was one man's reward, enduring study and respect the other's.

As a dramatic illustration of the difference, consider James's critique of Shaler's "probability argument" for God's benevolence from the fact of human evolution. James read Shaler's *The Individual,* and wrote a very warm, though critical, letter to his dear friend. He praised "the gravity and dignity and peacefulness" of Shaler's thoughts, but singled out the probability argument for special rebuttal.

James points out that the actual result of evolution is the only sample we have. We cannot compute a "probability" or even speak in such terms. Any result in a sample of one would appear equally miraculous when you consider the vast range of alternative possibilities. But something had to happen. We may only talk of odds if we could return to the beginning, list a million possible outcomes, and then lay cold cash upon one possibility alone:

> We never know what ends may have been kept from realization, for the dead tell no tales. The surviving witness would in any case, and whatever he were, draw the conclusion that the universe was planned to make him and the like of him succeed, for it actually did so. But your argument that it is millions to one that it didn't do so by chance doesn't apply. It would apply if the witness had preexisted in an independent form and framed his scheme, and then the world had realized it. Such a coincidence would prove the world to have a kindred mind to his. But there has been no such

coincidence. The world has come but once, the witness is there after the fact and simply approves. . . . Where only one fact is in question, there is no relation of "probability" at all. [James's letter is reprinted, in full, in *The Autobiography of Nathaniel Southgate Shaler*, 1909.]

Old, bad arguments never die (they don't fade away either), particularly when they match our hopes. Shaler's false probability argument is still a favorite among those who yearn to find a cosmic rationale for human importance. And James's retort remains as brilliant and as valid today as when he first presented the case to Shaler. We could save ourselves from a lot of current nonsense if every devotee of the anthropic principle (strong version), every fan of Teilhard's noosphere, simply read and understood James's letter to Shaler.

James then continues with the ultimate Darwinian riposte to Shaler's doctrine of cosmic hope and importance. Human intellect is a thing of beauty—truly awesome. But our evolution need not record any more than a Darwinian concatenation of improbabilities:

I think, therefore, that the excellence we have reached and now approve may be due to no general design, but merely to a succession of the short designs we actually know of, taking advantage of opportunity, and adding themselves together from point to point.

Which brings us back to Mr. Eli Grant. (I do hope, compassionate reader, that you have been worrying about this poor man's fate while I temporized in higher philosophical realms.) The young Shaler tried to cover his ass by exposing Grant's. Obviously, he succeeded, but what happened to the poor janitor, left to take the rap?

This story has a happy ending, based on two sources of evidence: one inferential, the other direct. Since Agassiz never found out, never saw the note, and since Mr. Hartt, like Godot, never arrived, we may assume that Grant's zealous accident eluded Agassiz's watchful eye. More directly, I am delighted to report that I found (in yet another drawer) a record book for the

Department of Invertebrate Paleontology in 1887. Mr. Eli Grant is still listed as janitor.

Was Mr. Grant meant to survive because he did? Does his tenure on the job indicate the workings of a benevolent and controlling mind? (Why not, for I can envisage 100 other scenarios, all plausible but less happy.) Or was Mr. Grant too small to fall under God's direct providence? But if so, by what hubris do we consider ourselves any bigger in a universe of such vastness? Such unprofitable, such unanswerable questions. Let us simply rejoice in the happy ending of a small tale, and give the last word to William James, still trying to set his friend Shaler straight:

> What if we did come where we are by chance, or by mere fact, with no one general design? What is gained, is gained, all the same. As to what may have been lost, who knows of it, in any case?

---

## Postscript: A Letter from Jimmy Carter

I had heard many stories of Jimmy Carter's personal kindness, and I had long admired him as the most intellectual of presidents since Roosevelt (the competition has not been too fierce of late). But I was delighted and surprised (to the point of shock) when I received a call, late one afternoon, from a woman who said, in a strong southern accent: "Please hold the line; President Carter would like to speak with you." My first reaction, undoubtedly impolitic, was to blurt out: "President Carter who?" (I did think of Jimmy, but his tenure had ended nearly ten years ago and I didn't realize that certain titles, like diamonds and sainthood, are forever.) She replied with more than a hint of indignation: "Former president Jimmy Carter of the United States." I allowed that I would hold.

He came on the line a minute later. My first reaction was surprise that the voice sounded so much like that of our president from 1977 to 1980. My second reaction was to chastise myself for such incredible stupidity since it was, after all, Mr. Carter on the line—and people do tend to sound like themselves (even basically competent folks can be mighty dimwitted when flustered). My third reaction was to wonder why, in heaven's name, he was calling me. So I listened and soon found out. Carter said that he had read and enjoyed several of my books. He then read of my bout with cancer in the preface to *The Flamingo's Smile.* He wanted, in this light, to express his best wishes for my health but hesitated to call, lest I might be too ill to be disturbed. So he phoned my publishers, found out that my next book was in production, and that I had recovered. Feeling, therefore, that he would not be intruding, he had decided to call—simply to express his good wishes and his hopes for my continued good health.

What a lovely man, and what a gracious and kind act. I sent him, as a most inadequate expression of thanks, a copy of the book, *Wonderful Life,* when it appeared a few months later. Not long thereafter, I received a letter in reply:

> I had a chance to read *Wonderful Life* while traveling to Kenya, Sudan, and Ethiopia recently. Rosalynn and I were spending three weeks mediating between the Ethiopian government and the Eritrean People's Liberation Front. . . . You may or may not be familiar with the horrendous wars in those countries. Between negotiating sessions, I found your book to be thoroughly enjoyable—perhaps your best so far.

But Carter then voiced a major criticism, quite disabling if valid. I argue in *Wonderful Life* that human evolution would almost surely not occur again if we could rewind the tape of life back to the early history of multicellular animals (erasing what actually happened of course) and let it play again from an identical starting point (too many initial possibilities relative to later survivors, with no reason to think that survivors prevailed for reasons of superiority or any other version of predictability, and too much randomness and contingency in the pathways of life's later history). But Carter made a brilliant riposte to this central claim of my book.

Jimmy Carter, if I understand his religious attitudes properly, upholds an unconventional point of view among Christian intellectuals on the issue of relationships between God and nature. Most theologians (in agreement with most scientists) now argue that facts of nature represent a domain different from the realm of religious attitudes and beliefs, and that these two worlds of equal value interact rather little. But Jimmy Carter is a twentieth-century natural theologian—that is, he accepts the older argument, popular before Darwin, that the state of nature should provide material for inferring the existence and character of God. Nineteenth-century versions of natural theology, as embodied in Paley's classic work (1802) of the same name, tended to argue that God's nature and benevolence were manifest in the excellent design of organisms and the harmony of ecosystems—in other words, in natural goodness. Such an attitude would be hard to maintain in a century that knew two world wars, Hiroshima, and the Holocaust. If natural theology is to be advanced in our times, a new style of argument must be developed—one that acknowledges the misfits, horrors, and improbabilities, but sees God's action as manifest nonetheless. I think that Carter developed a brilliant twentieth-century version of natural theology by criticizing my book in the next paragraph of his letter:

> You seem to be straining mightily to prove that everything that has happened prior to an evolutionary screening period was just an accident, and that if the tape of life was replayed in countless different ways it is unlikely that cognitive creatures would have been created or evolved. It may be that when you raise "one chance in a million" to the 4th or 5th power there comes a time when pure "chance" can be questioned. I presume that you feel more at ease with the luck of 1 out of 10 to the 30th power than with the concept of a creator who/that has done some orchestrating.

In other words, Carter asks, can't the improbability of our evolution become so great that the very fact of its happening must indicate some divine intent? One chance in ten can be true chance, but the realization of one chance in many billion might indicate intent. What else could a twentieth-century natural theologian do but locate God in the realization of improbability, rather than in the ineffable beauty of design! Carter's argument is fascinating but, I believe, wrong—and wrong for the same reason that James invokes against Shaler (the central point of this essay). In fact, Carter's argument is Shaler's updated and more sophisticated. Shaler claimed that God must have superintended our evolution because the derailment or disruption of any of thousands of links in our evolutionary chain would have canceled the possibility of our eventual appearance. James replied that we cannot read God in contingencies of history because a probability cannot even be calculated for a singular occurrence known only after the fact (whereas probabilities could be attached to predictions made at the beginning of a sequence). James might as well have been answering Carter when he wrote to Shaler:

> But your argument that it is millions to one that it didn't do so by chance doesn't apply. It would apply if the witness had preexisted in an independent form and framed his scheme, and then the world had realized it. Such a coincidence would prove the world to have a kindred mind to his. But there has been no such coincidence. The world has come but once, the witness is there after the fact and simply approves. . . . Where only one fact is in question, there is no relation of "probability" at all.

Such is our intellectual heritage, such our continuity, that fine thinkers can speak to each other across the centuries.

# 22 | Kropotkin Was No Crackpot

IN LATE 1909, two great men corresponded across oceans, religions, generations, and races. Leo Tolstoy, sage of Christian nonviolence in his later years, wrote to the young Mohandas Gandhi, struggling for the rights of Indian settlers in South Africa:

> God helps our dear brothers and co-workers in the Transvaal. The same struggle of the tender against the harsh, of meekness and love against pride and violence, is every year making itself more and more felt here among us also.

A year later, wearied by domestic strife, and unable to endure the contradiction of life in Christian poverty on a prosperous estate run with unwelcome income from his great novels (written before his religious conversion and published by his wife), Tolstoy fled by train for parts unknown and a simpler end to his waning days. He wrote to his wife:

> My departure will distress you. I'm sorry about this, but do understand and believe that I couldn't do otherwise. My position in the house is becoming, or has become, unbearable. Apart from anything else, I can't live any longer in these conditions of luxury in which I have been living, and I'm doing what old men of my age commonly do: leaving this worldly life in order to live the last days of my life in peace and solitude.

The great novelist Leo Tolstoy, late in life. THE BETTMANN ARCHIVE.

But Tolstoy's final journey was both brief and unhappy. Less than a month later, cold and weary from numerous long rides on Russian trains in approaching winter, he contracted pneumonia and died at age eighty-two in the stationmaster's home at the railroad stop of Astapovo. Too weak to write, he dictated his last letter on November 1, 1910. Addressed to a son and daughter who did not share his views on Christian nonviolence, Tolstoy offered a last word of advice:

The views you have acquired about Darwinism, evolution, and the struggle for existence won't explain to you the meaning of your life and won't give you guidance in your actions, and a life without an explanation of its meaning and importance, and without the unfailing guidance that stems from it is a pitiful existence. Think about it. I say it, probably on the eve of my death, because I love you.

Tolstoy's complaint has been the most common of all indictments against Darwin, from the publication of the *Origin of Species* in 1859 to now. Darwinism, the charge contends, undermines morality by claiming that success in nature can only be measured by victory in bloody battle—the "struggle for existence" or "survival of the fittest" to cite Darwin's own choice of mottoes. If we wish "meekness and love" to triumph over "pride and violence" (as Tolstoy wrote to Gandhi), then we must repudiate Darwin's vision of nature's way—as Tolstoy stated in a final plea to his errant children.

This charge against Darwin is unfair for two reasons. First, nature (no matter how cruel in human terms) provides no basis for our moral values. (Evolution might, at most, help to explain why we have moral feelings, but nature can never decide for us whether any particular action is right or wrong.) Second, Darwin's "struggle for existence" is an abstract metaphor, not an explicit statement about bloody battle. Reproductive success, the criterion of natural selection, works in many modes: Victory in battle may be one pathway, but cooperation, symbiosis, and mutual aid may also secure success in other times and contexts. In a famous passage, Darwin explained his concept of evolutionary struggle (*Origin of Species,* 1859, pp. 62–63):

> I use this term in a large and metaphorical sense including dependence of one being on another, and including (which is more important) not only the life of the individual, but success in leaving progeny. Two canine animals, in a time of dearth, may be truly said to struggle with each other which shall get food and live. But a plant on the edge of a desert is said to struggle for life against the drought. . . . As the mistletoe is disseminated by birds, its existence depends on birds; and it may metaphorically be said to struggle with

other fruit-bearing plants, in order to tempt birds to devour and thus disseminate its seeds rather than those of other plants. In these several senses, which pass into each other, I use for convenience sake the general term of struggle for existence.

Yet, in another sense, Tolstoy's complaint is not entirely unfounded. Darwin did present an encompassing, metaphorical definition of struggle, but his actual examples certainly favored bloody battle—"Nature, red in tooth and claw," in a line from Tennyson so overquoted that it soon became a knee-jerk cliché for this view of life. Darwin based his theory of natural selection on the dismal view of Malthus that growth in population must outstrip food supply and lead to overt battle for dwindling resources. Moreover, Darwin maintained a limited but controlling view of ecology as a world stuffed full of competing species—so balanced and so crowded that a new form could only gain entry by literally pushing a former inhabitant out. Darwin expressed this view in a metaphor even more central to his general vision than the concept of struggle—the metaphor of the wedge. Nature, Darwin writes, is like a surface with 10,000 wedges hammered tightly in and filling all available space. A new species (represented as a wedge) can only gain entry into a community by driving itself into a tiny chink and forcing another wedge out. Success, in this vision, can only be achieved by direct takeover in overt competition.

Furthermore, Darwin's own chief disciple, Thomas Henry Huxley, advanced this "gladiatorial" view of natural selection (his word) in a series of famous essays about ethics. Huxley maintained that the predominance of bloody battle defined nature's way as nonmoral (not explicitly immoral, but surely unsuited as offering any guide to moral behavior).

From the point of view of the moralist the animal world is about on a level of a gladiator's show. The creatures are fairly well treated, and set to fight—whereby the strongest, the swiftest, and the cunningest live to fight another day. The spectator has no need to turn his thumbs down, as no quarter is given.

But Huxley then goes further. Any human society set up along these lines of nature will devolve into anarchy and misery—Hobbes's brutal world of *bellum omnium contra omnes* (where *bellum* means "war," not beauty): the war of all against all. Therefore, the chief purpose of society must lie in mitigation of the struggle that defines nature's pathway. Study natural selection and do the opposite in human society:

> But, in civilized society, the inevitable result of such obedience [to the law of bloody battle] is the re-establishment, in all its intensity, of that struggle for existence—the war of each against all—the mitigation or abolition of which was the chief end of social organization.

This apparent discordance between nature's way and any hope for human social decency has defined the major subject for debate about ethics and evolution ever since Darwin. Huxley's solution has won many supporters—nature is nasty and no guide to morality except, perhaps, as an indicator of what to avoid in human society. My own preference lies with a different solution based on taking Darwin's metaphorical view of struggle seriously (admittedly in the face of Darwin's own preference for gladiatorial examples)—nature is sometimes nasty, sometimes nice (really neither, since the human terms are so inappropriate). By presenting examples of all behaviors (under the metaphorical rubric of struggle), nature favors none and offers no guidelines. The facts of nature cannot provide moral guidance in any case.

But a third solution has been advocated by some thinkers who do wish to find a basis for morality in nature and evolution. Since few can detect much moral comfort in the gladiatorial interpretation, this third position must reformulate the way of nature. Darwin's words about the metaphorical character of struggle offer a promising starting point. One might argue that the gladiatorial examples have been over-sold and misrepresented as predominant. Perhaps cooperation and mutual aid are the more common results of struggle for existence. Perhaps communion rather than combat leads to greater reproductive success in most circumstances.

The most famous expression of this third solution may be

Petr Kropotkin, a bearded but gentle anarchist. THE
BETTMANN ARCHIVE.

found in *Mutual Aid,* published in 1902 by the Russian revolutionary anarchist Petr Kropotkin. (We must shed the old stereotype of anarchists as bearded bomb throwers furtively stalking
about city streets at night. Kropotkin was a genial man, almost
saintly according to some, who promoted a vision of small
communities setting their own standards by consensus for the
benefit of all, thereby eliminating the need for most functions of
a central government.) Kropotkin, a Russian nobleman, lived in
English exile for political reasons. He wrote *Mutual Aid* (in English) as a direct response to the essay of Huxley quoted above,
"The Struggle for Existence in Human Society," published in *The
Nineteenth Century,* in February 1888. Kropotkin responded to

Huxley with a series of articles, also printed in *The Nineteenth Century* and eventually collected together as the book *Mutual Aid*.

As the title suggests, Kropotkin argues, in his cardinal premise, that the struggle for existence usually leads to mutual aid rather than combat as the chief criterion of evolutionary success. Human society must therefore build upon our natural inclinations (not reverse them, as Huxley held) in formulating a moral order that will bring both peace and prosperity to our species. In a series of chapters, Kropotkin tries to illustrate continuity between natural selection for mutual aid among animals and the basis for success in increasingly progressive human social organization. His five sequential chapters address mutual aid among animals, among savages, among barbarians, in the medieval city, and amongst ourselves.

I confess that I have always viewed Kropotkin as daftly idiosyncratic, if undeniably well meaning. He is always so presented in standard courses on evolutionary biology—as one of those soft and woolly thinkers who let hope and sentimentality get in the way of analytic toughness and a willingness to accept nature as she is, warts and all. After all, he was a man of strange politics and unworkable ideals, wrenched from the context of his youth, a stranger in a strange land. Moreover, his portrayal of Darwin so matched his social ideals (mutual aid naturally given as a product of evolution without need for central authority) that one could only see personal hope rather than scientific accuracy in his accounts. Kropotkin has long been on my list of potential topics for an essay (if only because I wanted to read his book, and not merely mouth the textbook interpretation), but I never proceeded because I could find no larger context than the man himself. Kooky intellects are interesting as gossip, perhaps as psychology, but true idiosyncrasy provides the worst possible basis for generality.

But this situation changed for me in a flash when I read a very fine article in the latest issue of *Isis* (our leading professional journal in the history of science) by Daniel P. Todes: "Darwin's Malthusian Metaphor and Russian Evolutionary Thought, 1859–1917." I learned that the parochiality had been mine in my ignorance of Russian evolutionary thought, not Kropotkin's in his isolation in England. (I can read Russian, but only painfully, and with a dictionary—which means, for all practical purposes, that I can't read the language.) I knew that Darwin had become a hero of the Russian intelligentsia and had influenced academic life in Russia perhaps more than in any other country. But virtually none of this Russian work has ever been translated or even discussed in English literature. The ideas of this school are unknown to us; we do not even recognize the names of the major protagonists. I knew Kropotkin because he had published in English and lived in England, but I never understood that he represented a standard, well-developed Russian critique of Darwin, based on interesting reasons and coherent national traditions. Todes's article does not make Kropotkin more correct, but it does place his writing into a general context that demands our respect and produces substantial enlightenment. Kropotkin was part of a mainstream flowing in an unfamiliar direction, not an isolated little arroyo.

This Russian school of Darwinian critics, Todes argues, based its major premise upon a firm rejection of Malthus's claim that competition, in the gladiatorial mode, must dominate in an ever more crowded world, where population, growing geometrically, inevitably outstrips a food supply that can only increase arithmetically. Tolstoy, speaking for a consensus of his compatriots, branded Malthus as a "malicious mediocrity."

Todes finds a diverse set of reasons behind Russian hostility to Malthus. Political objections to the dog-eat-dog character of Western industrial competition arose from both ends of the Russian spectrum. Todes writes:

> Radicals, who hoped to build a socialist society, saw Malthusianism as a reactionary current in bourgeois political economy. Conservatives, who hoped to preserve the communal virtues of tsarist Russia, saw it as an expression of the "British national type."

But Todes identifies a far more interesting reason in the immediate experience of Russia's land and natural history. We all have a tendency to spin universal theories from a limited domain of surrounding circumstance. Many geneticists read the entire world of evolution in the confines of a laboratory bottle filled with fruit flies. My own increasing dubiousness about universal adaptation arises in large part, no doubt, because I study a peculiar snail that varies so widely and capriciously across an apparently unvarying environment, rather than a bird in flight or some other marvel of natural design.

Russia is an immense country, under-populated by any nineteenth-century measure of its agricultural potential. Russia is also, over most of its area, a harsh land, where competition is more likely to pit organism against environment (as in Darwin's metaphorical struggle of a plant at the desert's edge) than organism against organism in direct and bloody battle. How could any Russian, with a strong feel for his own countryside, see Malthus's principle of overpopulation as a foundation for evolutionary theory? Todes writes:

> It was foreign to their experience because, quite simply, Russia's huge land mass dwarfed its sparse population. For a Russian to see an inexorably increasing population inevitably straining potential supplies of food and space required quite a leap of imagination.

If these Russian critics could honestly tie their personal skepticism to the view from their own backyard, they could also recognize that Darwin's contrary enthusiasms might record the parochiality of his different surroundings, rather than a set of necessarily universal truths. Malthus makes a far better prophet in a crowded, industrial country professing an ideal of open competition in free markets. Moreover, the point has often been made that both Darwin and Alfred Russel Wallace independently developed the theory of natural selection after primary experience with natural history in the tropics. Both claimed inspiration from Malthus, again independently; but if fortune favors the prepared mind, then their tropical experience probably predisposed both men to read Malthus with resonance and approval. No other area on earth is so packed with species, and therefore so replete with competition of body against body. An Englishman who had learned the ways of nature in the tropics was almost bound to view evolution differently from a Russian nurtured on tales of the Siberian wasteland.

For example, N. I. Danilevsky, an expert on fisheries and population dynamics, published a large, two-volume critique of Darwinism in 1885. He identified struggle for personal gain as the credo of a distinctly British "national type," as contrasted with old Slavic values of collectivism. An English child, he writes, "boxes one on one, not in a group as we Russians like to spar." Danilevsky viewed Darwinian competition as "a purely English doctrine" founded upon a line of British thought stretching from Hobbes through Adam Smith to Malthus. Natural selection, he wrote, is rooted in "the war of all against all, now termed the struggle for existence—Hobbes' theory of politics; on competition—the economic theory of Adam Smith. . . . Malthus applied the very same principle to the problem of population. . . . Darwin extended both Malthus' partial theory and the general theory of the political economists to the organic world." (Quotes are from Todes's article.)

When we turn to Kropotkin's *Mutual Aid* in the light of Todes's discoveries about Russian evolutionary thought, we must reverse the traditional view and interpret this work as mainstream Russian criticism, not personal crankiness. The central logic of Kropotkin's argument is simple, straightforward, and largely cogent.

Kropotkin begins by acknowledging that struggle plays a central role in the lives of organisms and also provides the chief impetus for their evolution. But Kropotkin holds that struggle must not be viewed as a unitary phenomenon. It must be divided into two fundamentally different forms with contrary evolutionary meanings. We must recognize, first of all, the struggle of organism against organism for limited resources—the theme that Malthus imparted to Darwin and that Huxley described as gladiatorial. This form of direct struggle does lead to competition for personal benefit.

But a second form of struggle—the style that Darwin called metaphorical—pits organism against the harshness of surrounding physical environments, not against other members of the same species. Organisms must struggle to keep warm, to survive the sudden and unpredictable dangers of fire and storm, to persevere through harsh periods of drought, snow, or pestilence. These forms of struggle between organism and environment are best waged by cooperation among members of the same species—by mutual aid. If the struggle for existence pits two lions against one zebra, then we shall witness a feline battle and an equine carnage. But if lions are struggling jointly against the harshness of an inanimate environment, then fighting will not remove the common enemy—while cooperation may overcome a peril beyond the power of any single individual to surmount.

Kropotkin therefore created a dichotomy within the general notion of struggle—two forms with opposite import: (1) organism against organism of the same species for limited resources, leading to competition; and (2) organism against environment, leading to cooperation.

> No naturalist will doubt that the idea of a struggle for life carried on through organic nature is the greatest generalization of our century. Life *is* struggle; and in that struggle the fittest survive. But the answers to the questions "by which arms is the struggle chiefly carried on?" and "who are the fittest in the struggle?" will widely differ according to the importance given to the two different aspects of the struggle: the direct one, for food and safety among separate individuals, and the struggle which Darwin described as "metaphorical"—the struggle, very often collective, against adverse circumstances.

Darwin acknowledged that both forms existed, but his loyalty to Malthus and his vision of nature chock-full of species led him to emphasize the competitive aspect. Darwin's less sophisticated votaries then exalted the competitive view to near exclusivity, and heaped a social and moral meaning upon it as well.

> They came to conceive of the animal world as a world of perpetual struggle among half-starved individuals, thirsting for one another's blood. They made modern literature resound with the war-cry of *woe to the vanquished,* as if it were the last word of modern biology. They raised the "pitiless" struggle for personal advantages to the height of a biological principle which man must submit to as well, under the menace of otherwise succumbing in a world based upon mutual extermination.

Kropotkin did not deny the competitive form of struggle, but he argued that the cooperative style had been underemphasized and must balance or even predominate over competition in considering nature as a whole.

> There is an immense amount of warfare and extermination going on amidst various species; there is, at the same time, as much, or perhaps even more, of mutual support, mutual aid, and mutual defense. . . . Sociability is as much a law of nature as mutual struggle.

As Kropotkin cranked through his selected examples, and built up steam for his own preferences, he became more and more convinced that the cooperative style, leading to mutual aid, not only predominated in general but also characterized the most advanced creatures in any group—ants among insects, mammals among vertebrates. Mutual aid therefore becomes a more important principle than competition and slaughter:

> If we . . . ask Nature: "who are the fittest: those who are continually at war with each other, or those who support one another?" we at once see that those animals which acquire habits of mutual aid are undoubtedly the fittest. They have more chances to survive, and they attain, in their respective classes, the highest development of intelligence and bodily organization.

If we ask why Kropotkin favored cooperation while most nineteenth-century Darwinians advocated competition as the predominant result of struggle in nature, two major reasons stand out. The first seems less interesting, as obvious under the slightly cynical but utterly realistic principle that true believers tend to read their social preferences into nature. Kropotkin, the anarchist who yearned to replace laws of central government with consensus of local communities, certainly hoped to locate a deep preference for mutual aid in the innermost evolutionary marrow of our being. Let mutual aid pervade nature and human cooperation becomes a simple instance of the law of life.

> Neither the crushing powers of the centralized State nor the teachings of mutual hatred and pitiless struggle which

came, adorned with the attributes of science, from obliging philosophers and sociologists, could weed out the feeling of human solidarity, deeply lodged in men's understanding and heart, because it has been nurtured by all our preceding evolution.

But the second reason is more enlightening, as a welcome empirical input from Kropotkin's own experience as a naturalist and an affirmation of Todes's intriguing thesis that the usual flow from ideology to interpretation of nature may sometimes be reversed, and that landscape can color social preference. As a young man, long before his conversion to political radicalism, Kropotkin spent five years in Siberia (1862–1866) just after Darwin published the *Origin of Species*. He went as a military officer, but his commission served as a convenient cover for his yearning to study the geology, geography, and zoology of Russia's vast interior. There, in the polar opposite to Darwin's tropical experiences, he dwelled in the environment least conducive to Malthus's vision. He observed a sparsely populated world, swept with frequent catastrophes that threatened the few species able to find a place in such bleakness. As a potential disciple of Darwin, he looked for competition, but rarely found any. Instead, he continually observed the benefits of mutual aid in coping with an exterior harshness that threatened all alike and could not be overcome by the analogues of warfare and boxing.

Kropotkin, in short, had a personal and empirical reason to look with favor upon cooperation as a natural force. He chose this theme as the opening paragraph for *Mutual Aid:*

Two aspects of animal life impressed me most during the journeys which I made in my youth in Eastern Siberia and Northern Manchuria. One of them was the extreme severity of the struggle for existence which most species of animals have to carry on against an inclement Nature; the enormous destruction of life which periodically results from natural agencies; and the consequent paucity of life over the vast territory which fell under my observation. And the other was, that even in those few spots where animal life teemed in abundance, I failed to find—although I was eagerly looking for it—that bitter struggle for the means of existence among animals belonging to the same species, which was considered by most Darwinists (though not always by Darwin himself) as the dominant characteristic of struggle for life, and the main factor of evolution.

What can we make of Kropotkin's argument today, and that of the entire Russian school represented by him? Were they just victims of cultural hope and intellectual conservatism? I don't think so. In fact, I would hold that Kropotkin's basic argument is correct. Struggle does occur in many modes, and some lead to cooperation among members of a species as the best pathway to advantage for individuals. If Kropotkin overemphasized mutual aid, most Darwinians in Western Europe had exaggerated competition just as strongly. If Kropotkin drew inappropriate hope for social reform from his concept of nature, other Darwinians had erred just as firmly (and for motives that most of us would now decry) in justifying imperial conquest, racism, and oppression of industrial workers as the harsh outcome of natural selection in the competitive mode.

I would fault Kropotkin only in two ways—one technical, the other general. He did commit a common conceptual error in failing to recognize that natural selection is an argument about advantages to individual organisms, however they may struggle. The result of struggle for existence may be cooperation rather than competition, but mutual aid must benefit individual organisms in Darwin's world of explanation. Kropotkin sometimes speaks of mutual aid as selected for the benefit of entire populations or species—a concept foreign to classic Darwinian logic (where organisms work, albeit unconsciously, for their own benefit in terms of genes passed to future generations). But Kropotkin also (and often) recognized that selection for mutual aid directly benefits each individual in its own struggle for personal success. Thus, if Kropotkin did not grasp the full implication of Darwin's basic argument, he did include the orthodox solution as his primary justification for mutual aid.

More generally, I like to apply a somewhat cynical rule of thumb in judging arguments about nature that also have overt social implications: When such claims imbue nature with just those properties that make us feel good or fuel our prejudices, be doubly suspicious. I am especially wary of arguments that find kindness, mutuality, synergism, harmony—the very elements that we strive mightily, and so often unsuccessfully, to put into our own lives—intrinsically in nature. I see no evidence for Teilhard's noosphere, for Capra's California style of holism, for Sheldrake's morphic resonance. Gaia strikes me as a metaphor, not a mechanism. (Metaphors can be liberating and enlightening, but new scientific theories must supply new statements about causality. Gaia, to me, only seems to reformulate, in different terms, the basic conclusions long achieved by classically reductionist arguments of biogeochemical cycling theory.)

There are no shortcuts to moral insight. Nature is not intrinsically anything that can offer comfort or solace in human terms—if only because our species is such an insignificant latecomer in a world not constructed for us. So much the better. The answers to moral dilemmas are not lying out there, waiting to be discovered. They reside, like the kingdom of God, within us—the most difficult and inaccessible spot for any discovery or consensus.

# 23 | Fleeming Jenkin Revisited

THE EUROPEAN REVOLUTIONS of 1848 played a pivotal role in several famous lives. Karl Marx, exiled from Germany, published the last issue of his *Neue Rheinische Zeitung* in red, then moved to England, where he constructed *Das Kapital* in the reading room of the British Museum. The young Richard Wagner, espousing an idealistic socialism that he would later reject with vigor, manned the barricades of Dresden, then fled from Germany to avoid a warrant for his arrest and missed the premiere of *Lohengrin*.

Another man, destined for a lesser but secure reputation, experienced a touch of the same excitement. In February 1848, Henry Charles Fleeming Jenkin, a fourteen-year-old boy from Scotland, found himself in Paris surrounded by rebellion. He wrote to a friend in Edinburgh: "Now then, Frank, what do you think of it? I in a revolution and out all day. Just think, what fun!"

In 1867, the same Fleeming Jenkin would taste revolution of a different kind—this time as a transient participant, not a mere observer. In his much revised fifth edition of the *Origin of Species*, Charles Darwin made a substantial concession by admitting that favorable variations arising in single individuals could not spread through entire populations. (In retrospect, Darwin need not have conceded. He based his admission on a false view of heredity. In a Mendelian world, unknown to Darwin, such favorable variations can spread—see subsequent discussion in this essay. Nonetheless, Darwin's concession represents a small but celebrated incident in the history of evolutionary thought.) Darwin wrote:

340

> I saw . . . that the preservation in a state of nature of any occasional deviation of structure . . . would be a rare event; and that, if preserved, it would generally be lost by subsequent intercrossing with ordinary individuals. Nevertheless, until reading an able and valuable article in the *North British Review* (1867), I did not appreciate how rarely single variations whether slight or strongly marked could be inherited.

Nearly every book in the history of evolution recounts the tale and refers to the author of this "able and valuable article" as "a Scottish engineer" or, more often, "an obscure Scottish engineer." The author was Fleeming (pronounced Flemming) Jenkin. Darwin, more explicit and vexed in private letters than in public texts, wrote to Joseph Hooker in 1869: "Fleeming Jenkin has given me much trouble. . . ."—and to Alfred Russel Wallace a few days later: "Fleeming Jenkin's arguments have convinced me."

All evolutionists recognize (and mispronounce by excessive literalism) Fleeming Jenkin as the man who forced an explicit, though unnecessary, concession from Darwin. But we know nothing about him and tend to assume that he rose from general obscurity for one small moment in our sun—a lamentable parochialism on our part.

My own career has included two fortuitous and peculiar intersections with Fleeming Jenkin—so I decided that I must write a column about him now, before a third encounter elevates coincidence to inescapable pattern. I was an undergraduate at Antioch College from 1958 to 1963. Antioch was (and is) a wonderful school in the finest American tradition of small liberal arts colleges. But it doesn't boast much in the way of library facilities for scholarship based on original sources. One day in 1960, I was browsing aimlessly through the stacks and found a crumbling run of the *North British Review* for the mid-nineteenth century. I recognized the name from Darwin's citation, and my heart skipped a beat as I hoped against hope that the volume for 1867 lay within the series. It did, and I then spent a more anxious minute convinced that I had the wrong title or that the issue for the right month would be missing. It wasn't. I found Jenkin's article and rushed to the pre-Xerox wet processor (anachronistically named smellox by a friend of mine several years later, in honor of the unpleasant chemical that left its signature even after drying). I fed dimes into the machine and soon had my precious copy of the original Fleeming Jenkin. What a prize I thought I had. I was sure that I possessed the only copy in the whole world. (Can you imagine what one peek at the Harvard library does to such naivete?) I have carried that copy with me ever since, assigning its properly Xeroxed offspring to classes now and again, but never dreaming that I would write anything about Jenkin.

Then, last month, I was browsing through a friend's Victorian literature collection, aimlessly running my eye along the titles of Robert Louis Stevenson's complete works. I found *Treasure Island, Kidnapped,* and all the other items of my old "Author's" card game. But my heart skipped another beat at the next title: *Memoir of Fleeming Jenkin.* The "obscure Scottish engineer" had achieved sufficient renown (in areas far from my own parish, where he only dabbled, however successfully) to win a full volume from Stevenson's pen. I kicked myself for sectarian assumptions in the granting of "importance," vowed to learn more about Jenkin (and to tell my fellow evolutionists), and raided the stacks of Widener Library, where I found several copies of Stevenson's memoir, amidst (no doubt) a liberal sprinkling of *North British Review*s for 1867 (which, *smelloci gratia,* I didn't need).

An interesting man, Fleeming Jenkin—and made all the more appealing by the strength of Stevenson's prose. Jenkin spent most of his life in Edinburgh, where he campaigned for the improvement of home sanitation, conducted some of Britain's first experiments with the phonograph, produced and directed amateur theatricals, hated golf (for a Scotsman, I suppose, about as bad as an American who barfs on apple pie), and became the first professor of engineering at the University of Edinburgh. Most important, he was a close friend and colleague of Lord Kelvin and spent most of his career designing and outfitting transoceanic cables with the great physicist.

Stevenson's book has a lovely, archaic charm. It describes a moral perfection that cannot be, and belongs to the genre of guiding homilies based on lives of the great. If Jenkin ever gazed at a woman other than his wife, if he ever raised his voice in anger or acted in even momentary pettiness, we are not told. Instead, we get glimpses of a simpler and formal world based on unquestioned certainties. In 1877, Jenkin writes to his absent wife about their son: "Frewen had to come up and sit in my room for company last night and I actually kissed him, a thing that has not occurred for years." The Captain, Jenkin's aged father, dies at age eighty-four but achieves solace in his last hour from a false report on the rescue of General "Chinese" Gordon at Khartoum: "He has been waiting with painful interest for news of Gordon and Khartoum; and by great good fortune, a false report reached him that the city was relieved, and the men of Sussex (his old neighbors) had been the first to enter. He sat up in bed and gave three cheers for the Sussex regiment."

Stevenson's memoir contains exactly one line on Jenkin's 1867 foray into evolutionary theory: "He had begun by this time to write. His paper on Darwin . . . had the merit of convincing on one point the philosopher himself." Evidently, Jenkin needed neither evolution nor the *North British Review* to merit Stevenson's extended attention. I felt a bit ashamed at my own previous parochialism. Do grocers know Thomas Jefferson (or was it Benjamin Franklin) only as the man who invented that thing that gets the cereal boxes down from the top shelf?

The backward reading of history has cruelly misserved many fine thinkers, Jenkin included. (Professionals refer to this unhappy tactic as "Whiggish history" in dubious memory of those Whig historians who evaluated predecessors exclusively by their adherence to ideals of Whig politics unknown in their own times.)

Jenkin has suffered because commentators extract from his 1867 article just the one small point that provoked Darwin's concession—and then analyze his argument in modern terms by pointing out that a twentieth-century Darwin could stick to his Mendelian guns. No modern evolutionary biologist, to my knowledge, has ever considered Jenkin's treatise as a whole and appreciated its force, despite its errors in modern terms. I shall attempt this rescue but bow first to the constraints of history and discuss the point that secured Jenkin's slight renown in evolutionary circles.

Darwin and Jenkin accepted the usual notion of heredity prevalent in their times—a concept called blending inheritance. Under blending inheritance, the offspring of two parents tend to lie halfway between for inherited characters. Jenkin pointed out to Darwin, or so the usual and quite inadequate story goes, that blending inheritance would challenge natural selection because any favorable variant would be swamped out by back-breeding with the predominant parental forms. Jenkin's own example will make his argument clear. It also serves as a sad reminder of unquestioned racism in Victorian England—and as an indication that, for all our pressing problems, we have improved somewhat during the past century:

> Suppose a white man to have been wrecked on an island inhabited by negroes. . . . Suppose him to possess the physical strength, energy, and ability of a dominant white race . . . grant him every advantage which we can conceive a white to possess over the native. . . . Yet from all these admissions, there does not follow the conclusion that, after a limited or unlimited number of generations, the inhabitants of the island will be white. Our shipwrecked hero would probably become king; he would kill a great many blacks in the struggle for existence; he would have a great many wives and children, while many of his subjects would live and die as bachelors. . . . In the first generation there will be some dozens of intelligent young mulattoes, much superior in average intelligence to the negroes. We might expect the throne for some generations to be occupied by a more or less yellow king; but can any one believe that the whole island will gradually acquire a white, or even a yellow population . . . for if a very highly favored white cannot blanch a nation of negroes, it will hardly be contended that a comparatively dull mulatto has a good chance of producing a tawny tribe.

In other words, by blending inheritance, the offspring of the first generation will be only half white. Most of these mulattoes, since full blacks so greatly predominate (and following prohibitions against incest), will marry full blacks, and their offspring of the second generation will be one-quarter white. By the same argument, the proportion of white blood will dilute to one-eighth in the third generation and soon dwindle to oblivion, despite supposed advantages.

Darwin, or so the story goes, saw the strength of this argument and retreated in frustrated impotence toward the Lamarckian views that he had previously rejected. Whiggery then comes to the rescue. Inheritance is Mendelian, or "particulate," not blending (though Darwin died long before the rediscovery of Mendel's laws in 1900). Traits based on genetic mutations do not dilute; genes that determine such traits are entities or particles that do not degrade by mixing with genes of the other parent in offspring. Indeed, if recessive, a favorable trait will appear in *no* offspring of the first generation (in matings between the favored mutant and ordinary partners carrying the dominant gene). But the trait does not dilute to oblivion. In the second generation, one-quarter of the offspring between mixed parents will carry two doses of the advantageous recessive gene and will express the favored trait. Any subsequent matings between these double recessives will pass the favored trait to all offspring—and it can spread through the population if concentrated by natural selection. (Skin color and height seem to blend because they are determined by such a large number of particulate genes. The average effect may be a blend, yet the genes remain intact and subject to selection.)

But this usual story fails when we properly locate Jenkin's point about blending in the wider context of an argument that pervades the entire essay—and do not simply extract the item as a kernel deserving modern notice while discarding the rest as chaff. As historian Peter J. Vorzimmer notes in his excellent book, *Charles Darwin: The Years of Controversy* (1970), Jenkin presented his arguments about blending in discussing only one particular kind of variation—*single* favorable variants *substantially different* from parental forms.

Darwin was no fool. He had thought about variation as deeply as any man. His longest book, the two-volume *Variation of Animals and Plants Under Domestication* (1868), summarizes everything he and almost everyone else knew about the subject. Can we seriously believe that he had never thought about problems that blending posed for natural selection—that he needed a prod from an engineer to recognize the difficulty? As Vorzimmer shows, Darwin had pondered long and hard about problems provoked by blending. Jenkin did not introduce Darwin to this basic problem of inheritance; rather, he made a distinction between the kinds of variation that blending affects, and Darwin welcomed the argument because it reinforced and sharpened one of his favorite views. Darwin did not retreat before Jenkin's onslaught, but rather felt more secure in his preferred belief—hence his expressed gratitude to Jenkin and hence (I assume) Stevenson's single comment that Jenkin had convinced "on one point the philosopher himself." Stevenson, the novelist, understood. We have forgotten.

The real issue has been lost in a terminology understood in Darwin's time but no longer familiar. Let us return to Darwin's letter to Wallace, quoting the passage this time in full: "I always thought individual differences more important than single variations, but now I have come to the conclusion that they are of paramount importance, and in this I believe I agree with you. Fleeming Jenkin's arguments have convinced me."

In Darwin's time, "individual differences" referred to recurrent variations of small scale, while "single variations" identified unique changes of large scope and import—often called "sports." Debate had focused on whether small-scale and continuous, or occasional and larger, variations supplied the raw material for evolutionary change. Darwin, the quintessential continuationist of this or any other age, had long preferred recurrent small-scale changes but had continued to flirt (largely by weight of tradition) with larger sports. Now, the simple point of Jenkin's argument: Note that he speaks of *one* white man identified (in the racist tradition) as vastly superior to the natives—in other words, a single sport. Jenkin's famous blending argument refers only to single, marked variations—not to the continuous recurrent variations that Darwin preferred. By accepting Jenkin's view, Darwin could finally rid himself of a form of variation that he had never favored.

As for recurrent, small-scale variation (individual differences, in Darwin's terminology), blending posed no insurmountable problem, and Darwin had resolved the issue in his own mind long before reading Jenkin. A blending variation can still establish itself in a population under two conditions: first, if the favorable variation continues to arise anew so that any dilution by blending can be balanced by reappearances, thus keeping the trait visible to natural selection; second, if individuals bearing the favored trait can recognize each other and mate preferentially—a process known as assortative mating in evolutionary jargon. Assortative mating can arise for several reasons, including aesthetic preference for mates of one's own appearance and simple isolation of the favored variants from normal individuals. Darwin recognized both recurrent appearance and isolation as the primary reasons for natural selection's continued power in the face of blending.

With this background, we can finally exhume the real point and logic of Jenkin's essay—an issue still very much alive, and discussed (through all his factual errors) in a most interesting and perceptive way by Fleeming Jenkin. Jenkin's essay is a critique of Darwin's continuationist perspective—his distinctive claim, still maintained by the evolutionary orthodoxy—that all large-scale phenomena of evolution may be rendered by accumulating, through vast amounts of time, the tiny changes that we observe in modern populations. I call this conventional view the "extrapolationist" argument; I also share Jenkin's opinion (but for different reasons) that this traditional mode of thinking cannot explain all of evolution. I find it supremely ironic that the one small section of Jenkin's article *not* about Darwin's claim for continuity (his argument that single sports will be swamped by blending) has become the only part that we remember—and, to make matters worse, usually misinterpret. Such, however, is the usual fate of Whig heroes and villains.

A simple, almost pedantic, précis of Jenkin's argument should rescue his larger point. Jenkin's essay proceeds in four parts. The first, on limits of variation, admits that Darwin's favored style of recurrent, continuous variation does occur and can be manipulated by natural selection to change the average form of a species. But, Jenkin argues, such variations always fiddle in minor ways with parts already present; they cannot construct anything new. Thus, natural selection can make dogs big, small, blocky, or elongate—but cannot change a dog into something else. Jenkin expresses this argument in his powerful metaphor of the "sphere of variation." Natural selection may move the average form anywhere within the sphere, but not beyond its fixed limits:

> A given animal or plant appears to be contained, as it were, within a sphere of variation; one individual lies near one portion of the surface, another individual, of the same species, near another part of the surface; the average animal at the center.

Common experience, Jenkin affirms, supports his view. Artificial selection practiced by breeders proceeds rapidly at first but soon reaches frustrating limits. Jenkin writes of racehorses:

> Hundreds of skillful men are yearly breeding thousands of racers. Wealth and honor await the man who can breed one horse to run one part in five thousand faster than his fellows. As a matter of experience, have our racers improved in speed by one part in a thousand during the last twenty generations?

Darwin, Jenkin claims, maintains an unwarranted faith in the power of simple time to overcome these barriers:

> The difference between six years and six myriads, blending by a confused sense of immensity, leads men to say hastily that if six or sixty years can make a pouter out of a common pigeon, six myriads may change a pigeon to something like a thrush; but this seems no more accurate than to conclude that because we observe that a cannon-ball has traversed a mile in a minute, therefore in an hour it will be sixty miles off, and in the course of ages that it will reach the fixed stars.

Darwin might argue, Jenkin admits, that once a species reaches the limit of its glass sphere, time will eventually reconstitute this edge as a new center and produce a new sphere around the previously peripheral point. Jenkin also rejects this argument:

> The average or original race . . . will [in Darwin's view] spontaneously lose the tendency to relapse and acquire a tendency to vary outside the sphere. What is to produce this change? Time simply, apparently. . . . This seems rather like the idea that keeping a bar of iron hot or cold for a long time would leave it permanently hot or cold at the end of the period when the heating or cooling agent was withdrawn.

Jenkin's second section, on types of variation, begins by admitting Darwin's point that small-scale recurrent variations will not be destroyed by blending. But these are the very variations subject to strict limits by the previous argument about rigid spheres. What kind of variation might then induce the evolution of something substantially new? Single sports might seem promising, but these are the rare events that will be swamped by blending—and readers may now note the point that Jenkin himself wished to make with the only part of his argument that we remember. But perhaps some kinds of sports do not blend and do perpetuate their kind. Fine, Jenkin admits. Perhaps such creatures do occasionally arise and produce new species. But such a process is not Darwinian evolution, for Darwin insisted that natural selection acts as a creative force by gradually accumulating favorable variants. Indeed, would such a process be very different from what the vernacular calls "creation"?

The third part argues that even if Darwin could find (which he can't) some way to accumulate small-scale recurrent variations into something new, geology does not supply enough time for such a slow process. Here Jenkin relied on the false arguments of his dearest friend, Lord Kelvin, about the earth's relatively young age (see Essay 8, on Kelvin, in *The Flamingo's Smile*).

The last section presents a powerful (and I think entirely correct) argument about the difficulty of inferring historical pathways from current situations. Jenkin contends that nearly any current situation can arise via several historical routes; thus the situation by itself cannot specify the pathway. Jenkin points out that Darwin bases much of his argument upon the lack of definite boundaries in nature—the intergradation of species into species, or geographic region into region. Darwin's continuationism predicts just such an absence of boundaries since species are gradually and imperceptibly changed into their descendants, while a creator should leave gaps between his incarnated objects. But Jenkin argues that many natural items come as continua, yet clearly do not arise by a process of historical transformation. Arguing, as he does throughout, by metaphor and analogy, Jenkin writes:

> Legal difficulties furnish another illustration. Does a particular case fall within a particular statute? Is it ruled by this or that precedent? The number of statutes or groups is limited; the number of possible combinations of events almost unlimited.

Taken as an entirety, Jenkin's argument possesses a kind of relentless logic. The critique of Darwin's extrapolationism serves as its unifying theme. Part one argues that small-scale variation cannot extend beyond fixed limits. Part two claims that no style of variation can make something substantially new in a Darwinian world. Part three proposes that even if Darwin could find a way, geology does not permit enough time. Finally, part four holds that we cannot infer historical transformations from the admitted continua of nature.

I don't want to fall into a Whiggish pit and judge Jenkin by current standards (everything that has gone before in this essay adequately discharges, I hope, my dues to anti-Whiggery). But old arguments usually repay our close attention because we often stop discussing the fundamentals once an orthodoxy triumphs, and we need to consult the original debates in order to rediscover the largest issues—perhaps never really resolved but merely swept under a rug of concord. Much of Jenkin's argument fails today, for few things last a century in science. He was clearly wrong about blending in part two and about time in part three. But I believe that he was right about continua in part four and that we are still plagued by a tendency to make an almost automatic inference about history when we fail to find clear boundaries.

The first argument about limits of variation has also risen again in current debates about evolutionary processes. I don't accept Jenkin's metaphor of the sphere, because small quantitative changes can accumulate to qualitative effects or leaps (contrary to Jenkin's position), and because I accept Darwin's argument that new spheres can be reconstituted about previously peripheral points.

But neither (probably) is a species the kind of almost equipotential sphere that strict Darwinians envisaged—unconstrained and capable of rolling anywhere that natural selection pushes. Constraints imposed by genetics and development have emerged as a central topic in contemporary evolutionary debate—and Fleeming Jenkin did present an insight worth considering.

In short, Darwin's strictly extrapolationist vision may not describe large-scale evolution very well—small, local adaptations built in the refiner's fire of Darwinian competition among organisms struggling for reproductive success may not, by extension, explain trends that persist for millions of years or relays of changing diversity that mass extinctions produce. Jenkin, who presented the most logical dissection of Darwin's continuationist vision in 1867, does reach across a century to set us thinking, however superannuated his specific claims.

We may give the last word to Jenkin, via Robert Louis Stevenson. One day as a young man, Stevenson reports, Fleeming Jenkin argued bitterly with two young women about a pressing issue of Victorian hypermorality: Can a misdeed against moral codes ever be condoned, whatever the circumstances—stealing a knife to prevent a murder, for example? (Jenkin, to his credit, argued the affirmative.) As he left the house, his anger mellowed. He realized that even the most apparently peculiar belief deserves respect if argued honorably and if properly constructed upon a set of basic premises different from those usually cherished:

> From such passages-at-arms, many retire mortified and ruffled; but Fleeming had no sooner left the house than he fell into delighted admiration of the spirit of his adversaries. From that it was but a step to ask himself "what truth was sticking in their heads"; for even the falsest form of words (in Fleeming's life-long opinion) reposed upon some truth.

---

## Postscript

My parochialism and ignorance in the case of Fleeming Jenkin were even deeper than I had realized. Having corrected the most blatant omission of failing to recognize the importance of his mainline career in engineering, I discovered, after publishing this essay, that I had also missed a tangential foray equal in importance to Jenkin's critique of Darwin. Several professors of economics wrote to inform me that Jenkin had made cogent contributions to the "dismal science" as well.

Robert B. Ekelund, Jr., of Auburn University, stated:

> Jenkin, an engineer by training, was the first English economist to draw and clearly understand supply and demand curves, the most familiar staple in all of economics. In two

amazing essays, published in 1868 and 1870, Jenkin developed demand and supply theory, applied it to labor markets, and introduced an innovative combination of stock and flow concepts for analyzing market fluctuations.

Christopher Bell of Davidson College then sent me an article from *Oxford Economics Papers* (Volume 15, 1963) by A. D. Brownlie and M. F. Lloyd Prichard entitled "Professor Fleeming Jenkin, 1833–1885, Pioneer in Engineering and Political Economy." This fascinating article cites the opinion of the great economist J. A. Schumpeter (1883–1950), who regarded Jenkin as "an economist of major importance, whose main papers . . . form an obvious stepping stone between J. S. Mill and Marshall."

In a time of great industrial strife, and considerable opposition to trade unionism, Jenkin used his quantitative analysis of supply and demand to defend, as practical and necessary, the rights of workers to form associations for collective bargaining. He wrote that "the total abolition of trade unions is out of the question as impolitic, undeserved, and impossible . . . [But] we must insist that the great power granted to the bodies of workmen shall be administered under stringent regulations."

Jenkin, scarcely a radical in politics, favored no massive redistribution of wealth, but only some minor tinkering for greater satisfaction and productivity of workers. He wrote: "Great inequality is necessary and desirable (observations seem to show that trade will extend faster with large profits and small wages than with small profits and large wages)." For, basically, Jenkin held firm to the ideals of the laissez-faire system so strongly identified with the intellectual history of his nation, particularly with Adam Smith in his own home city of Edinburgh. Jenkin wrote:

> We cannot deny that each man, acting rationally for his own advantage, will conduce to the good of all; and if the motive be not the highest, it is one which at least can always be counted on.

Yet Jenkin tempered the harshness of pure laissez-faire with a realization that the central argument, practically applied by people in power, almost always acted as a rationale for unfairness toward workers. He wrote:

They [laborers] think it monstrous that one of two parties to a bargain should be told to shut his eyes and open his hands and take the wages fixed by Political Economy, which allegorical personage looks very like an employer on pay day.

A wonderful irony pervades all these themes, one that only an evolutionary biologist could fully identify and appreciate. Brownlie and Lloyd Prichard point out that Jenkin's economic writings were consigned to oblivion, largely because the two great opinion makers of later nineteenth-century English economics, Jevons and Marshall, "treated him shabbily to say the least." (Both Jevons and Marshall sensed that the "amateur" Jenkin had anticipated some of the "original" work that served as a basis of their own reputations. They therefore sought to disparage and discredit, and then to ignore, this gifted thinker, who did only limited work in economics and did not really threaten their turf, or even their prestige, in any large sense—an act, all too characteristic, alas, of conventional academic ungenerosity.)

Now, the irony: Jenkin was, basically, a proponent of the laissez-faire school. Darwin, as I have often argued in these essays, established his central theory of natural selection by importing the structure of Adam Smith's economic arguments into nature (with organisms struggling for individual reproductive success as the analogue of "each man, acting rationally for his own advantage" in Jenkin's quotation—and with organic progress and balance of nature arising as a result, just as "the good of all" supposedly emerges from concatenated selfishness in Adam Smith's system). How ironic then that Jenkin was belittled and disparaged for truly original work in the parent discipline of economics—but, thanks to Darwin's greater geniality and sense of fairness, honored and acknowledged for similarly cogent contributions to a field that had so benefited, just a little before (in 1859 when Darwin published the *Origin*), from generous consideration of economic theories.

# 24 | The Passion of Antoine Lavoisier

GALILEO AND LAVOISIER have more in common than their brilliance. Both men are focal points in a cardinal legend about the life of intellectuals—the conflict of lonely and revolutionary genius with state power. Both stories are apocryphal, however inspiring. Yet they only exaggerate, or encapsulate in the epitome of a bon mot, an essential theme in the history of thinking and its impact upon society.

Galileo, on his knees before the Inquisition, abjures his heretical belief that the earth revolves around a central sun. Yet, as he rises, brave Galileo, faithful to the highest truth of factuality, addresses a stage whisper to the world: *eppur se muove*—nevertheless, it does move. Lavoisier, before the revolutionary tribunal during the Reign of Terror in 1794, accepts the inevitable verdict of death, but asks for a week or two to finish some experiments. Coffinhal, the young judge who has sealed his doom, denies his request, stating, *La république n'a pas besoin de savants* (the Republic does not need scientists).

Coffinhal said no such thing, although the sentiments are not inconsistent with emotions unleashed in those frightening and all too frequent political episodes so well characterized by Marc Antony in his lamentation over Caesar: "O judgment! thou are fled to brutish beasts, And men have lost their reason." Lavoisier, who had been under arrest for months, was engaged in no experiments at the time. Moreover, as we shall see, the charges leading to his execution bore no relationship to his scientific work.

But if Coffinhal's chilling remark is apocryphal, the second most famous quotation surrounding the death of Lavoisier is ac-

# DISCOURS

## D'OUVERTURE ET DE CLÔTURE

DU

## COURS DE ZOOLOGIE

Donné dans le Muséum national d'Histoire naturelle,
l'an IX de la République,

PAR LE C^EN LACEPÈDE,

Membre du Sénat, et de l'Institut national de France; l'un des
Professeurs du Muséum d'Histoire naturelle; membre de l'Institut
national de la République Cisalpine; de la société d'Arragon; de
celle des Curieux de la Nature, de Berlin; des sociétés d'Histoire
naturelle, des Pharmaciens, Philotechnique, Philomatique, et des
Observateurs de l'homme, de Paris; de celle d'Agriculture d'Agen;
de la société des Sciences et Arts de Montauban; du Lycée
d'Alençon, etc.

A PARIS;

CHEZ PLASSAN, IMPRIMEUR-LIBRAIRE;

L'AN IX DE LA REPUBLIQUE.

Title page for Lacépède's opening and closing addresses for the zoology course at the Natural History Museum in 1801–1802—but identified only as "year 9 of the Republic."

curate and well attested. The great mathematician Joseph Louis Lagrange, upon hearing the news about his friend's execution, remarked bitterly: "It took them only an instant to cut off that head, but France may not produce another like it in a century."

The French revolution had been born in hope and expansiveness. At the height of enthusiasm for new beginnings, the revolutionary government suppressed the old calendar, and started time all over again, with Year I beginning on September 22,

1792, at the founding of the French republic. The months would no longer bear names of Roman gods or emperors, but would record the natural passage of seasons—as in *brumaire* (foggy), *ventose* (windy), *germinal* (budding), and to replace parts of July and August, originally named for two despotic caesars, *thermidor*. Measures would be rationalized, decimalized, and based on earthly physics, with the meter defined as one ten-millionth of a quarter meridian from pole to equator. The metric system is our enduring legacy of this revolutionary spirit, and Lavoisier himself played a guiding role in devising the new weights and measures.

But initial optimism soon unraveled under the realities of internal dissension and external pressure. Governments tumbled one after the other, and Dr. Guillotin's machine, invented to make execution more humane, became a symbol of terror by sheer frequency of public use. Louis XVI was beheaded in January 1793 (Year I of the republic). Power shifted from the Girondins to the Montagnards, as the Terror reached its height and the war with Austria and Prussia continued. Finally, as so often happens, the architect of the terror, Robespierre himself, paid his visit to Dr. Guillotin's device, and the cycle played itself out. A few years later, in 1804, Napoleon was crowned as emperor, and the First Republic ended. Poor Lavoisier had been caught in the midst of the cycle, dying for his former role as tax collector on May 8, 1794, less than three months before the fall of Robespierre on July 27 (9 Thermidor, Year II).

Old ideals often persist in vestigial forms of address and writing, long after their disappearance in practice. I was reminded of this phenomenon when I acquired, a few months ago, a copy of the opening and closing addresses for the course in zoology at the Muséum d'Histoire naturelle of Paris for 1801–1802. The democratic fervor of the revolution had faded, and Napoleon had already staged his *coup d'état* of 18 Brumaire (November 9, 1799), emerging as emperor de facto, although not crowned until 1804. Nonetheless, the author of these addresses, who would soon resume his full name Bernard-Germain-Etienne de la Ville-sur-Illon, comte de Lacépède, is identified on the title page only as C<sup>en</sup> Lacépède (for *citoyen*, or "citizen"—the democratic form adopted by the revolution to abolish all distinctions of address). The long list of honors and memberships, printed in small type below Lacépède's name, is almost a parody on the ancient forms; for instead of the old affiliations that always included "member of the royal academy of this or that" and "counsellor to the king or count of here or there," Lacépède's titles are rigorously egalitarian—including "one of the professors at the museum of natural history," and member of the society of pharmacists of Paris, and of agriculture of Agen. As for the year of publication, we have to know the history detailed above—for the publisher's date is given, at the bottom, only as "l'an IX de la République."

Lacépède was one of the great natural historians in the golden age of French zoology during the late eighteenth and early nineteenth centuries. His name may be overshadowed in retrospect by the illustrious quartet of Buffon, Lamarck, Geoffroy, and Cuvier, but Lacépède—who was chosen by Buffon to complete his life's work, the multivolumed *Histoire naturelle*—deserves a place with these men, for all were *citoyens* of comparable merit. Although Lacépède supported the revolution in its moderate first phases, his noble title bred suspicion and he went into internal exile during the Terror. But the fall of Robespierre prompted his return to Paris, where his former colleagues persuaded the government to establish a special chair for him at the Muséum, as zoologist for reptiles and fishes.

By tradition, the opening and closing addresses for the zoology course at the Muséum were published in pamphlet form each year. The opening address for Year IX, "Sur l'histoire des races ou principales variétés de l'espèce humaine" (On the history of races and principal varieties of the human species), is a typical statement of the liberality and optimism of Enlightenment thought. The races, we learn, may differ in current accomplishments, but all are capable of greater and equal achievement, and all can progress.

But the bloom of hope had been withered by the Terror. Progress, Lacépède asserts, is not guaranteed, but is possible only if untrammeled by the dark side of human venality. Memories of dire consequences for unpopular thoughts must have been fresh, for Lacépède cloaked his criticism of revolutionary excesses in careful speech and foreign attribution. Ostensibly, he was only describing the evils of the Indian caste system in a passage that must be read as a lament about the Reign of Terror:

Hypocritical ambition, . . . abusing the credibility of the multitude, has conserved the ferocity of the savage state in the midst of the virtues of civilization. . . . After having reigned by terror [*regné par la terreur*], submitting even monarchs to their authority, they reserved the domain of science and art to themselves [a reference, no doubt, to the suppression of the independent academies by the revolutionary government in 1793, when Lacépède lost his first post at the Muséum], and surrounded themselves with a veil of mystery that only they could lift.

At the end of his address, Lacépède returns to the familiar theme of political excesses and makes a point, by no means original of course, that I regard as the central structural tragedy in the working of any complex system, including organisms and social institutions—the crushing asymmetry between the need for slow and painstaking construction and the potential for almost instantaneous destruction:

Thus, the passage from the semisavage state to civilization occurs through a great number of insensible stages, and requires an immense amount of time. In moving slowly through these successive stages, man fights painfully against his habits; he also battles with nature as he climbs, with great effort, up the long and perilous path. But it is not the same with the loss of the civilized state; this is almost sudden. In this morbid fall, man is thrown down by all his ancient tendencies; he struggles no longer, he gives up, he does not battle obstacles, he abandons himself to the burdens that surround him. Centuries are needed to nurture the tree of science and make it grow, but one blow from the hatchet of destruction cuts it down.

The chilling final line, a gloss on Lagrange's famous statement about the death of Lavoisier, inspired me to write about the founder of modern chemistry, and to think a bit more about the tragic asymmetry of creation and destruction.

Antoine-Laurent Lavoisier, born in 1743, belonged to the nobility through a title purchased by his father (standard practice for boosting the royal treasury during the *ancien régime*). As a

Lavoisier and his wife as painted by the great artist David, who later became a fervent supporter of the revolution. THE METROPOLITAN MUSEUM OF ART, PURCHASE, MR. AND MRS. CHARLES WRIGHTSMAN GIFT, 1977.

leading liberal and rationalist of the Enlightenment (a movement that attracted much of the nobility, including many wealthy intellectuals who had purchased their titles to rise from the bourgeoisie), Lavoisier fitted an astounding array of social and scientific services into a life cut short by the headsman at age fifty-one.

We know him best today as the chief founder of modern chemistry. The textbook one-liners describe him as the discoverer (or at least the namer) of oxygen, the man who (though anticipated by Henry Cavendish in England) recognized water as a compound of the gases hydrogen and oxygen, and who correctly described combustion, not as the liberation of a hypothetical substance called phlogiston, but as the combination of burning material with oxygen. But we can surely epitomize his contribution more accurately by stating that Lavoisier set the basis for modern chemistry by recognizing the nature of elements and compounds—by finally dethroning the ancient taxonomy of air, water, earth, and fire as indivisible elements; by identifying gas, liquid, and solid as states of aggregation for a single substance subjected to different degrees of heat; and by developing quantitative methods for defining and identifying true elements. Such a brief statement can only rank as a caricature of Lavoisier's scientific achievements, but this essay treats his other life in social service, and I must move on.

Lavoisier, no shrinking violet in the game of self-promotion, openly spoke of his new chemistry as "a revolution." He even published his major manifesto, *Traité élémentaire de chimie,* in 1789, starting date of the other revolution that would seal his fate.

Lavoisier, liberal child of the Enlightenment, was no opponent of the political revolution, at least in its early days. He supported the idea of a constitutional monarchy, and joined the most moderate of the revolutionary societies, the Club of '89. He served as an alternate delegate in the States General, took his turn as a *citoyen* at guard duty, and led several studies and commissions vital to the success of the revolution—including a long stint as *régisseur des poudres* (director of gunpowder, where his brilliant successes produced the best stock in Europe, thus providing substantial help in France's war against Austria and Prussia). He worked on financing the revolution by *assignats* (paper money backed largely by confiscated church lands), and he served on the commission of weights and measures that formulated the metric system. Lavoisier rendered these services to all governments, including the most radical, right to his death, even hoping at the end that his crucial work on weights and measures might save his life. Why, then, did Lavoisier end up in two pieces on the *place de la Révolution* (long ago renamed, in pleasant newspeak, *place de la Concorde*)?

The fateful move had been made in 1768, when Lavoisier joined the infamous Ferme Générale, or Tax Farm. If you regard the IRS as a less than benevolent institution, just consider taxation under the *ancien régime* and count your blessings. Taxation was regressive with a vengeance, as the nobility and clergy were entirely exempt, and poor people supplied the bulk of the royal treasury through tariffs on the movement of goods across provincial boundaries, fees for entering the city of Paris, and taxes on such goods as tobacco and salt. (The hated *gabelle*, or "salt tax," was applied at iniquitously differing rates from region to region, and was levied not on actual consumption but on presumed usage—thus, in effect, forcing each family to buy a certain quantity of taxed salt each year.)

Moreover, the government did not collect taxes directly. They set the rates and then leased (for six-year periods) the privilege of collecting taxes to a private finance company, the Ferme Générale. The Tax Farm operated for profit like any other private business. If they managed to collect more than the government levy, they kept the balance; if they failed to reach the quota, they took the loss. The system was not only oppressive in principle; it was also corrupt. Several shares in the Tax Farm were paid for no work as favors or bribes; many courtiers, even the king himself, were direct beneficiaries. Nonetheless, Lavoisier chose this enterprise for the primary investment of his family fortune, and he became, as members of the firm were called, a *fermier-général,* or "farmer-general."

(Incidentally, since I first read the sad story of Lavoisier some twenty-five years ago, I have been amused by the term farmer-general, for it conjures up a pleasantly rustic image of a country yokel, dressed in his Osh Kosh b'Gosh overalls, and chewing on a stalk of hay while trying to collect the *gabelle*. But I have just learned from the *Oxford English Dictionary* that my image is not only wrong, but entirely backward. A farm, defined as a piece of agricultural land, is a derivative term. In usage dating to Chaucer, a farm, from the medieval Latin *firma*, "fixed payment," is "a fixed yearly sum accepted from a person as a composition for taxes or other moneys which he is empowered to collect." By extension, to farm is to lease anything for a fixed rent. Since most leases applied to land, agricultural plots become "farms," with a first use in this sense traced only to the sixteenth century; the leasers of such land then became "farmers." Thus, our modern phrase "farming out" records the original use, and has no agricultural connotation. And Lavoisier was a farmer-general in the true sense, with no mitigating image of bucolic innocence.)

I do not understand why Lavoisier chose the Ferme Générale

for his investment, and then worked so assiduously in his role as tax farmer. He was surely among the most scrupulous and fair-minded of the farmers, and might be justifiably called a reformer. (He opposed the overwatering of tobacco, a monopoly product of the Ferme, and he did, at least in later years, advocate taxation upon all, including the radical idea that nobles might pay as well.) But he took his profits, and he provoked no extensive campaign for reform as the money rolled in. The standard biographies, all too hagiographical, tend to argue that he regarded the Ferme as an investment that would combine greatest safety and return with minimal expenditure of effort—all done to secure a maximum of time for his beloved scientific work. But I do not see how this explanation can hold. Lavoisier, with his characteristic energy, plunged into the work of the Ferme, traveling all over the country, for example, to inspect the tobacco industry. I rather suspect that Lavoisier, like many modern businessmen, simply jumped at a good and legal investment without asking too many ethical questions.

But the golden calf of one season becomes the shattered idol of another. The farmers-general were roundly hated, in part for genuine corruption and iniquity, in part because tax collectors are always scapegoated, especially when the national treasury is bankrupt and the people are starving. Lavoisier's position was particularly precarious. As a scheme to prevent the loss of taxes from widespread smuggling of goods into Paris, Lavoisier advocated the building of a wall around the city. Much to Lavoisier's distress, the project, financed largely (and involuntarily) through taxes levied upon the people of Paris, became something of a boondoggle, as millions were spent on fancy ornamental gates. Parisians blamed the wall for keeping in fetid air and spreading disease. The militant republican Jean-Paul Marat began a campaign of vilification against Lavoisier that only ended when Charlotte Corday stabbed him to death in his bath. Marat had written several works in science and had hoped for election to the Royal Academy, then run by Lavoisier. But Lavoisier had exposed the emptiness of Marat's work. Marat fumed, bided his time, and waited for the season when patriotism would become a good refuge for scoundrels. In January 1791, he launched his attack in *l'Ami du Peuple* (*The Friend of the People*):

I denounce you, Coryphaeus of charlatans, Sieur Lavoisier [coryphaeus, meaning highest, is the leader of the chorus in a classical Greek drama] Farmer-general, Commissioner of Gunpowders. . . . Just to think that this contemptible little man who enjoys an income of forty thousand livres has no other claim to fame than that of having put Paris in prison with a wall costing the poor thirty millions. . . . Would to heaven he had been strung up to the nearest lamppost.

The breaching of the wall by the citizens of Paris on July 12, 1789, was the prelude to the fall of the Bastille two days later.

Lavoisier began to worry very early in the cycle. Less than seven months after the fall of the Bastille, he wrote to his old friend Benjamin Franklin:

After telling you about what is happening in chemistry, it would be well to give you news of our Revolution. . . . Moderate-minded people, who have kept cool heads during the general excitement, think that events have carried us too far . . . we greatly regret your absence from France at this time; you would have been our guide and you would have marked out for us the limits beyond which we ought not to go.

But these limits were breached, just as Lavoisier's wall had fallen, and he could read the handwriting on the remnants. The Ferme Générale was suppressed in 1791, and Lavoisier played no further role in the complex sorting out of the farmers' accounts. He tried to keep his nose clean with socially useful work on weights and measures and public education. But time was running out for the farmers-general. The treasury was bankrupt, and many thought (quite incorrectly) that the iniquitously hoarded wealth of the farmers-general could replenish the nation. The farmers were too good a scapegoat to resist; they were arrested *en masse* in November 1793, commanded to put their accounts in order, and to reimburse the nation for any ill-gotten gains.

The presumed offenses of the farmers-general were not capital under revolutionary law, and they hoped initially to win their personal freedom, even though their wealth and possessions might be confiscated. But they had the misfortune to be in the wrong place (jail) at the worst time (as the Terror intensified).

Eventually, capital charges of counterrevolutionary activities were drummed up, and in a mock trial lasting only part of a day, the farmers-general were condemned to the guillotine.

Lavoisier's influential friends might have saved him, but none dared (or cared) to speak. The Terror was not so inexorable and efficient as tradition holds. Fourteen of the farmers-general managed to evade arrest, and one was saved by the intervention of Robespierre. Madame Lavoisier, who lived to a ripe old age, marrying and divorcing Count Rumford, and reestablishing one of the liveliest salons in Paris, never allowed any of these men over her doorstep again. One courageous (but uninfluential) group offered brave support in Lavoisier's last hours. A deputation from the Lycée des Arts came to the prison to honor Lavoisier and crown him with a wreath. We read in the minutes of that organization: "Brought to Lavoisier in irons, the consolation of friendship . . . to crown the head about to go under the ax."

It is a peculiar attribute of human courage that when no option remains but death, criteria of judgment shift to the manner of dying. Chronicles of the revolution are filled with stories about who died with dignity—and who went screaming to the knife. Antoine Lavoisier died well. He wrote a last letter to his cousin, in apparent calm, not without humor, and with an intellectual's faith in the supreme importance of mind.

> I have had a fairly long life, above all a very happy one, and I think that I shall be remembered with some regrets and perhaps leave some reputation behind me. What more could I ask? The events in which I am involved will probably save me from the troubles of old age. I shall die in full possession of my faculties.

Lavoisier's rehabilitation came almost as quickly as his death. In 1795, the Lycée des Arts held a first public memorial service, with Lagrange himself offering the eulogy and unveiling a bust of Lavoisier inscribed with the words: "Victim of tyranny, respected friend of the arts, he continues to live; through genius he still serves humanity." Lavoisier's spirit continued to inspire, but his head, once filled with great thoughts as numerous as the unwritten symphonies of Mozart, lay severed in a common grave.

Many people try to put a happy interpretation upon Lacépède's observation about the asymmetry of painstaking creation and instantaneous destruction. The collapse of systems, they argue, may be a prerequisite to any future episode of creativity—and the antidote, therefore, to stagnation. Taking the longest view, for example, mass extinctions do break up stable ecosystems and provoke episodes of novelty further down the evolutionary road. We would not be here today if the death of dinosaurs had not cleared some space for the burgeoning of mammals.

I have no objection to this argument in its proper temporal perspective. If you choose a telescope and wish to peer into an evolutionary future millions of years away, then a current episode of destruction may be read as an ultimate spur. But if you care for the here and now, which is (after all) the only time we feel and have, then massive extinction is only a sadness and an opportunity lost forever. I have heard people argue that our current wave of extinctions should not inspire concern because the earth will eventually recover, as so oft before, and perhaps with pleasant novelty. But what can a conjecture about ten million years from now possibly mean to our lives—especially since we have the power to blow up our planet long before then, and rather little prospect, in any case, of surviving so long ourselves (since few vertebrate species live for 10 million years).

The argument of the "long view" may be correct in some meaninglessly abstract sense, but it represents a fundamental mistake in categories and time scales. Our only legitimate long view extends to our children and our children's children's children—hundreds or a few thousands of years down the road. If we let the slaughter continue, they will share a bleak world with rats, dogs, cockroaches, pigeons, and mosquitoes. A potential recovery millions of years later has no meaning at our appropriate scale. Similarly, others could do the unfinished work of Lavoisier, if not so elegantly; and political revolution did spur science into some interesting channels. But how can this mitigate the tragedy of Lavoisier? He was one of the most brilliant men ever to grace our history, and he died at the height of his powers and health. He had work to do, and he was not guilty.

My title, "The Passion of Antoine Lavoisier," is a double entendre. The modern meaning of *passion*, "overmastering zeal or enthusiasm," is a latecomer. The word entered our language from the Latin verb for suffering, particularly for suffering physical pain. The Saint Matthew and Saint John Passions of J. S. Bach are musical dramas about the suffering of Jesus on the cross. This essay, therefore, focuses upon the final and literal passion of Lavoisier. (Anyone who has ever been disappointed in love—that is, nearly all of us—will understand the intimate connection between the two meanings of passion.)

But I also wanted to emphasize Lavoisier's passion in the modern meaning. For this supremely organized man—farmer-general; commissioner of gunpowder; wall builder; reformer of prisons, hospitals, and schools; legislative representative for the nobility of Blois; father of the metric system; servant on a hundred government committees—really had but one passion amidst this burden of activities for a thousand lifetimes. Lavoisier loved science more than anything else. He awoke at six in the morning and worked on science until eight, then again at night from seven until ten. He devoted one full day a week to scientific experiments and called it his *jour de bonheur* (day of happiness). The letters and reports of his last year are painful to read, for Lavoisier never abandoned his passion—his conviction that reason and science must guide any just and effective social order. But those who received his pleas, and held power over him, had heard the different drummer of despotism.

Lavoisier was right in the deepest, almost holy, way. His passion harnessed feeling to the service of reason; another kind of passion was the price. Reason cannot save us and can even persecute us in the wrong hands; but we have no hope of salvation without reason. The world is too complex, too intransigent; we cannot bend it to our simple will. Bernard Lacépède was probably thinking of Lavoisier when he wrote a closing flourish following his passage on the great asymmetry of slow creation and sudden destruction:

Ah! Never forget that we can only stave off that fatal degradation if we unite the liberal arts, which embody the sacred fire of sensibility, with the sciences and the useful arts, without which the celestial light of reason will disappear.

The Republic needs scientists.

# 25 | The Godfather of Disaster

LEMUEL GULLIVER, marooned by pirates on a small Pacific island, lamented his apparently inevitable fate: "I considered how impossible it was to preserve my life, in so desolate a place; and how miserable my end must be." But then the floating island of Laputa appeared and he rode up on a chain to safety.

The Laputans, Gulliver soon discovered, were an odd lot, with an ethereal turn of mind well suited to their abode. Their thoughts, he noted, "are so taken up with intense speculation" that they can neither speak nor hear the words of others unless explicitly roused. Thus, each Laputan of status employs a "flapper" who gently strikes the ear or mouth of his master with an inflated bladder full of small pebbles whenever his lordship must either attend or answer.

The Laputans are not catholic in their distractions; only music and mathematics incite their unworldly concentration. Gulliver finds that their mathematical obsession extends to all spheres of life; he obtains for his first meal "a shoulder of mutton, cut into an equilateral triangle; a piece of beef into rhomboides; and a pudding into a cycloid."

But mathematics has its negative side, at least psychologically. The Laputans are not lost in a blissful reverie about the perfection of circles or the infinitude of pi. They are scared. Their calculations have taught them that "the earth very narrowly escaped a brush from the tail of the last comet . . . and that the next, which they have calculated for one and thirty years hence, will probably destroy [them]." The Laputans live in fear: "When they meet an acquaintance in the morning, the first question is about the sun's

367

health; how he looked at his setting and rising, and what hopes they have to avoid the stroke of the approaching comet."

Jonathan Swift, as usual, was not writing abstract humor in reciting the Laputans' fear of comets. He was satirizing the influential theory of a political and religious enemy, William Whiston, handpicked successor to Isaac Newton as Lucasian Professor of Mathematics at Cambridge. In 1696, Whiston had published the first edition of a work destined for scientific immortality of the worst sort—as a primer of how not to proceed. Whiston called his treatise *A New Theory of the Earth from its Original to the Consummation of all Things, Wherein the Creation of the World in Six Days, the Universal Deluge, and the General Conflagration, as laid down in the Holy Scriptures, are shewn to be perfectly agreeable to Reason and Philosophy.*

Whiston has descended through history as the worst example of religious superstition viewed as an impediment to science. Whiston, we are told, was so wed to the few thousand years of Moses' chronology that he had to postulate absurd catastrophes via cometary collisions in order to encompass the earth's history in so short a time. This dismissal is no modern gloss but an old tradition in scientific rhetoric. Charles Lyell, conventional father of modernity in geological thought, poured contempt upon Whiston's extraterrestrial and catastrophic theories because they foreclosed proper attention to gradual, earth-based causes. Lyell wrote in 1830:

> [Whiston] retarded the progress of truth, diverting men from the investigation of the laws of sublunary nature, and inducing them to waste time in speculations on the power of comets to drag the waters of the ocean over the land—on the condensation of the vapors of their tails into water, and other matters equally edifying.

But Whiston did not only suffer the abuse of posthumous reputation; he became an object of ridicule in his own time as well (as Swift's satire indicates). His contemporary troubles did not stem from his cometary theory (which resembled several others of his day and did not strike fellow intellectuals as outré) but from his religious heterodoxies. Whiston's public support of the Arian heresy (a denial of the Trinity, and the consubstantiality of Christ with God the Father) led to dismissal from his Cambridge professorship (as Newton, his erstwhile champion, and a quieter, more measured exponent of the same heresy, remained conspicuously silent). Resettled in London, Whiston was tried twice for heresy and, though not formally convicted, lost most of his previous prestige and lived the rest of his long life (he died in 1752 at the age of eighty-four) as an independent intellectual, viewed as a prophet by some and as a crank by most. In the eighth plate of Hogarth's *The Rake's Progress,* set in the mental hospital of Bedlam, an inmate covers the wall with a sketch of Whiston's scheme for measuring longitude.

Despite continual rejection of Whiston, from his own time to ours, we must still grant him a major role in the history of science. The French historian Jacques Roger ended his article on Whiston (in the *Dictionary of Scientific Biography*) with these words:

> His writings were much disputed but also widely read throughout the eighteenth century, and not just in England. For example, Buffon, who summarized Whiston's theory in order to ridicule it, borrowed more from him than he was willing to admit. . . . It may be said that all the cosmogonies based on the impact of celestial bodies, including that of Jeans, owed something, directly or indirectly, to Whiston's inventions.

Moreover, we must not forget the early acclaim of his contemporaries. The greatest figure in all the history of science, Isaac Newton, personally chose Whiston as his successor. In my copy of Whiston's *New Theory* (the second edition of 1708), a Mr. Nathaniel Hancock, who bought the book in 1723, has inscribed on its title page, in a beautiful, flowing hand, the following judgment of Whiston and his book by John Locke:

> I have not heard any one of my acquaintance speak of it, but with great commendations (as I think it deserves). . . . He is one of those sort of writers that I always fancy should be most encouraged; I am always for the builders.

Comets were in the air in late seventeenth century Britain. In 1680, a great comet brightened the skies of Europe, followed two years later by a smaller object that sent Edmond Halley to the drawing boards of history and mathematics. Moreover, the seventeenth century had been a time of extraordinary change and tension in Britain—the execution of Charles I, Cromwell's Protectorate, the Restoration, and the Glorious Revolution to mention just a few of the tumultuous events of Whiston's age. These happenings fostered a revival of millennial thought—a scrutiny of the prophecies in Daniel and Revelation, leading to a conclusion that the end of this world lay in sight, and that the blessed millennium, or thousand-year reign of Christ, would soon begin. Since comets had long been viewed as harbingers or signals of great transitions and disasters (literally, "evil stars"), Whiston chose a propitious time to implicate comets as the prime movers of our planet's history.

Whiston's *New Theory* tried, above all, to establish a consistency between the two great sources of truth, as defined by his countrymen: the infallibility of Scripture and the mathematical beauty of the cosmos, so recently revealed by Newton. Whiston began his account of our planet's history by summarizing his method of inquiry in a single page, entitled *Postulata*. The first two statements illustrate his attempt to join Moses with Newton:

1. The obvious or literal sense of scripture is the true and real one, where no evident reason can be given to the contrary.
2. That which is clearly accountable in a natural way, is not, without reason, to be ascribed to a Miraculous Power.

Comets became Whiston's *deus ex machina* for rendering the cataclysmic events of Genesis with the forces of Newton's universe.

Consider Whiston's descriptions of the earth, from cradle to grave, with each of its five principal events tied to cometary causes:

1. *The Hexameron, or Moses' six days of creation.* Whiston prefaced the body of his work with a ninety-four-page "Discourse Concerning the Nature, Stile, and Extent of the Mosaick History of the Creation." Here, he attempts to preserve the literal sense of Scripture (first postulate above) in the light of Newton's nearly infinite universe. How could all this vastness be made in six days, and how could our earth, one tiny speck in one corner of the cosmos, be the focus of such infinitude? Whiston devotes his

preface to a single argument: Moses described the origin of the earth alone, not the entire universe; moreover, he tailored his words to describe not the abstract properties of nature's laws, but the visual appearance of events as an untutored observer might have witnessed them on the congealing surface of our planet. With these provisos, everything happened exactly as Genesis proclaims.

The earth began as a comet, and the chaos described in Genesis 1 ("and the earth was without form and void") represents the original swirling atmosphere. Whiston's contemporaries did not know the true size of comets, and many assumed, as he did, that comets might be of planetary dimensions, and therefore suitable for transformation into a planet. Whiston wrote:

Tis very reasonable to believe, that a planet is a comet formed into a regular and lasting constitution, and placed at a proper distance from the sun . . . and a comet is a chaos, i.e., a planet unformed or in its primaeval state, placed in a very eccentrical [orbit].

To transform this comet, with its highly elliptical pathway, into a planet, God needs to render its orbit more nearly circular. The chaotic atmosphere will then clear and precipitate to form the solid surface of a planet. Whiston's attitude toward miracles (temporary suspension by God of his own natural laws) remained ambiguous. His second postulate stated a preference for natural explanations, but only when possible. He never did resolve whether the change in orbit that converted our cometary ancestor into the present earth had been a true miracle (accomplished by the immediate agency of God's own hand) or a natural event (the result of gravitational influences exerted by another body moving through the heavens according to Newton's laws). But since Newton's laws are God's laws, Whiston attached only limited importance to the distinction—for the transition from comet to planet occurred either by God's direct action or by laws that God had established in full knowledge of the later, desired result.

In any case, once the comet's orbit had been adjusted to its planetary pathway, the events of Genesis 1 would proceed naturally, as viewed by an observer on earth. The creation of light on the first day represents an initial clarification of a formerly opaque atmosphere (so that a brightness always present could finally be perceived). Similarly, the "creation" of the sun and moon records a further lightening of atmosphere.

> This fourth day is therefore the very time when . . . these heavenly bodies, which were in being before, but so as to be wholly strangers to a spectator on earth, were rendered visible.

Meanwhile, the products of this former atmosphere settled out by order of density into a series of concentric layers—solid at the center, water above, and a solid froth on top—to form the earth.

If all this activity still seems a bit much to compress into a mere six days, Whiston added an argument to increase our confidence. The original earth underwent no diurnal rotation on its axis but maintained a constant position as it revolved around the sun. The nearly equatorial Eden therefore experienced a year divided into halves: one of day; the second of night. Since we define a "day" as a single alternation of light and darkness, the days of Genesis 1 were all a year long—not a vast span for the work accomplished, but a big step in the right direction.

2. *The Fall, and expulsion of Adam and Eve from Eden.* The pristine earth stood bolt upright with no seasons, tides, or winds to disturb its primeval bliss. But "as soon as Man had sinned . . . and as God Almighty had pronounced a curse on the ground, and its production, presently the earth began a new and strange motion, and revolved from west to east on its own axis." This axis tilted to its present inclination of some 21 degrees, and the earth began its diurnal rotation, with days, nights, winds, and seasons. Whiston ascribed this change to a cometary collision:

> Now the only assignable cause is that of the impulse of a comet with little or no atmosphere, or of a central solid hitting obliquely upon the earth along some parts of its present equator.

3. *Noah's flood.* All the great works of this late seventeenth century vogue for "theories of the earth" (notably Burnet's *Sacred Theory of the Earth* and Woodward's *Essay Towards a Natural History of the Earth*) regarded an explanation of the Deluge as their cen-

tral test and focus. Events of the Creation were too distant and shrouded in mystery, phenomena of the coming millennium too tentative. But the Flood was a relatively recent incident, begun (or so Whiston deduced) precisely "on the 17th day of the 2nd month from the autumnal equinox . . . in the 2349th year before the Christian era." Any proper theory of the earth must, above all, render this cardinal and precisely specified event of a history remembered and recorded in the ancient chronicles.

The comet that unleashed the Flood did not strike the earth directly but passed close enough for two great effects that combined to produce the Deluge. First, the earth passed (for about two hours) directly through the "vaporous tail" of the comet, thus absorbing by gravity enough water to unleash forty days and nights of rain. Second, the tides generated by close passage of such an enormous body stretched the round earth into an oblate spheroid and eventually cracked the solid surface, allowing the underlying layer of water to rise and contribute to the great flood (Genesis, remember, speaks not only of rain from above, but also of upwelling from the "fountains of the deep").

(In a rather uncomfortable bit of special pleading, even in his own terms, Whiston argued that the cometary impact at the Fall had not unleashed a similar flood because this previous comet had no atmosphere. If we then ask why this earlier impact, more direct after all than the near miss that made the Flood, did not tear the surface and raise the waters from the abyss, Whiston responds that such a fracturing requires not only the gravitational force of the comet itself but also the pressing weight of waters from its tail.)

Above all, Whiston took delight in his cometary theory because it had resolved this cardinal event in our history as a consequence of nature's divinely appointed laws, and had thereby removed the need for a special, directly miraculous explanation:

> Whatever difficulties may hitherto have rendered this most noted catastrophe of the old world, that it was destroyed by waters, very hard, if not wholly inexplicable without an Omnipotent Power, and Miraculous Interposition: since the theory of comets, with their atmospheres and tails is discovered, they must vanish of their own accord. . . . We shall easily see that a deluge of waters is by no means an impossi-

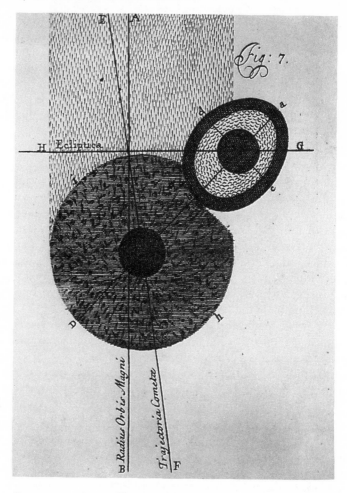

Cometary action as illustrated by Whiston in 1696. A passing comet (large object in the center) induces Noah's flood. The earth (upper right), entering the comet's tail, will receive its 40 days and nights of rain. The comet's gravity is stretching the earth into a spheroid. Under this gravitational tug, the earth's outer surface will soon crack, releasing water from below (the light, middle layer) to contribute to the deluge.

ble thing; and in particular that such an individual deluge
. . . which Moses describes, is no more so, but fully accountable that it might be, nay almost demonstrable that it really
was.

4. *The coming conflagration.* The prophetic books of Daniel and
Revelation speak of a worldwide fire that will destroy the current
earth, but in a purifying way that will usher in the millennium.
Whiston proposed (as the Laputans feared) that a comet would
instigate this conflagration for a set of coordinated reasons. This
comet would strip off the earth's cooling atmosphere, raise the
molten material at the earth's core, and contribute its own fiery
heat. Moreover, the passage of this comet would slow the earth's
rotation, thus initiating an orbit so elliptical that the point of
closest approach to the sun would be sufficient to ignite our
planet's surface. Thus, Whiston writes, "the theory of comets"
can provide "almost as commensurate and complete an account
of the future burning, as it already has done of the ancient drown-
ing of the earth."

5. *The consummation.* As prophecy relates, the thousand-year
reign of Christ will terminate with a final battle between the just
and the forces of evil led by the giants Gog and Magog. Thereafter, the bodies of the just shall ascend to heaven, those of the
damned shall sink in the other direction—and the earth's appointed role shall be over. This time a comet shall make a direct
hit—no more glancing blows for diurnal rotation or near misses
for floods—and knock the earth either clear out of the solar system or into an orbit so elliptical that it will become, as it was in the
beginning, a comet.

Our conventional, modern reading of Whiston as an impediment to true science arises not only from the fatuous character of
this particular reconstruction but also, and primarily, from our
recognition that Whiston invoked the laws of nature only to validate a predetermined goal—the rendering of biblical history—
and not, as modern ideals proclaim, to chart with objectivity, and
without preconception, the workings of the universe. Consider,
for example, Whiston's reverie on how God established the laws
of nature so that a comet would instigate a flood just when human
wickedness deserved such a calamity.

That Omniscient Being, who foresaw when the degeneracy of human nature would be arrived at an unsufferable degree of wickedness . . . and when consequently his vengeance ought to fall upon them; predisposed and preadapted the orbits and motions of both the comet and the earth, so that at the very time, and only at that time, the former should pass close by the latter and bring that dreadful punishment upon them.

Yet such an assessment of Whiston seems singularly unfair and anachronistic. How can we justify a judgment of modern taxonomies that didn't exist in the seventeenth century? We dismiss Whiston because he violated ideals of science as we now define the term. But, in Whiston's time, science did not exist as a separate domain of inquiry; the word itself had not been coined. No matter how we may view such an enterprise today, Whiston's mixture of natural events and scriptural traditions defined a primary domain at the forefront of scholarship in his time. We have since defined Whiston's *New Theory* as a treatise in the history of science because we remain intrigued with his use of astronomical arguments but have largely lost both context for and interest in his exegesis of millennarian prophecy. But Whiston would not have accepted such a categorization; he would not even have recognized our concerns and divisions. He did not view his effort as a work of science, but as a treatise in an important contemporary tradition for using all domains of knowledge—revelations of Scripture, history of ancient chronicles, and knowledge of nature's laws—to reconstruct the story of human life on our planet. The *New Theory* contains—and by Whiston's explicit design—far more material on theological principles and biblical exegesis than on anything that would now pass muster as science.

Moreover, although Whiston later achieved a reputation as a crank in his own time, he wrote the *New Theory* at the height of his conventional acceptability. He showed the manuscript to Christopher Wren and won the hearty approval of this greatest among human architects. He then gave (and eventually dedicated) the work to Newton himself, and so impressed Mr. Numero Uno in our current pantheon of scientific heroes that he ended up as Newton's handpicked successor at Cambridge.

In fact, Whiston's arguments in the *New Theory* are neither marginal nor oracular, but preeminently Newtonian in both spirit and substance. In reading the *New Theory*, I was particularly struck by feature of organization, a conceit really, that most commentators pass over. Whiston ordered his book in a manner that strikes us as peculiar (and ultimately quite repetitious). He presents the entire argument as though it could be laid out in a mathematical and logical framework, combining sure knowledge of nature's laws with clear strictures of a known history in order to deduce the necessity of cometary action as a primary cause.

Whiston begins with the page of *Postulata*, or general principles of explanation cited previously. He then lists eighty-five "lemmata," or secondary postulates derived directly from laws of nature. The third section discusses eleven "hypotheses"—not "tentative explanations" in the usual, modern sense of the word, but known facts of history assumed beforehand and used as terms in later deductions. Whiston then pretends that he can combine these lemmas and facts to deduce the proper explanation of our planet's history. The next section lists 101 "phaenomena," or particular facts that require explanation. The final chapter on "solutions" runs through these facts again to supply cometary (and other) explanations based on the lemmas and hypotheses. (Whiston then ends the book with four pages of "corollaries" extolling God's power and scriptural authority.)

I call this organization a conceit because it bears the form, but not the substance, of deductive necessity. The lemmas are not an impartial account of consequences from Newton's laws but a tailored list designed beforehand to yield the desired results. The hypotheses are not historical facts in the usual sense of verified, direct observations but inferences based on a style of biblical exegesis not universally followed even in Whiston's time. The solutions are not deductive necessities but possible readings that do not include other alternatives (even if one accepts the lemmas and hypotheses).

Still, we must not view Whiston's *New Theory* as a caricature of Newtonian methodology (if only from the direct evidence that Newton himself greatly admired the book). The Newton of our pantheon is a sanitized and modernized version of the man himself, as abstracted from his own time for the sake of glory, as Whiston has been for the sake of infamy. Newton's thinking combined the same interests in physics and prophecy, although an almost conspiratorial silence among scholars has, until recently, foreclosed discussion of Newton's voluminous religious writings, most of which remain unpublished. (James Force's excellent study, *William Whiston, Honest Newtonian*, 1985, should be consulted on this issue.) Newton and Whiston were soul mates, not master and jester. Whiston's perceived oddities arose directly from his Newtonian convictions and his attempt to use Newtonian methods (in both scientific and religious argument) to resolve the earth's history.

I have, over the years, written many essays to defend maligned figures in the traditional history of science. I usually proceed, as I have so far with Whiston, by trying to place an unfairly denigrated man into his own time and to analyze the power and interest of his arguments in their own terms. I have usually held that judgment by modern standards is the pitfall that led to our previous, arrogant dismissal—and that we should suppress our tendency to justify modern interest by current relevance.

Yet I would also hold that old arguments can retain a special meaning and importance for modern scientific debates. Some issues are so broad and general that they transcend all social contexts to emerge as guiding themes in scientific arguments across the centuries (see my book *Time's Arrow, Time's Cycle* for such a discussion about metaphors of linear and cyclical time in geology). In these situations, old versions can clarify and instruct our current research because they allow us to tease out the generality from its overlay of modern prejudices and to grasp the guiding power of a primary theme through its application to a past world that we can treat more abstractly, and without personal stake.

Whiston's basic argument about comets possesses this character of instructive generality. We must acknowledge, first of all, the overt and immediate fact that one of the most exciting items in contemporary science—the theory of mass extinction by extraterrestrial impact—calls upon the same agency (some versions even cite comets as the impacting bodies). Evidence continues to accumulate for the hypothesis that a large extraterrestrial object struck the earth some 65 million years ago and triggered, or at least greatly promoted, the late Cretaceous mass extinction (the sine qua non of our own existence, since the death of dinosaurs cleared ecological space for the evolution of large mammals). Intensive research is now under way to test the generality of this claim by searching for evidence of similar impacts during other episodes of mass extinction. We await the results with eager anticipation.

But theories of mass extinction do not provide the main reason why we should pay attention to Whiston today. After all, the similarities may be only superficial: Whiston made a conjecture to render millennarian prophecy; the modern theory has mustered some surprising facts to explain an ancient extinction. Guessing right for the wrong reason does not merit scientific immortality. No, I commend Whiston to modern attention for a different and more general cause—because the form and structure of his general argument embody a powerful abstraction that we need to grasp today in our search to understand the roles of stability, gradual change, and catastrophe in the sciences of history.

Whiston turned to comets for an interesting reason rooted in his Newtonian perspective, not capriciously as an easy way out for the salvation of Moses. Scientists who work with the data of history must, above all, develop general theories about how substantial change can occur in a universe governed by invariant natural laws. In Newton's (and Whiston's) world view, immanence and stability are the usual consequences of nature's laws: The cosmos does not age or progress anywhere. Therefore, if substantial changes did occur, they must be rendered by rapid and unusual events that, from time to time, interrupt the ordinary world of stable structure. In other words, Whiston's catastrophic theory of change arose primarily from his belief in the general stability of nature. Change must be an infrequent fracture or rupture. He wrote:

> We know no other natural causes that can produce any great and general changes in our sublunary world, but such bodies as can approach to the earth, or, in other words, but comets.

A major intellectual movement began about a century after Whiston wrote and has persisted to become the dominant ideology of our day. Whiston's notion of stability as the ordinary state of things yielded to the grand idea that change is intrinsic to the workings of nature. The poet Robert Burns wrote:

> Look abroad through nature's range
> Nature's mighty law is change.

This alternative idea of gradual and progressive change as inherent in nature's ways marked a major reform in scientific thinking and led to such powerful theories as Lyellian geology and Darwinian evolution. But this notion of slow, intrinsic alteration also established an unfortunate dogma that fostered an amnesia about other legitimate styles of change and often still leads us to restrict our hypotheses to one favored style falsely viewed as preferable (or even true) a priori. For example, the *New York Times* recently suggested that impact theories be disregarded on general principles:

> Terrestrial events, like volcanic activity or change in climate or sea level, are the most immediate possible cause of mass extinctions. Astronomers should leave to astrologers the task of seeking the causes of earthly events in the stars [editorial, April 2, 1985].

Perhaps they will now grant this paleontologist equal power of judgment over their next price increase.

The world is too complex for subsumption under any general theory of change. Whiston's model of stability, punctuated now and then by changes of great magnitude that induce new steady states, did not possess the generality that he or Newton supposed. But neither does Lyellian gradualism explain the entire course of our planet's history (and Lyell will have to eat his words about Whiston, just as the editors of the *Times* must now feast on theirs about the theory of mass extinction by extraterrestrial impact). Whiston's general style of argument—change as an interruption of usual stability—is on the ascendancy again as a worthy alternative to a way of thinking that has become too familiar, too automatic.

On the wall of Preservation Hall in New Orleans hangs a tattered and greasy sign, but the most incisive I have ever seen. It gives a price scale for requests by the audience to the aged men of the band who play jazz in the old style:

| Traditional Requests | $1 |
| Others | $2 |
| The Saints | $5 |

Preservation Hall guards against too frequent repetition of the most familiar with the usual currency of our culture—currency itself. Scholars must seek other, more active tactics. We must have gadflies—and historical figures may do posthumous service—to remind us constantly that our usual preferences, channels, and biases are not inevitable modes of thought. I nominate William Whiston to the first rank of reminders as godfather to punctuational theories of change in geology.

Funny, isn't it? Whiston longed "to be in that number, when the Saints go marching in"; in fact, he wrote the *New Theory* largely to suggest that cometary impact would soon usher in this blessed millenium. Yet he is now a soul mate to those who wish to hear a different drummer.

# 8 | Evolution and Creation

# 26 | Knight Takes Bishop?

I HAVE NOT THE SLIGHTEST doubt that truth possesses inestimable moral value. In addition, as Mr. Nixon once found to his sorrow, truth represents the only way to keep a complex story straight, for no one can remember all the details of when he told what to whom unless his words have an anchor in actual occurrence.

> Oh, what a tangled web we weave,
> When first we practice to deceive!

Yet, for a scholar, there is nothing quite like falsehood. Lies are pinpoints—identifiable historical events that can be traced. Falsehoods also have motivations—points of departure for our ruminations on the human animal. Truth, on the other hand, simply happens. Its accurate report teaches us little beyond the event itself.

In this light, we should note with interest that the most famous story in all the hagiography of evolution is, if not false outright, at least grossly distorted by biased reconstruction long after the fact. I speak of Thomas Henry Huxley's legendary encounter with the bishop of Oxford, "Soapy Sam" Wilberforce, at the 1860 meeting of the British Association for the Advancement of Science, held in His Lordship's own see.

Darwin had published the *Origin of Species* in November 1859. Thus, when the British Association for the Advancement of Science met at Oxford in the summer of 1860, this greatest of all debates received its first prominent public airing. On Saturday,

June 30, more than 700 people wedged themselves into the largest room of Oxford's Zoological Museum to hear what was, by all accounts, a perfectly dreadful hour-long peroration by an American scholar, Dr. Draper, on the "intellectual development of Europe considered with reference to the views of Mr. Darwin." Leonard Huxley wrote, in *Life and Letters of Thomas Henry Huxley:*

> The room was crowded to suffocation. . . . The very windows by which the room was lighted down the length of its west side were packed with ladies, whose white handkerchiefs, waving and fluttering in the air at the end of the Bishop's speech, were an unforgettable factor in the acclamation of the crowd.

The throng, as Leonard Huxley notes, had not come to hear Dr. Draper drone on about Europe. Word had circulated widely that "Soapy Sam" Wilberforce, the silver-tongued bishop of Oxford, would attend with the avowed purpose of smashing Mr. Darwin in the discussion to follow Draper's paper.

The story of Wilberforce's oration and Huxley's rejoinder has been enshrined among the half-dozen greatest legends of science—surely equal to Newton beaned by an apple or Archimedes jumping from his bath and shouting "Eureka!" through the streets of Syracuse. We have read the tale from comic book to novel to scholarly tome. We have viewed the scene, courtesy of the BBC, in our living rooms. The story has an "official version" codified by Darwin's son Francis, published in his *Life and Letters of Charles Darwin,* and expanded in Leonard Huxley's biography of his father. This reconstruction has become canonical, copied from source to later source hundreds of times, and rarely altered even by jot or tittle. Consider just one of countless retellings, chosen as an average and faithful version (from Ruth Moore's *Charles Darwin,* Hutchinson, 1957):

> For half an hour the Bishop spoke savagely ridiculing Darwin and Huxley, and then he turned to Huxley, who sat with him on the platform. In tones icy with sarcasm he put his famous question: was it through his grandfather or his grandmother that he claimed descent from an ape? . . . At the Bishop's question, Huxley had clapped the knee of the

surprised scientist beside him and whispered: "The Lord hath delivered him into mine hands." . . . [Huxley] tore into the arguments Wilberforce had used. . . . Working himself up to his climax, he shouted that he would feel no shame in having an ape as an ancestor, but that he would be ashamed of a brilliant man who plunged into scientific questions of which he knew nothing. In effect, Huxley said that he would prefer an ape to the Bishop as an ancestor, and the crowd had no doubt of his meaning.

The room dissolved into an uproar. Men jumped to their feet, shouting at this direct insult to the clergy. Lady Brewster fainted. Admiral Fitzroy, the former Captain of the Beagle, waved a Bible aloft, shouting over the tumult that it, rather than the viper he had harbored in his ship, was the true and unimpeachable authority. . . .

The issue had been joined. From that hour on, the quarrel over the elemental issue that the world believed was involved, science versus religion, was to rage unabated.

We may list as the key, rarely challenged features of this official version the following claims:

1. Wilberforce directly bearded and taunted Huxley by pointedly asking, in sarcastic ridicule, whether he claimed descent from an ape on his grandfather's or grandmother's side.

2. Huxley, before rising to the challenge, mumbled his famous mock-ecclesiastical sarcasm about the Lord's aid in his coming rhetorical victory.

3. Huxley than responded to Wilberforce's arguments in loud, clear, and forceful tones.

4. Huxley ended his speech with a devastatingly effective parry to the bishop's taunt.

5. Although Huxley said only that he would prefer an ape to a man who used skills of oratory to obfuscate rather than to seek truth, many took him to mean (and some thought he had said) that he would prefer an ape to a bishop as an ancestor. (Huxley, late in life, disavowed this stronger version about apes and bishops. When Wilberforce's son included it in a biography of his father, Huxley protested and secured a revision.)

6. Huxley's riposte inspired an uproar. The meeting ended forthwith and in tumult.

7. Although Moore, to her credit, does not make this claim, we are usually told that Huxley had scored an unambiguous and decisive victory—a key incident in Darwin's triumph.

8. This debate focused the world's attention on the real and deep issue of Darwin's century—science versus religion. Huxley's victory was a pivotal moment in the battle for science and reason against superstition and dogma.

I have had a strong interest in this story ever since, as an assistant professor on sabbatical leave at Oxford in 1970, I occupied a dingy office in the back rooms of the Zoological Museum, now crammed with cabinets of fossils and subdivided into cubicles, but then the large and open room where Huxley and Wilberforce fell to blows. For six months, I sat next to a small brass plaque announcing that the great event had occurred on my very spot. I also felt strong discomfort about the official tale for two definite reasons. First, it is all too pat—the victor and the vanquished, good triumphing over evil, reason over superstition. So few heroic tales in the simplistic mode turn out to be true. Huxley was a brilliant orator, but why should Wilberforce have failed so miserably? Much as I dislike the man, he was no fool. He was as gifted an orator as Huxley and a dominant intellectual force among conservative Anglicans.

Second, I knew from preliminary browsings that the official tale was a reconstruction, made by Darwin's champions some quarter century after the fact. Amazingly enough (for all its later fame), no one bothered to record the event in any detail at the time itself. No stenographer was present. The two men exchanged words to be sure, but no one knows what they actually said, and the few sketchy reports of journalists and letter writers contain important gaps and contradictions. Ironically, the official version has been so widely accepted and unchallenged not because we know its truth by copious documentation, but rather because so little data exist for a potential challenge.

For years, this topic has been about number fifty in my list of one hundred or so potential essays (sorry folks, but, the Lord and editors willing, you may have me to kick around for some time to come). Yet for want of new data about my suspicions, it remained well back in my line of processing, until I received a letter from my friend and distinguished Darwin scholar Sam Schweber of Brandeis University. Schweber wrote: "I came across a letter from Balfour Stewart to David Forbes commenting on the BAAS meeting he just attended at which he witnessed the Huxley-Wilberforce debate. It is probably the most accurate statement of what transpired." I read Stewart's letter and sat bolt upright with attention and smiles. Stewart wrote, describing the scene along the usual lines, thus vouching for the basic outline:

> There was an animated discussion in a large room on Saturday last at Oxford on Darwin's theory where the Bishop of Oxford and Prof. Huxley fell to blows. . . . There was one good thing I cannot help mentioning. The Bishop said he had been informed that Prof. Huxley had said he didn't care whether his grandfather was an ape [*sic* for punctuation] now he [the bishop] would not like to go to the Zoological Gardens and find his father's father or his mother's mother in some antiquated ape. To which Prof. Huxley replied that he would rather have for his grandfather an honest ape low in the scale of being than a man of exalted intellect and high attainments who used his power to pervert the truth.

Colorful, though nothing new so far. But I put an ellipsis early in the quotation, and I should now like to restore the missing words. Stewart wrote: "I think the Bishop had the best of it." Score one big point for my long-held suspicions. Balfour Stewart was no benighted cleric, but a distinguished scientist, Fellow of the Royal Society, and director of the Kew Observatory. Balfour Stewart also thought that Wilberforce had won the debate!

This personal discovery sent me to the books (I thank my research assistant, Ned Young, for tracking down all the sources, no mean job for so many obscure bits and pieces). We gathered all the eyewitness accounts (damned few) and found a half dozen or so modern articles, mostly by literary scholars, on aspects of the debate. (See Janet Browne, 1978; Sheridan Gilley, 1981; J. R. Lucas, 1979. I especially commend Browne's detective work on Francis Darwin's construction of the official version, and Gilley's incisive and well-written account of the debate.) I confess disappointment in finding that Stewart's letter was no new discovery. Yet I remain surprised that its key value—the claim by an important scientist that Wilberforce had won—has received so little attention. So far as I know, Stewart's letter has never been quoted *in extenso,* and no reference gives it more than a passing sentence. But I was delighted to find that the falsity of the official version is common knowledge among a small group of scholars. All the more puzzling, then, that the standard, heroic account continues to hold sway.

What is so wrong with the official tale, as epitomized in my eight points above? We should begin by analyzing the very few eyewitness accounts recorded right after the event itself.

Turning to reports by journalists, we must first mark the outstanding negative evidence. In a nation with a lively press, and with traditions for full and detailed reporting (so hard to fathom from our age of television and breathless paragraphs for the least common denominator), the great debate stands out for its nonattention. *Punch,* Wilberforce's frequent and trenchant critic, ignored the exchange but wrote poem and parody aplenty on another famous repartee about evolution from the same meeting—Huxley versus Owen on the brains of humans and gorillas. The *Athenaeum,* in one of but two accounts (the other from *Jackson's Oxford Journal*), presents a straightforward report that, in its barest outline, already belies the standard version in two or three crucial respects. On July 7, the reporter notes Oxford's bucolic charms: "Since Friday, the air has been soft, the sky sunny. A sense of sudden summer has been felt in the meadows of Christ Church and in the gardens of St. John's; many a dreamer of dreams, tempted by the summer warmth . . . and stealing from section A or B [of the meeting] has consulted his ease and taken a boat." But we then learn of a contrast between fireworks inside and punting lazily downstream while taking one's *dolce far niente.*

> The Bishop of Oxford came out strongly against a theory which holds it possible that man may be descended from an ape. . . . But others—conspicuous among these, Prof. Huxley—have expressed their willingness to accept, for themselves, as well as for their friends and enemies, all actual truths, even the last humiliating truth of a pedigree not registered in the Herald's College. The dispute has at least made Oxford uncommonly lively during the week.

The next issue, July 14, devotes a full page of tiny type to Dr. Draper and his aftermath—the longest eyewitness account ever penned. The summary of Wilberforce's remarks indicates that his half-hour oration was not confined to gibe and rhetoric, but primarily presented a synopsis of the competent (if unoriginal) critique of the *Origin* that he later published in the *Quarterly Review.* The short paragraph allotted to Huxley's reply does not mention the famous repartee—an omission of no great import in a press that, however detailed, could be opaquely discreet. But the account of Huxley's words affirms what all letter writers (see below) also noted—that Huxley spoke briefly and presented no detailed refutation of the bishop's arguments. Instead, he focused his remarks on the logic of Darwin's argument, asserting that evolution was no mere speculation, but a theory supported by copious evidence even if the process of transmutation could not be directly observed.

By the standard account, chaos should now break out, FitzRoy should jump up raving, and Henslow should gavel the meeting closed. No such thing; the meeting went on. FitzRoy took the podium in his turn. Two other speakers followed. And then, the true climax—not entirely omitted in Francis Darwin's "official" version so many years later, but so relegated to a few lines of afterthought that the incident simply dropped out of most later accounts—leading to the popular impression that Huxley's riposte had ended the meeting. Henslow turned to Joseph Hooker, the botanist of Darwin's inner circle, and asked him "to state his view of the botanical aspect of the question."

The *Athenaeum* gave Hooker's remarks four times the coverage awarded to Huxley. It was Hooker who presented a detailed refutation of Wilberforce's specific arguments. It was Hooker who charged directly that the bishop had distorted and misunderstood Darwin's theory. We get some flavor of Hooker's force and effectiveness from a section of the *Athenaeum*'s report:

> In the first place, his Lordship, in his eloquent address, had as it appeared to him [Hooker], completely misunderstood Mr. Darwin's hypothesis: his Lordship intimated that this maintained the doctrine of the transmutation of existing species one into another, and had confounded this with that of the successive development of species by variation and natural selection. The first of these doctrines was so wholly opposed to the facts, reasonings and results of Mr. Darwin's

work, that he could not conceive how any one who had read it could make such a mistake—the whole book, indeed, being a protest against that doctrine.

Moreover, it was Hooker who presented the single most effective debating point against Wilberforce (according to several eyewitness accounts) by stating publicly that he had long opposed evolution but had been led to the probable truth of Darwin's claim by so many years of direct experience with the form and distribution of plants. The bishop did not respond, and Henslow closed the meeting after Hooker's successful speech.

When we turn to the few letters of eyewitnesses, we find the *Athenaeum* account affirmed, the official story further compromised, and some important information added—particularly on the exchange about apes and ancestors. We must note, first of all, that the three letters most commonly cited—those of Green, Fawcett, and Hooker himself—were all written by participants or strong partisans of Darwin's side. For example, future historian J. R. Green, source of the standard version for Huxley's actual words, began his account (to the geologist W. Boyd Dawkins) with a lovely Egyptian metaphor of fealty to Darwin:

> On Saturday morning I met Jenkins going to the Museum. We joined company, and he proposed going to Section D, the Zoology, etc. "to hear the Bishop of Oxford smash Darwin." "Smash Darwin! Smash the Pyramids," said I in great wrath. . . .

(These one-sided sources make Balfour Stewart's neglected letter all the more important—for he was the only uncommitted scientist who reported his impressions right after the debate.)

We may draw from these letters, I believe, three conclusions that further refute the official version. First, Huxley's words may have rung true, but his oratory was faulty. He was ill at ease (his great career as a public speaker lay in the future). He did not project; many in the audience did not hear what he said. Hooker wrote to Darwin on July 2:

> Well, Sam Oxon [short for *Oxoniensis,* Latin for "of Oxford," Wilberforce's ecclesiastical title] got up and spouted

for half an hour with inimitable spirit, ugliness and empti-ness and unfairness. . . . Huxley answered admirably and turned the tables, but he could not throw his voice over so large an assembly, nor command the audience; and he did not allude to Sam's weak points nor put the matter in a form or way that carried the audience.

The chemist A. G. Vernon-Harcourt could not recall Huxley's famous words many years later because he had not heard them over the din. He wrote to Leonard Huxley: "As the point became clear, there was a great burst of applause, which mostly drowned the end of the sentence."

Second, for all the admitted success of Huxley's great moment, Hooker surely made the more effective rebuttal—and the meet-ing ended with his upbeat. I hesitate to take Hooker's own ac-count at face value, but he was so scrupulously modest and self-effacing, and so willing to grant Huxley all the credit later on as the official version congealed, that I think we may titrate the adrenaline of his immediate joy with the modesty of his general bearing and regard his account to Darwin as pretty accurate:

> My blood boiled, I felt myself a dastard; now I saw my ad-vantage; I swore to myself that I would smite that Amale-kite, Sam, hip and thigh. . . . There and then I smashed him amid rounds of applause. I hit him in the wind and then proceeded to demonstrate in a few words: (1) that he could never have read your book, and (2) that he was absolutely ignorant of the rudiments of Bot [botanical] Science. I said a few more on the subject of my own experience and con-version, . . . Sam was shut up—had not one word to say in reply, and the meeting *was dissolved forthwith* [Hooker's ital-ics].

Third, and most important, we do not really know what either man said in the famous exchange about apes and ancestors. Hux-ley's retort is not in dispute. The eyewitness versions differ sub-stantially in wording, but all agree in content. We might as well cite Green's version, if only because it became canonical when Huxley himself "approved" it for Francis Darwin's biography of his father:

I asserted, and I repeat—that a man has no reason to be ashamed of having an ape for his grandfather. If there were an ancestor whom I should feel shame in recalling, it would rather be a *man,* a man of restless and versatile intellect, who, not content with an equivocal success in his own sphere of activity, plunges into scientific questions with which he has no real acquaintance, only to obscure them by an aimless rhetoric, and distract the attention of his hearers from the real points at issue by eloquent digressions and skilled appeals to religious prejudice.

Huxley later demurred only about the word "equivocal," asserting that he would not have besmirched the bishop's competence in matters of religion.

Huxley's own, though lesser-known version (in a brief letter written to his friend Dyster on September 9, 1860) puts the issue more succinctly, but to the same effect:

If then, said I, the question is put to me would I rather have a miserable ape for a grandfather or a man highly endowed by nature and possessed of great means of influence and yet who employs those faculties and that influence for the mere purpose of introducing ridicule into a grave scientific discussion—I unhesitatingly affirm my preference for the ape.

But what had Wilberforce said to incur Huxley's wrath? Quite astonishingly, on this pivotal point of the entire legend, we have nothing but a flurry of contradictory reports. No two accounts coincide. All mention apes and grandfathers, but beyond this anchor of agreement, we find almost every possible permutation of meaning.

We don't know, first of all, whether or not Wilberforce committed that most dubious imposition upon Victorian sensibilities by daring to mention *female* ancestry from apes—that is, did he add grandmothers or speak only of grandfathers? Several versions cite only the male parent, as in Green's letter: "He [Wilberforce] had been told that Professor Huxley had said that he didn't see that it mattered much to a man whether his grandfather was an ape or not. Let the learned professor speak for himself." Yet, I am inclined to the conclusion that Wilberforce must have said something about grandmothers. The distaff side of descent occurs in several versions, Balfour Stewart's neglected letter in particular (see earlier citation), by disinterested observers or partisans of Wilberforce. I can understand why opponents might have delighted in such an addition ("merely corroborative detail, intended to give artistic verisimilitude to an otherwise bald and unconvincing narrative," as Pooh-Bah liked to say). But why should sympathetic listeners remember such a detail if the bishop had not included it himself?

But, far more important, it seems most unlikely that the central claim of the official version can be true—namely, that Wilberforce taunted Huxley by asking him pointedly whether he could trace his personal ancestry from grandparents back to apes (made all the worse if the bishop really asked whether he could trace it on his mother's side). No contemporary account puts the taunt quite so baldly. The official version cites a letter from Lyell (who was not there) since the anonymous eyewitness (more on him later) who supplied Francis Darwin's account could not remember the exact words. Lyell wrote: "The Bishop asked whether Huxley was related by his grandfather's or grandmother's side to an ape." The other common version of this taunt was remembered by Isabel Sidgwick in 1898: "Then, turning to his antagonist with a smiling insolence, he begged to know, was it through his grandfather or his grandmother that he claimed his descent from a monkey?"

We will never know for sure, but the memories of Canon Farrar seem so firm and detailed, and ring so true to me, that I shall place my money on his version. Farrar was a liberal clergyman who once organized a meeting for Huxley to explain Darwinism to fellow men of the cloth. His memories, written in 1899 to Leonard Huxley, are admittedly forty years old, but his version makes sense of many puzzles and should be weighted well on that account—especially since he regarded Huxley as the victor and did not write to reconstruct history in the bishop's cause. Farrar wrote, taking the official version of Wilberforce's taunt to task:

> His words are quite misquoted by you (which your father refuted). They did not appear vulgar, nor insolent nor per-

sonal, but flippant. He had been talking of the perpetuity of
species in birds [a correct memory since all agree that Wilberforce criticized Darwin on the breeds of pigeons in exactly this light]: and then denying *a fortiori* the derivation of
the species Man from Ape, he rhetorically invoked the help
of feeling: and said (I swear to the sense and form of the
sentence, if not to the words) "If anyone were to be willing
to trace his descent through an ape as his grandfather,
would he be willing to trace his descent similarly on the side
of his grandmother." It was (you see) the arousing of antipathy about degrading women to the Quadrumana [four-
footed apes]. It was not to the point, but it was the purpose.
It did not sound insolent, but unscientific and unworthy of
the zoological argument which he had been sustaining. It
was a bathos. Your father's reply . . . showed that there was a
vulgarity as well as a folly in the Bishop's words; and the
impression distinctly was, that the Bishop's party as they left
the room, felt abashed; and recognized that the Bishop had
forgotten to behave like a gentleman.

Farrar's analysis of Huxley's victory includes an interesting comment on Victorian sensibilities:

The victory of your father, was not the ironical dexterity
shown by him, but the fact that he had got a victory in respect of manners and good breeding. You must remember
that the whole audience was made up of gentlefolk, who
were not prepared to endorse anything vulgar.

Finally, Farrar affirms the other major falsity of the official version by acknowledging the superiority of Hooker's reply:

The speech which really left its mark scientifically on the
meeting, was the short one of Hooker. . . . I should say that
to fair minds, the intellectual impression left by the discussion was that the Bishop had stated some facts about the
perpetuity of species, but that no one had really contributed
any valuable point to the opposite side except Hooker . . .
but that your father had scored a victory over Bishop Wilberforce in the question of good manners.

And so, in summary, we may conclude that the heroic legend of the official version fails badly in two crucial points—our ignorance of Wilberforce's actual words and the near certainty that the forgotten Hooker made a better argument than Huxley. What, then, can we conclude, based on such poor evidence, about such a key event in the hagiography of science? Huxley did not debate Wilberforce at Oxford in 1860; rather, they both spoke, one after the other, in a prolonged discussion of Draper's paper. They had one short and wonderful exchange of rhetorical barbs on a totally nonintellectual point prompted by a whimsical remark, perhaps even a taunt, that Wilberforce made about apes and ancestry, though no one remembered precisely what he said. Huxley made a sharp and effective retort. Everyone enjoyed the incident immensely and recalled it in a variety of versions. Some thought Huxley had won the exchange; others credit Wilberforce. Huxley hardly dealt with Wilberforce's case against Darwin. Hooker, however, made an effective reply in Darwin's behalf, and the meeting ended.

All events before the codification of the official version support this ambiguous and unheroic account. In particular, Wilberforce seemed not a bit embarrassed by the incident. Disraeli spoke about it in his presence. Wilberforce reprinted his review of Darwin's *Origin,* the basis of his remarks that fateful day, in an 1874 collection of his works. His son recounted the tale with credit in Wilberforce's biography. Moreover, Darwin and Wilberforce remained on good terms. The ever genial Darwin wrote to Asa Gray that he found Wilberforce's review "uncommonly clever, not worth anything scientifically, but quizzes me in splendid style. I chuckled with laughter at myself." Wilberforce, told by the vicar of Downe about Darwin's reaction, said: "I am glad he takes it in this way. He is such a capital fellow."

Moreover, though I don't believe that self-justification provides much evidence for anything, we do have a short testimony from Wilberforce himself. He wrote to Sir Charles Anderson just three days after the event: "On Saturday Professor Henslow who presided over the Zoological Section called on me by name to address the Section on Darwin's Theory. So I could not escape and had quite a long fight with Huxley. I think I thoroughly beat him." This letter, now housed in the Bodleian Library of Oxford University, escaped all notice until 1978, when Josef L. Altholz

cited it in the *Journal of the History of Medicine.* I would not exaggerate the importance of this document because it smacks of insin-
cerity at least once—so why not in its last line as well? We know
that 700 people crammed the Museum's largest room to witness
the proceedings. They didn't come to hear Dr. Draper on the
intellectual development of Europe. Wilberforce was on the dais,
and if he didn't know that he would speak, how come everyone
else did?

Why then, and how, did the official version so color this event
as a primal victory for evolution? The answer largely lies with
Huxley himself, who successfully promoted, in retrospect, a version that suited his purposes (and had probably, by then, dis-
placed the actual event in his memory). Huxley, though not
antireligious, was uncompromisingly and pugnaciously anticleri-
cal. Moreover, he despised Wilberforce and his mellifluous soph-
istries. When Wilberforce died in 1873, from head injuries
sustained in a fall from his horse, Huxley remarked (as the story
goes): "For once, reality and his brains came into contact and the
result was fatal."

Janet Browne has traced the construction of the official version
in Francis Darwin's biography of his father. The story is told
through an anonymous eyewitness, but Browne proves that
Hooker himself wrote the account, volunteering for the task with
direct purpose (writing to Francis): "Have you any account of the
Oxford meeting? If not, I will, if you like, see what I can do towards vivifying it (and vivisecting the Bishop) for you." Hooker
dredged his memory with pain and uncertainty. He had forgotten
his letter to Darwin and admitted, "It is impossible to be sure of
what one heard, or of impressions formed, after nearly thirty
years of active life." And further, "I have been driven wild formulating it from memory." Huxley then vetted Hooker's account
and the official story was set.

The tale was then twice embellished—first, in 1892, when
Francis published a shorter biography of Charles Darwin, and
Huxley contributed a letter, now remembering for the first time
(more than thirty years later) his *sotto voce* crack, "The Lord hath
delivered him into mine hands"; second, in 1900, when Leonard
Huxley wrote the life of his father. Thus, dutiful sons presented
the official version as constructed by a committee of two—the
chief participants Huxley and Hooker—from memories colored

by thirty years of battle. We can only agree with Sheridan Gilley, who writes:

> The standard account is a wholly one-sided effusion from the winning side, put together long after the event, uncritically copied from book to book, and shaped by the hagiographic conventions of the Victorian life and letters.

So much for correcting a moment of history. But why should we care today? Does the heroic version do any harm? And does its rectification have any meaning beyond our general preference for accuracy? Stories do not become primary legends simply because they tell rip-roaring narratives; they must stand as exemplars, particular representations of something deeper and far more general. The official version of Huxley versus Wilberforce is an archetype for a common belief about the nature of science and its history. The fame and meaning of the official version lie in this wider context. Yet this common belief is not only wrong (or at least seriously oversimplified) but ultimately harmful to science. Thus, in debunking the official version of Huxley versus Wilberforce, we might make a helpful correction for science itself.

Ruth Moore captured the general theme in her version of the standard account: "From that hour on, the quarrel over the elemental issue that the world believed was involved, science versus religion, was to rage unabated." The story has archetypal power because Huxley and Wilberforce, in the official version, are not mere men but symbols, or synecdoches, for a primal struggle: religion versus science, reaction versus enlightenment, dogma versus truth, darkness versus light.

All men have blind spots, however broad their vision. Thomas Henry Huxley was the most eloquent spokesman that evolution has ever known. But his extreme anticlericalism led him to an uncompromising view of organized religion as the enemy of science. Huxley could envision no allies among the official clergy. Conservatives like Wilberforce were enemies pure and simple; liberals lacked the guts to renounce what fact and logic had falsified, as they struggled to marry the irreconcilable findings of science with their supernatural vision. He wrote in 1887 of those "whose business seems to be to mix the black of dogma and the white of science into the neutral tint of what they call liberal theology." Huxley did view his century as a battleground between science and organized religion—and he took great pride in the many notches on his own gun.

This cardboard dichotomy seems favorable for science at first (and superficial) glance. It enshrines science as something pure and apart from the little quirks and dogmas of daily life. It exalts science as a disembodied method for discovering truth at all costs, while social institutions—religion in particular—hold fast to antiquated superstition. Comfort and social stability resist truth, and science must therefore fight a lonely battle for enlightenment. Its heroes, in bad times, are true martyrs—Bruno at the stake, Galileo before the Inquisition—or, in better times, merely irritated, as Huxley was, by ecclesiastical stupidity.

But no battle exists between science and religion—the two most separate spheres of human need. A titanic struggle occurs, always has, always will, between questioning and authority, free inquiry and frozen dogma—but the institutions representing these poles are not science and religion. These struggles occur *within* each field, not primarily across disciplines. The general ethic of science leads to greater openness, but we have our fossils, often in positions of great power. Organized religion, as an arm of state power so frequently in history, has tended to rigidity—but theologies have also spearheaded social revolution. Official religion has not opposed evolution as a monolith. Many prominent evolutionists have been devout, and many churchmen have placed evolution at the center of their personal theologies. Henry Ward Beecher, America's premier pulpiteer during Darwin's century, defended evolution as God's way in a striking commercial metaphor: "Design by wholesale is grander than design by retail"—better, that is, to ordain general laws of change than to make each species by separate fiat.

The struggle of free inquiry against authority is so central, so pervasive that we need all the help we can get from every side. Inquiring scientists must join hands with questioning theologians if we wish to preserve that most fragile of all reeds, liberty itself. If scientists lose their natural allies by casting entire institutions as enemies, and not seeking bonds with soul mates on other paths, then we only make a difficult struggle that much harder.

Huxley had not planned to enter that famous Oxford meeting.

He was still inexperienced in public debate, not yet Darwin's bull-dog. He wrote: "I did not mean to attend it—did not see the good of giving up peace and quietness to be episcopally pounded." But his friends prevailed upon him, and Huxley, savoring victory, left the meeting with pleasure and resolution:

> Hooker and I walked away from the meeting together, and I remember saying to him that this experience had changed my opinion as to the practical value of the art of public speaking, and that from that time forth I should carefully cultivate it, and try to leave off hating it.

So Huxley became the greatest popular spokesman for science in his century—as a direct result of his famous encounter with Wilberforce. He waded into the public arena and struggled for three decades to breach the boundaries between science and the daily life of ordinary people. And yet, ironically, his Manichean view of science and religion—abetted so strongly by the official version, his own construction in part, of the debate with Wilberforce—harmed his greatest hope by establishing boundaries to exclude natural allies and, ultimately, by encircling science as something apart from other human passions. We may, perhaps, read one last document of the great Oxford debate in a larger metaphorical context as a plea, above all, for solidarity among people of like minds and institutions of like purposes. Darwin to Hooker upon receiving his account of the debate: "Talk of fame, honor, pleasure, wealth, all are dirt compared with affection."

# 27 | Genesis and Geology

HERBERT HOOVER produced a fine translation, still in use, of Agricola's sixteenth-century Latin treatise on mining and geology. In the midst of his last presidential campaign, Teddy Roosevelt published a major monograph on the evolutionary significance of animal coloration (see Essay 14). Woodrow Wilson was no intellectual slouch, and John F. Kennedy did aptly remark to a group of Nobel laureates assembled at the White House that the building then contained more intellectual power than at any moment since the last time Thomas Jefferson dined there alone.

Still, when we seek a political past of intellectual eminence in the midst of current emptiness, we cannot do better than the helm of Victorian Britain. High ability may not have prevailed generally, as the wise Private Willis, guard to the House of Commons, reminds us in Gilbert and Sullivan's *Iolanthe:*

> When in that House M.P.'s divide,
> If they've a brain and cerebellum, too,
> They've got to leave that brain outside,
> And vote just as their leaders tell 'em to.
> But then the prospect of a lot
> Of dull M.P.'s in close proximity
> All thinking for themselves is what
> No man can face with equanimity.

But the men at the top—the Tory leader Benjamin Disraeli and his Liberal counterpart W. E. Gladstone—were formidable in many various ways. Disraeli maintained an active career as a respected romantic novelist, publishing the three-volume *Endymion* in 1880, at the height of his prestige and just a year before his death. Gladstone, a distinguished Greek scholar, wrote his three-volume *Studies on Homer and the Homeric Age* (1858) while temporarily out of office.

In 1885, following a series of setbacks including the death of General Gordon at Khartoum, Gladstone's government fell, and he resigned as prime minister. He did not immediately proceed to unwind with his generation's rum swizzle on a Caribbean beach (Chivas Regal on the links of Saint Andrews, perhaps). Instead, he occupied his enforced leisure by writing an article on the scientific truth of the book of Genesis—"Dawn of Creation and of Worship," published in *The Nineteenth Century,* in November 1885. Thomas Henry Huxley, who invented the word *agnostic* to describe his own feelings, read Gladstone's effort with disgust and wrote a response to initiate one of the most raucous, if forgotten, free-for-alls of late nineteenth century rhetoric. (Huxley disliked Gladstone and once described him as suffering from "severely copious chronic glossorrhoea.")

But why bring up a forgotten and musty argument, even if the protagonists were two of the most colorful and brilliant men of the nineteenth century? I do so because current events have brought their old subject—the correlation of Genesis with geology—to renewed attention.

Our legislative victory over "creation science" (Supreme Court in *Edwards* v. *Aguillard,* June 1987) ended an important chapter in American social history, one that stretched back to the Scopes trial of 1925. (Biblical literalism will never go away, so long as cash flows and unreason retains its popularity, but the legislative strategy of passing off dogma as creation science and forcing its instruction in classrooms has been defeated.) In this happy light, we are now free to ask the right question once again: In what *helpful* ways may science and religion coexist?

Ever since the Edwards decision, I have received a rash of well-meaning letters suggesting a resolution very much like Gladstone's. These letters begin by professing pleasure at the defeat of fundamentalism. Obviously, six days of creation and circa 6,000 years of biblical chronology will not encompass the earth's history. But, they continue, once we get past the nonsense of literalism, are we not now free to read Genesis 1 as factual in a more general sense? Of course the days of creation can't be twenty-four hours long. Of course the origin of light three days before the creation of the sun poses problems. But aren't the general order and story consistent with modern science, from the big bang to Darwinian theory? After all, plants come first in Genesis, then creatures of the sea, then land animals, and finally humans. Well, isn't this right? And, if so, then isn't Genesis true in the broad sense? And if true, especially since the scribes of Genesis could not have understood the geological evidence, must not the words be divinely inspired? This sequence of claims forms the core of Gladstone's article. Huxley's words therefore deserve a resurrection.

Huxley's rebuttal follows the argument that most intellectuals—scientists and theologians alike—make today. First, while the broadest brush of the Genesis sequence might be correct—plants first, people last—many details are dead wrong by the testimony of geological evidence from the fossil record. Second, this lack of correlation does not compromise the power and purpose of religion or its relationship with the sciences. Genesis is not a treatise on natural history.

Gladstone wrote his original article as a response to a book by Professor Alfred Réville of the Collège de France—*Prolegomena to the History of Religions* (1884). Gladstone fancied himself an expert on Homer, and he had labored for thirty years to show that common themes of the Bible and the most ancient Greek texts could be harmonized to expose the divine plan revealed by the earliest historical records of different cultures. Gladstone was most offended by Réville's dismissal of his Homeric claims, but his article focuses on the veracity of Genesis.

Gladstone did not advocate the literal truth of Genesis; science had foreclosed this possibility to any Victorian intellectual. He accepted, for example, the standard argument that the "days" of creation are metaphors for periods of undetermined length separating the major acts of a coherent sequence. But Gladstone then insisted that these major acts conform precisely to the order best specified by modern science—the cosmological events of the first four days (Genesis 1:1–19) to Laplace's "nebular hypothesis" for the origin of the sun and planets, and the biological events of "days" five and six (Genesis 1:20–31) to the geological record of fossils and Darwin's theory of evolution. He placed special emphasis on a fourfold sequence in the appearance of animals: the "water population" followed by the "air population" on the fifth day, and the "land population" and its "consummation in man" on the sixth day:

> And God said, Let the waters bring forth abundantly the moving creature that hath life, and fowl that may fly above the earth in the open firmament of heaven [Verse 20]. . . . And God said, Let the earth bring forth the living creature after its kind, cattle, and creeping thing, and beast of the earth after its kind; and it was so [Verse 24]. . . . And God said, Let us make man in our image [Verse 26].

Gladstone then caps his argument with the claim still echoed by modern reconcilers: This order, too good to be guessed by writers ignorant of geological evidence, must have been revealed by God to the scribes of Genesis:

> Then, I ask, how came . . . the author of the first chapter of Genesis to know that order, to possess knowledge which natural science has only within the present century for the first time dug out of the bowels of the earth? It is surely impossible to avoid the conclusion, first, that either this writer was gifted with faculties passing all human experience, or else his knowledge was divine.

In a closing flourish, Gladstone enlarged his critique in a manner sure to inspire Huxley's wrath. He professed himself satisfied as to the possibility of physical evolution, even by Darwin's mechanism. But the spirit, the soul, the "mind of man" must be divine in origin, thereby dwarfing to insignificance anything in the merely material world. Gladstone chided Darwin for reaching too far, for trying to render the ethereal realm by his crass and heartless mechanism. He ridiculed the idea "that natural selection and the survival of the fittest, all in the physical order, exhibit to us the great *arcanum* of creation, the sun and center of life, so that mind and spirit are dethroned from their old supremacy, are no longer sovereign by right, but may find somewhere by charity a place assigned them, as appendages, perhaps only as excrescences, of the material creation."

Ending on a note of deep sadness, Gladstone feared for our equanimity, our happiness, our political stability, our hopes for a moral order, should the festering sore of agnosticism undermine our assurance of God's existence and benevolence—"this belief, which has satisfied the doubts and wiped away the tears, and found guidance for the footsteps of so many a weary wanderer on earth, which among the best and greatest of our race has been so cherished by those who had it, and so longed and sought for by those who had it not." If science could now illustrate God by proving that he knew his stuff when he whispered into Moses' ear, then surely that sore could be healed.

Huxley, who had formally retired just a few months before, and who had forsworn future controversy of exactly this kind, responded with an article in the December issue of *The Nineteenth Century*—"The Interpreters of Genesis and the Interpreters of Nature." Obviously pleased with himself, and happy with his return to fighting form, he wrote to Herbert Spencer: "Do read my polishing off of the G.O.M. [Gladstone was known to friends and enemies alike as the "Grand Old Man"]. I am proud of it as a work of art, and as evidence that the volcano is not yet exhausted."

Huxley begins by ridiculing the very notion that harmonizing Genesis with geology has any hope of success or intellectual potential to illustrate anything meaningful. He places Gladstone among "those modern representatives of Sisyphus, the reconcilers of Genesis with science." (Sisyphus, king of Corinth, tried to cheat death and was punished in Hades with the eternal task of repeatedly rolling a large stone to the top of a hill, only to have it roll down again just as it reached the top.)

Huxley arranged his critique by citing four arguments against Gladstone's insistence that Genesis specified an accurate "fourfold order" of creation—water population, air population, land population, and man. Huxley wrote:

> If I know anything at all about the results attained by the natural sciences of our time, it is a demonstrated conclusion and established fact that the fourfold order given by Mr. Gladstone is not that in which the evidence at our disposal tends to show that the water, air and land populations of the

globe have made their appearance. . . . The facts which demolish his whole argument are of the commonest notoriety. [Huxley uses "notoriety" not in its current, pejorative meaning, but in the old sense of "easily and evidently known to all."]

He then presents his arguments in sequence:

1. Direct geological evidence shows that land animals arose before flying creatures. This reversal of biblical sequence holds whether we view the Genesis text as referring only to vertebrates (for terrestrial amphibians and reptiles long precede birds) or to all animals (for such terrestrial arthropods as scorpions arise before flying insects).

2. Even if we didn't know, or chose not to trust, the geological sequence, we could deduce on purely anatomical grounds that flying creatures must have evolved from preexisting terrestrial ancestors. Structures used in flight are derived modifications of terrestrial features:

> Every beginner in the study of animal morphology is aware that the organization of a bat, of a bird, or of a pterodactyle, presupposes that of a terrestrial quadruped, and that it is intelligible only as an extreme modification of the organization of a terrestrial mammal or reptile. In the same way, winged insects (if they are to be counted among the "air-population") presuppose insects which were wingless, and therefore as "creeping things," which were part of the land-population.

3. Whatever the order of first appearances, new species within all groups—water, air, and land dwellers—have continued to arise throughout subsequent time, whereas Genesis implies that God made *all* the sea creatures, then all the denizens of the air, and so on.

4. However we may wish to quibble about the order of animals, Gladstone should not so conveniently excise plants from his discussion. Genesis pushes their origin back to the third day, before the origin of any animal. But plants do not precede animals in the fossil record; and the terrestrial flowering plants specifically men-

tioned in Genesis (grass and fruit tree) arise very late, long after the first mammals.

Huxley then ends his essay with a powerful statement—every bit as relevant today as 100 years ago at its composition—on the proper domains and interactions of science and religion. Huxley expresses no antipathy for religion, properly conceived, and he criticizes scientists who overstep the boundaries and possibilities of their discipline as roundly as he condemns an antiquated and overextended role for the biblical text:

> The antagonism between science and religion, about which we hear so much, appears to me to be purely factitious, fabricated on the one hand by short-sighted religious people, who confound . . . theology with religion; and on the other by equally short-sighted scientific people who forget that science takes for its province only that which is susceptible of clear intellectual comprehension.

The moral precepts for our lives, Huxley argues, have been developed by great religious thinkers, and no one can improve on the Prophet Micah's statement: ". . . what doth the Lord require of thee, but to do justly, and to love mercy, and to walk humbly with thy God." Nothing that science might discover about the factual world could possibly challenge, or even contact, this sublime watchword for a proper life:

> But what extent of knowledge, what acuteness of scientific criticism, can touch this, if anyone possessed of knowledge or acuteness could be absurd enough to make the attempt? Will the progress of research prove that justice is worthless and mercy hateful? Will it ever soften the bitter contrast between our actions and our aspirations, or show us the bounds of the universe, and bid us say, "Go to, now we comprehend the infinite"?

Conflicts develop not because science and religion vie intrinsically, but when one domain tries to usurp the proper space of the other. In that case, a successful defense of home territory is not only noble per se, but a distinct benefit to honorable people in both camps:

The antagonism of science is not to religion, but to the heathen survivals and the bad philosophy under which religion herself is often well-nigh crushed. And, for my part, I trust that this antagonism will never cease, but that to the end of time true science will continue to fulfill one of her most beneficent functions, that of relieving men from the burden of false science which is imposed upon them in the name of religion.

Gladstone responded with a volley of rhetoric. He began from the empyrean heights, pointing out that after so many years of parliamentary life he was a tired (if still grand) old man, and didn't know if he could muster the energy for this sort of thing anymore—particularly for such a nettlesome and trivial opponent as the merely academic Huxley:

As I have lived for more than half a century in an atmosphere of contention, my stock of controversial fire has perhaps become abnormally low; while Professor Huxley, who has been inhabiting the Elysian regions of science . . . may be enjoying all the freshness of an unjaded appetite.

(Much of the fun in reading through this debate lies not in the forcefulness of arguments or in the mastery of prose by both combatants, but in the sallying and posturing of two old gamecocks [Huxley was sixty, Gladstone seventy-six in 1885] pulling out every trick from the rhetorical bag—the musty and almost shameful, the tried and true, and even a novel flourish here and there.)

But once Gladstone got going, that old spark fanned quite a flame. His denunciations spanned the gamut. Huxley's words, on the one hand, were almost too trivial to merit concern—one listens "to his denunciations . . . as one listens to distant thunders, with a sort of sense that after all they will do no great harm." On the other hand, Huxley's attack could not be more dangerous. "I object," Gladstone writes, "to all these exaggerations . . . as savoring of the spirit of the Inquisition, and as restraints on literary freedom."

Yet, when Gladstone got down to business, he could muster only a feeble response to Huxley's particulars. He did effectively combat Huxley's one weak argument—the third charge that all groups continue to generate new species, whatever the sequence of their initial appearance. Genesis, Gladstone replies, only discusses the order of origin, not the patterns of subsequent history:

> If we arrange the schools of Greek philosophy in numerical order, according to the dates of their inception, we do not mean that one expired before another was founded. If the archaeologist describes to us as successive in time the ages of stone, bronze and iron, he certainly does not mean that no kinds of stone implement were invented after bronze began.

But Gladstone came to grief on his major claim—the veracity of the Genesis sequence: water population, air population, land population, and humans. So he took refuge in the oldest ploy of debate. He made an end run around his disproved argument and changed the terms of discussion. Genesis doesn't refer to all animals, but "only to the formation of the objects and creatures with which early man was conversant." Therefore, toss out all invertebrates (although I cannot believe that cockroaches had no foothold, even in the Garden of Eden) and redefine the sequence of water, air, land, and mentality as fish, bird, mammal, and man. At least this sequence matches the geological record. But every attempt at redefinition brings new problems. How can the land population of the sixth day—"every living thing that creepeth upon the earth"—refer to mammals alone and exclude the reptiles that not only arose long before birds but also provided the dinosaurian lineage of their ancestry. This problem backed Gladstone into a corner, and he responded with the weak rejoinder that reptiles are disgusting and degenerate things, destined only for our inattention (despite Eve and the serpent): "Reptiles are a family fallen from greatness; instead of stamping on a great period of life its leading character, they merely skulked upon the earth." Yet Gladstone sensed his difficulty and admitted that while reptiles didn't disprove his story, they certainly didn't help him either: "However this case may be regarded, of course I cannot draw from it any support to my general contention."

Huxley, smelling victory, moved in for the kill. He derided Gladstone's slithery argument about reptiles and continued to highlight the evident discrepancies of Genesis, read literally, with geology ("Mr. Gladstone and Genesis," *The Nineteenth Century*, 1896).

> However reprehensible, and indeed contemptible, terrestrial reptiles may be, the only question which appears to me to be relevant to my argument is whether these creatures are or are not comprised under the denomination of "everything that creepeth upon the ground."

Contrasting the approved tactics of Parliament and science, Huxley obliquely suggested that Gladstone might emulate the wise cobbler and stick to his last. Invoking reptiles once again, he wrote:

> Still, the wretched creatures stand there, importunately demanding notice; and, however different may be the practice in that contentious atmosphere with which Mr. Gladstone expresses and laments his familiarity, in the atmosphere of science it really is of no avail whatever to shut one's eyes to facts, or to try to bury them out of sight under a tumulus of rhetoric.

Gladstone's new sequence of fish, bird, mammal, and man performs no better than his first attempt in reconciling Genesis and geology. The entire enterprise, Huxley asserts, is misguided, wrong, and useless: "Natural science appears to me to decline to have anything to do with either [of Gladstone's two sequences]; they are as wrong in detail as they are mistaken in principle." Genesis is a great work of literature and morality, not a treatise on natural history:

> The Pentateuchal story of the creation is simply a myth [in the literary, not pejorative, sense of the term]. I suppose it to be a hypothesis respecting the origin of the universe which some ancient thinker found himself able to reconcile with his knowledge, or what he thought was knowledge, of

the nature of things, and therefore assumed to be true. As such, I hold it to be not merely an interesting but a venerable monument of a stage in the mental progress of mankind.

Gladstone, who was soon to enjoy a fourth stint as prime minister, did not respond. The controversy then flickered, shifting from the pages of *The Nineteenth Century* to the letters column of the *Times*. Then it died for a while, only to be reborn from time to time ever since.

I find something enormously ironical in this old battle, fought by Huxley and Gladstone a century ago and by much lesser lights even today. It doesn't matter a damn because Huxley was right in asserting that correspondence between Genesis and the fossil record holds no significance for religion or for science. Still, I think that Gladstone and most modern purveyors of his argument have missed the essence of the kind of myth that Genesis 1 represents. Nothing could possibly be more vain or intemperate than a trip on these waters by someone lacking even a rudder or a paddle in any domain of appropriate expertise. Still, I do feel that when read simply for its underlying metaphor, the story of Genesis 1 does contradict Gladstone's fundamental premise. Gladstone's effort rests upon the notion that Genesis 1 is a tale about *addition* and linear sequence—God makes this, then this, and then this in a sensible order. Since Gladstone also views evolution and geology as a similar story of progress by accretion, reconciliation becomes possible. Gladstone is quite explicit about this form of story:

> Evolution is, to me, a series with development. And like series in mathematics, whether arithmetical or geometrical, it establishes in things an unbroken progression; it places each thing . . . in a distinct relation to every other thing, and makes each a witness to all that have preceded it, a prophecy of all that are to follow it.

But I can't read Genesis 1 as a story about linear addition at all. I think that its essential theme rests upon a different metaphor—*differentiation* rather than accretion. God creates a chaotic and formless *totality* at first, and then proceeds to make divisions within—to precipitate islands of stability and growing complexity from the vast, encompassing potential of an initial state. Consider the sequence of "days."

On day one, God makes two primary and orthogonal divisions: He separates heaven from earth, and light from darkness. But each category only represents a diffuse potential, containing no differentiated complexity. The earth is "without form and void"; and no sun, moon, or stars yet concentrate the division of light from darkness. On the second day, God consolidates the separation of heaven and earth by creating the firmament and calling it heaven. The third day is then devoted to differentiating the chaotic earth into its stable parts—land and sea. Land then develops further by bringing forth plants. (Does this indicate that the writer of Genesis treated life under a taxonomy very different from ours? Did he see plants as essentially of the earth and animals as something separate? Would he have held that plants have closer affinity with soil than with animals?) The fourth day does for the firmament what the third day accomplished for earth: heaven differentiates and light becomes concentrated into two great bodies, the sun and moon.

The fifth and sixth days are devoted to the creation of animal life, but again the intended metaphor may be differentiation rather than linear addition. On the fifth day, the sea and then the air bring forth their intended complexity of living forms. On the sixth day, the land follows suit. The animals are not simply placed by God in their proper places. Rather, the places themselves "bring forth" or differentiate their appropriate inhabitants at the appointed times.

The final result is a candy box of intricately sculpted pieces, with varying degrees of complexity. But how did the box arise? Did the candy maker just add items piece by piece, according to a prefigured plan—Gladstone's model of linear addition? Or did he start with the equivalent of a tray of fudge, and then make smaller and smaller divisions with his knife, decorating each piece as he cut by sculpting wondrous forms from the potential inherent in the original material? I read the story in this second manner. And if differentiation be the more appropriate metaphor, then Genesis cannot be matching Gladstone's linear view

414 | BULLY FOR BRONTOSAURUS

of evolution. The two stories rest on different premises of organization—addition and differentiation.*

But does life's history really match either of these two stories? Addition and differentiation are not mutually exclusive truths inherent in nature. They are schemata of organization for human thought, two among a strictly limited number of ways that we have devised to tell stories about nature's patterns. Battles have been fought in their names many times before, sometimes strictly within biology. Consider, for example, the early nineteenth century struggle in German embryology between one of the greatest of all natural scientists, Karl Ernst von Baer, who viewed development as a process of differentiation from general forms to specific structures, and the *Naturphilosophen* (nature philosophers) with their romantic conviction that all developmental processes (including embryology) must proceed by linear addition of complexity as spirit struggles to incarnate itself in the highest, human form (see Chapter 2 of my book *Ontogeny and Phylogeny,* Harvard University Press, 1977).

My conclusion may sound unexciting, even wishy-washy, but I think that evolution just says yes to both metaphors for different parts of its full complexity. Yes, truly novel structures do arise in temporal order—first fins, then legs, then hair, then language—and additive models describe part of the story well. Yes, the coding rules of DNA have not changed, and all of life's history differentiates from a potential inherent from the first. Is the history of Western song a linear progression of styles or the construction of more and more castles for a kingdom fully specified in original blueprints by notes of the scale and rules of composition?

Finally, and most important, the bankruptcy of Gladstone's effort lies best exposed in this strictly limited number of deep metaphors available to our understanding. Gladstone was wrong in critical detail, as Huxley so gleefully proved. But what if he had been entirely right? What if the Genesis sequence had been generally accurate in its broad brush? Would such a correspondence

---

*After writing this essay, I visited the cathedral of San Marco in Venice, and was pleased to note that the early medieval mosaics of the great creation dome (in the south end of the narthex) picture the events of the first six days as an explicit sequence of divisions with differentiation.

mean that God had dictated the Torah word by word? Of course not. How many possible stories can we tell? How many can we devise beyond addition and differentiation? Simultaneous creation? Top down appearance? Some, perhaps, but not many. So what if the Genesis scribe wrote his beautiful myth in one of the few conceivable and sensible ways—and if later scientific discoveries then established some fortuitous correspondences with his tale? Bats didn't know about extinct pterodactyls, but they still evolved wings that work in similar ways. The strictures of flying don't permit many other designs—just as the limited pathways from something small and simple to something big and complex don't allow many alternatives in underlying metaphor. Genesis and geology happen not to correspond very well. But it wouldn't mean much if they did—for we would only learn something about the limits to our storytelling, not even the whisper of a lesson about the nature and meaning of life or God.

Genesis and geology are sublimely different. William Jennings Bryan used to dismiss geology by arguing that he was only interested in the rock of ages, not the age of rocks. But in our tough world—not cleft for us, and offering no comfortable place to hide—I think we had better pay mighty close attention to both.

# 28 | William Jennings Bryan's Last Campaign

I HAVE SEVERAL REASONS for choosing to celebrate our legal victory over "creation science" by trying to understand with sympathy the man who forged this long and painful episode in American history—William Jennings Bryan. In June 1987, the Supreme Court voided the last creationist statute by a decisive 7–2 vote, and then wrote their decision in a manner so clear, so strong, and so general that even the most ardent fundamentalists must admit the defeat of their legislative strategy against evolution. In so doing, the Court ended William Jennings Bryan's last campaign, the cause that he began just after World War I as his final legacy, and the battle that took both his glory and his life in Dayton, Tennessee, when, humiliated by Clarence Darrow, he died just a few days after the Scopes trial in 1925.

My reasons range across the domain of Bryan's own character. I could invoke rhetorical and epigrammatic expressions, the kind that Bryan, as America's greatest orator, laced so abundantly into his speeches—Churchill's motto for World War II, for example: "In victory: magnanimity." But I know that my main reason is personal, even folksy, the kind of one-to-one motivation that Bryan, in his persona as the Great Commoner, would have applauded. Two years ago, a colleague sent me an ancient tape of Bryan's voice. I expected to hear the pious and polished shoutings of an old stump master, all snake oil and orotund sophistry. Instead, I heard the most uncanny and friendly sweetness, high pitched, direct, and apparently sincere. Surely this man could not simply be dismissed, as by H. L. Mencken, reporting the Scopes

416

trial for the Baltimore *Sun:* as "a tinpot Pope in the Coca-Cola belt."

I wanted to understand a man who could speak with such warmth, yet talk such yahoo nonsense about evolution. I wanted, above all, to resolve a paradox that has always cried out for some answer rooted in Bryan's psyche. How could this man, America's greatest populist reformer, become, late in life, her arch reactionary?

For it was Bryan who, just one year beyond the minimum age of thirty-five, won the Democratic presidential nomination in 1896 with his populist rallying cry for abolition of the gold standard: "You shall not press down upon the brow of labor this crown of thorns. You shall not crucify mankind upon a cross of gold." Bryan who ran twice more, and lost in noble campaigns for reform, particularly for Philippine independence and against American imperialism. Bryan, the pacifist who resigned as Wilson's secretary of state because he sought a more rigid neutrality in the First World War. Bryan who stood at the forefront of most progressive victories in his time: women's suffrage, the direct election of senators, the graduated income tax (no one loves it, but can you think of a fairer way?). How could this man have then joined forces with the cult of biblical literalism in an effort to purge religion of all liberality, and to stifle the same free thought that he had advocated in so many other contexts?

This paradox still intrudes upon us because Bryan forged a living legacy, not merely an issue for the mists and niceties of history. For without Bryan, there never would have been antievolution laws, never a Scopes trial, never a resurgence in our day, never a decade of frustration and essays for yours truly, never a Supreme Court decision to end the issue. Every one of Bryan's progressive triumphs would have occurred without him. He fought mightily and helped powerfully, but women would be voting today and we would be paying income tax if he had never been born. But the legislative attempt to curb evolution was his baby, and he pursued it with all his legendary demoniac fury. No one else in the ill-organized fundamentalist movement had the inclination, and surely no one else had the legal skill or political clout. Ironically, fundamentalist legislation against evolution is the only truly distinctive and enduring brand that Bryan placed

William Jennings Bryan on the stump. Taken during the presidential campaign of 1896. THE BETTMANN ARCHIVE.

upon American history. It was Bryan's movement that finally bit the dust in Washington in June of 1987.

The paradox of shifting allegiance is a recurring theme in literature about Bryan. His biography in the *Encyclopaedia Britannica* holds that the Scopes trial "proved to be inconsistent with many progressive causes he had championed for so long." One prominent biographer located his own motivation in trying to discover "what had transformed Bryan from a crusader for social and economic reform to a champion of anachronistic rural evangelism, cheap moral panaceas, and Florida real estate" (L. W. Levine, 1965).

Two major resolutions have been proposed. The first, clearly the majority view, holds that Bryan's last battle was inconsistent with, even a nullification of, all the populist campaigning that had gone before. Who ever said that a man must maintain an unchanging ideology throughout adulthood; and what tale of human psychology could be more familiar than the transition from crusading firebrand to diehard reactionary. Most biographies treat the Scopes trial as an inconsistent embarrassment, a sad and unsettling end. The title to the last chapter of almost every book about Bryan features the word "retreat" or "decline."

The minority view, gaining ground in recent biographies and clearly correct in my judgment, holds that Bryan never transformed or retreated, and that he viewed his last battle against evolution as an extension of the populist thinking that had inspired his life's work (in addition to Levine, cited previously, see Paolo E. Coletta, 1969, and W. H. Smith, 1975).

Bryan always insisted that his campaign against evolution meshed with his other struggles. I believe that we should take him at his word. He once told a cartoonist how to depict the harmony of his life's work: "If you would be entirely accurate you should represent me as using a double-barreled shotgun, firing one barrel at the elephant as he tries to enter the treasury and another at Darwinism—the monkey—as he tries to enter the schoolroom." And he said to the Presbyterian General Assembly in 1923: "There has not been a reform for 25 years that I did not support. And I am now engaged in the biggest reform of my life. I am trying to save the Christian Church from those who are trying to destroy her faith."

But how can a move to ban the teaching of evolution in public schools be deemed progressive? How did Bryan link his previous efforts to this new strategy? The answers lie in the history of Bryan's changing attitudes toward evolution.

Bryan had passed through a period of skepticism in college. (According to one story, more than slightly embroidered no doubt, he wrote to Robert G. Ingersoll for ammunition but, upon receiving only a pat reply from his secretary, reverted immediately to orthodoxy.) Still, though Bryan never supported evolution, he did not place opposition high on his agenda; in fact, he evinced a positive generosity and pluralism toward Darwin. In "The Prince of Peace," a speech that ranked second only to the "Cross of Gold" for popularity and frequency of repetition, Bryan said:

> I do not carry the doctrine of evolution as far as some do; I am not yet convinced that man is a lineal descendant of the lower animals. I do not mean to find fault with you if you want to accept the theory. . . . While I do not accept the Darwinian theory I shall not quarrel with you about it.

(Bryan, who certainly got around, first delivered this speech in 1904, and described it in his collected writings as "a lecture delivered at many Chautauqua and religious gatherings in America, also in Canada, Mexico, Tokyo, Manila, Bombay, Cairo, and Jerusalem.")

He persisted in this attitude of laissez-faire until World War I, when a series of events and conclusions prompted his transition from toleration to a burning zeal for expurgation. His arguments did not form a logical sequence, and were dead wrong in key particulars; but who can doubt the passion of his feelings?

We must acknowledge, before explicating the reasons for his shift, that Bryan was no intellectual. Please don't misconstrue this statement. I am not trying to snipe from the depth of Harvard elitism, but to understand. Bryan's dearest friends said as much. Bryan used his first-rate mind in ways that are intensely puzzling to trained scholars—and we cannot grasp his reasons without mentioning this point. The "Prince of Peace" displays a profound ignorance in places, as when Bryan defended the idea of miracles by stating that we continually break the law of gravity: "Do we not suspend or overcome the law of gravitation every day? Every time we move a foot or lift a weight we temporarily overcome one of the most universal of natural laws and yet the world is not disturbed." (Since Bryan gave this address hundreds of times, I assume that people tried to explain to him the difference between laws and events, or reminded him that without gravity, our raised foot would go off into space. I must conclude that he didn't care because the line conveyed a certain rhetorical oomph.) He also explicitly defended the suppression of understanding in the service of moral good:

> If you ask me if I understand everything in the Bible, I answer no, but if we will try to live up to what we do understand, we will be kept so busy doing good that we will not have time to worry about the passages which we do not understand.

This attitude continually puzzled his friends and provided fodder for his enemies. One detractor wrote: "By much talking and little thinking his mentality ran dry." To the same effect, but with kindness, a friend and supporter wrote that Bryan was "almost unable to think in the sense in which you and I use that word. Vague ideas floated through his mind but did not unite to form any system or crystallize into a definite practical position."

Bryan's long-standing approach to evolution rested upon a threefold error. First, he made the common mistake of confusing the fact of evolution with the Darwinian explanation of its mechanism. He then misinterpreted natural selection as a martial theory of survival by battle and destruction of enemies. Finally, he made the logical error of arguing that Darwinism implied the moral virtuousness of such deathly struggle. He wrote in the *Prince of Peace* (1904):

> The Darwinian theory represents man as reaching his present perfection by the operation of the law of hate—the merciless law by which the strong crowd out and kill off the weak. If this is the law of our development then, if there is any logic that can bind the human mind, we shall turn backward toward the beast in proportion as we substitute the law of love. I prefer to believe that love rather than hatred is the law of development.

And to the sociologist E. A. Ross, he said in 1906 that "such a conception of man's origin would weaken the cause of democracy and strengthen class pride and the power of wealth." He persisted in this uneasiness until World War I, when two events galvanized him into frenzied action. First, he learned that the martial view of Darwinism had been invoked by most German intellectuals and military leaders as a justification for war and future domination. Second, he feared the growth of skepticism at home, particularly as a source of possible moral weakness in the face of German militarism.

Bryan united his previous doubts with these new fears into a campaign against evolution in the classroom. We may question the quality of his argument, but we cannot deny that he rooted his own justifications in his lifelong zeal for progressive causes. In this crucial sense, his last hurrah does not nullify, but rather continues, all the applause that came before. Consider the three principal foci of his campaign, and their links to his populist past:

1. For peace and compassion against militarism and murder. "I learned," Bryan wrote, "that it was Darwinism that was at the basis of that damnable doctrine that might makes right that had spread over Germany."

2. For fairness and justice toward farmers and workers and against exploitation for monopoly and profit. Darwinism, Bryan argued, had convinced so many entrepreneurs about the virtue of personal gain that government now had to protect the weak and poor from an explosion of anti-Christian moral decay: "In the United States," he wrote,

> pure-food laws have become necessary to keep manufacturers from poisoning their customers; child labor laws have become necessary to keep employers from dwarfing the bodies, minds and souls of children; anti-trust laws have become necessary to keep overgrown corporations from strangling smaller competitors, and we are still in a death grapple with profiteers and gamblers in farm products.

3. For absolute rule of majority opinion against imposing elites. Christian belief still enjoyed widespread majority support in America, but college education was eroding a consensus that once ensured compassion within democracy. Bryan cited studies showing that only 15 percent of college male freshmen harbored doubts about God, but that 40 percent of graduates had become skeptics. Darwinism, and its immoral principle of domination by a selfish elite, had fueled this skepticism. Bryan railed against this

insidious undermining of morality by a minority of intellectuals, and he vowed to fight fire with fire. If they worked through the classroom, he would respond in kind and ban their doctrine from the public schools. The majority of Americans did not accept human evolution, and had a democratic right to proscribe its teaching.

Let me pass on this third point. Bryan's contention strikes at the heart of academic freedom, and I have often treated this subject in previous essays. Scientific questions cannot be decided by majority vote. I merely record that Bryan embedded his curious argument in his own concept of populism. "The taxpayers," he wrote,

> have a right to say what shall be taught . . . to direct or dismiss those whom they employ as teachers and school authorities. . . . The hand that writes the paycheck rules the school, and a teacher has no right to teach that which his employers object to.

But what of Bryan's first two arguments about the influence of Darwinism on militarism and domestic exploitation? We detect the touch of the Philistine in Bryan's claims, but I think we must also admit that he had identified something deeply troubling— and that the fault does lie partly with scientists and their acolytes.

Bryan often stated that two books had fueled his transition from laissez-faire to vigorous action: *Headquarters Nights,* by Vernon L. Kellogg (1917), and *The Science of Power,* by Benjamin Kidd (1918). I fault Harvard University for many things, but all are overbalanced by its greatest glory—its unparalleled resources. Half an hour after I needed these obscure books if I ever hoped to hold the key to Bryan's activities, I had extracted them from the depths of Widener Library. I found them every bit as riveting as Bryan had, and I came to understand his fears, even to agree in part (though not, of course, with his analysis or his remedies).

Vernon Kellogg was an entomologist and perhaps the leading teacher of evolution in America (he held a professorship at Stanford and wrote a major textbook, *Evolution and Animal Life,* with his mentor and Darwin's leading disciple in America, David Starr Jordan, ichthyologist and president of Stanford University). During the First World War, while America maintained official neutrality, Kellogg became a high official in the international, nonpartisan effort for Belgian relief, a cause officially "tolerated" by Germany. In this capacity, he was posted at the headquarters of the German Great General Staff, the only American on the premises. Night after night, he listened to dinner discussions and arguments, sometimes in the presence of the Kaiser himself, among Germany's highest military officers. *Headquarters Nights* is Kellogg's account of these exchanges. He arrived in Europe as a pacifist, but left committed to the destruction of German militarism by force.

Kellogg was appalled, above all, at the justification for war and German supremacy advanced by these officers, many of whom had been university professors before the war. They not only proposed an evolutionary rationale but advocated a particularly crude form of natural selection, defined as inexorable, bloody battle:

> Professor von Flussen is Neo-Darwinian, as are most German biologists and natural philosophers. The creed of the *Allmacht* ["all might" or omnipotence] of a natural selection based on violent and competitive struggle is the gospel of the German intellectuals; all else is illusion and anathema.
> . . . This struggle not only must go on, for that is the natural law, but it should go on so that this natural law may work out in its cruel, inevitable way the salvation of the human species. . . . That human group which is in the most advanced evolutionary stage . . . should win in the struggle for existence, and this struggle should occur precisely that the various types may be tested, and the best not only preserved, but put in position to impose its kind of social organization—its *Kultur*—on the others, or, alternatively, to destroy and replace them. This is the disheartening kind of argument that I faced at Headquarters. . . . Add the additional assumption that the Germans are the chosen race, and that German social and political organization the chosen type of human community life, and you have a wall of logic and conviction that you can break your head against but can never shatter—by headwork. You long for the muscles of Samson.

Kellogg, of course, found in this argument only "horrible academic casuistry and . . . conviction that the individual is nothing, the state everything." Bryan conflated a perverse interpretation with the thing itself and affirmed his worst fears about the polluting power of evolution.

Benjamin Kidd was an English commentator highly respected in both academic and lay circles. His book *Social Evolution* (1894) was translated into a dozen languages and as widely read as anything ever published on the implications of evolution. In *The Science of Power* (1918), his posthumous work, Kidd constructs a curious argument that, in a very different way from Kellogg's, also fueled Bryan's dread. Kidd, a philosophical idealist, believed that life must move toward progress by rejecting material struggle and individual benefit. Like the German militarists, but to excoriate rather than to praise, Kidd identified Darwinism with these impediments to progress. In a chapter entitled "The Great Pagan Retrogression," Kidd presented a summary of his entire thesis:

1. Darwin's doctrine of force rekindled the most dangerous of human tendencies—our pagan soul, previously (but imperfectly) suppressed for centuries by Christianity and its doctrines of love and renunciation:

> The hold which the theories of the *Origin of Species* obtained on the popular mind in the West is one of the most remarkable incidents in the history of human thought. . . . Everywhere throughout civilization an almost inconceivable influence was given to the doctrine of force as the basis of legal authority. . . .
>
> For centuries the Western pagan had struggled with the ideals of a religion of subordination and renunciation coming to him from the past. For centuries he had been bored almost beyond endurance with ideals of the world presented to him by the Churches of Christendom. . . . But here was a conception of life which stirred to its depths the inheritance in him from past epochs of time. . . . This was the world which the masters of force comprehended. The pagan heart of the West sang within itself again in atavistic joy.

2. In England and America, Darwinism's worst influence lay in its justification for industrial exploitation as an expression of natural selection ("social Darwinism" in its pure form):

> The prevailing social system, born as it had been in struggle, and resting as it did in the last resort on war and on the toil of an excluded proletariat, appeared to have become clothed with a new and final kind of authority.

3. In Germany, Darwin's doctrine became a justification for war:

> Darwin's theories came to be openly set out in political and military textbooks as the full justification for war and highly organized schemes of national policy in which the doctrine of force became the doctrine of Right.

4. Civilization can only advance by integration: The essence of Darwinism is division by force for individual advantage. Social progress demands the "subordination of the individual to the universal" via "the iron ethic of Renunciation."

5. Civilization can only be victorious by suppressing our pagan soul and its Darwinian justification:

> It is the psychic and spiritual forces governing the social integration in which the individual is being subordinated to the universal which have become the winning forces in evolution.

This characterization of evolution has been asserted in many contexts for nearly 150 years—by German militarists, by Kidd, by hosts of the vicious and the duped, the self-serving and the well-meaning. But it remains deeply and appallingly wrong for three basic reasons.

1. Evolution means only that all organisms are united by ties of genealogical descent. This definition says nothing about the mechanism of evolutionary change: In principle, externally directed upward striving might work as well as the caricatured straw man of bloody Darwinian battle to the death. The objections, then, are to Darwin's theory of natural selection, not to evolution itself.

2. Darwin's theory of natural selection is an abstract argument about a metaphorical "struggle" to leave more offspring in subsequent generations, not a statement about murder and mayhem. Direct elimination of competitors is one pathway to Darwinian advantage, but another might reside in cooperation through social ties within a species or by symbiosis between species. For every act of killing and division, natural selection can also favor cooperation and integration in other circumstances. Nineteenth-century interpreters did generally favor a martial view of selection, but to every militarist, we may counterpose a Prince Kropotkin (see Essay 22), urging that the "real" Darwinism be recognized as a doctrine of integration and "mutual aid."

3. Whatever Darwinism represents on the playing fields of nature (and by representing both murder and cooperation at different times, it upholds neither as nature's principal way), Darwinism implies nothing about moral conduct. We do not find our moral values in the actions of nature. One might argue, as Thomas Henry Huxley did in his famous essay "Evolution and Ethics," that Darwinism embodies a law of battle, and that human morality must be defined as the discovery of an opposite path. Or one might argue, as grandson Julian did, that Darwinism is a law of cooperation and that moral conduct should follow nature. If two such brilliant and committed Darwinians could come to such opposite opinions about evolution and ethics, I can only conclude that Darwinism offers no moral guidance.

But Bryan made this common threefold error and continually characterized evolution as a doctrine of battle and destruction of the weak, a dogma that undermined any decent morality and deserved banishment from the classroom. In a rhetorical flourish near the end of his "Last Evolution Argument," the final speech that he prepared with great energy, but never had an opportunity to present at the Scopes trial, Bryan proclaimed:

> Again force and love meet face to face, and the question "What shall I do with Jesus?" must be answered. A bloody, brutal doctrine—Evolution—demands, as the rabble did nineteen hundred years ago, that He be crucified.

I wish I could stop here with a snide comment on Bryan as yahoo and a ringing defense for science's proper interpretation of Darwinism. But I cannot, for Bryan was right in one crucial way. Lord only knows, he understood precious little about science, and he wins no medals for logic of argument. But when he said that Darwinism had been widely protrayed as a defense of war, domination, and domestic exploitation, he was right. Scientists would not be to blame for this if we had always maintained proper caution in interpretation and proper humility in resisting the extension of our findings into inappropriate domains. But many of these insidious and harmful misinterpretations had been promoted by scientists. Several of the German generals who traded arguments with Kellogg had been university professors of biology.

Just one example from a striking source. In his "Last Evolution Argument," Bryan charged that evolutionists had misused science to present moral opinions about the social order as though they represented facts of nature.

> By paralyzing the hope of reform, it discourages those who labor for the improvement of man's condition. . . . Its only program for man is scientific breeding, a system under which a few supposedly superior intellects, self-appointed, would direct the mating and the movements of the mass of mankind—an impossible system!

I cannot fault Bryan here. One of the saddest chapters in all the history of science involves the extensive misuse of data to support biological determinism, the claim that social inequalities based on race, sex, or class cannot be altered because they reflect the innate and inferior genetic endowments of the disadvantaged (see my book *The Mismeasure of Man*). It is bad enough when scientists misidentify their own social preferences as facts of nature in their technical writings and even worse when writers of textbooks, particularly for elementary- and high-school students, promulgate these (or any) social doctrines as the objective findings of science.

Two years ago, I obtained a copy of the book that John Scopes used to teach evolution to the children of Dayton, Tennessee—*A Civic Biology*, by George William Hunter (1914). Many writers have looked into this book to read the section on evolution that Scopes taught and Bryan quoted. But I found something disturbing in another chapter that has eluded previous commentators—an egregious claim that science holds the moral answer to questions about mental retardation, or social poverty so misinterpreted. Hunter discusses the infamous Jukes and Kallikaks, the "classic," and false, cases once offered as canonical examples of how bad heredity runs in families. Under the heading "Parasitism and Its Cost to Society—the Remedy," he writes:

> Hundreds of families such as those described above exist today, spreading disease, immorality and crime to all parts of this country. The cost to society of such families is very severe. Just as certain animals or plants become parasitic on other plants or animals, these families have become parasitic on society. They not only do harm to others by corrupting, stealing or spreading disease, but they are actually protected and cared for by the state out of public money. Largely for them the poorhouse and the asylum exist. They take from society, but they give nothing in return. They are true parasites.
>
> If such people were lower animals, we would probably kill them off to prevent them from spreading. Humanity will not allow this, but we do have the remedy of separating the sexes in asylums or other places and in various ways preventing intermarriage and the possibilities of perpetuating such a low and degenerate race.

Bryan had the wrong solution, but he had correctly identified a problem!

Science is a discipline, and disciplines are exacting. All maintain rules of conduct and self-policing. All gain strength, respect, and acceptance by working honorably within their bounds and knowing when transgression upon other realms counts as hubris or folly. Science, as a discipline, tries to understand the factual state of nature and to explain and coordinate these data into general theories. Science teaches us many wonderful and disturbing things—facts that need weighing when we try to develop standards of conduct and ponder the great questions of morals and aesthetics. But science cannot answer these questions alone and cannot dictate social policy.

Scientists have power by virtue of the respect commanded by

the discipline. We may therefore be sorely tempted to misuse that power in furthering a personal prejudice or social goal—why not provide that extra oomph by extending the umbrella of science over a personal preference in ethics or politics? But we cannot, lest we lose the very respect that tempted us in the first place.

If this plea sounds like the conservative and pessimistic retrenching of a man on the verge of middle age, I reply that I advocate this care and restraint in order to demonstrate the enormous power of science. We live with poets and politicians, preachers and philosophers. All have their ways of knowing, and all are valid in their proper domains. The world is too complex and interesting for one way to hold all the answers. Besides, highfalutin morality aside, if we continue to overextend the boundaries of science, folks like Bryan will nail us properly for their own insidious purposes.

We should give the last word to Vernon Kellogg, the great teacher who understood the principle of strength in limits, and who listened with horror to the ugliest misuses of Darwinism. Kellogg properly taught in his textbook (with David Starr Jordan) that Darwinism cannot provide moral answers:

> Some men who call themselves pessimists because they cannot read good into the operations of nature forget that they cannot read evil. In morals the law of competition no more justifies personal, official, or national selfishness or brutality than the law of gravitation justifies the shooting of a bird.

Kellogg also possessed the cardinal trait lacked both by Bryan and by many of his evolutionary adversaries: humility in the face of our profound ignorance about nature's ways, combined with that greatest of all scientific privileges, the joy of the struggle to know. In his greatest book, *Darwinism Today* (1907), Kellogg wrote:

> We are ignorant, terribly, immensely ignorant. And our work is, to learn. To observe, to experiment, to tabulate, to induce, to deduce. Biology was never a clearer or more inviting field for fascinating, joyful, hopeful work.

Amen, brother!

## *Postscript*

As I was writing this essay, I learned of the untimely death from cancer (at age forty-seven) of Federal Judge William R. Overton of Arkansas. Judge Overton presided and wrote the decision in *McLean* v. *Arkansas* (January 5, 1982), the key episode that led to our final victory in the Supreme Court in June 1987. In this decision, he struck down the Arkansas law mandating equal time for "creation science." This precedent encouraged Judge Duplantier to strike down the similar Louisiana law by summary judgment (without trial). The Supreme Court then affirmed this summary judgment in their 1987 decision. (Since Arkansas and Louisiana had passed the only anti-evolution statutes in the country, these decisions close the issue.) Judge Overton's brilliant and beautifully crafted decision is the finest legal document ever written about this question—far surpassing anything that the Scopes trial generated, or any document arising from the two Supreme Court cases (*Epperson* v. *Arkansas* of 1968, striking down Scopes-era laws that banned evolution outright, and the 1987 decision banning the "equal time" strategy). Judge Overton's definitions of science are so cogent and clearly expressed that we can use his words as a model for our own proceedings. *Science,* the leading journal of American professional science, published Judge Overton's decision verbatim as a major article.

I was a witness in *McLean* v. *Arkansas* (see Essay 21 in *Hen's Teeth and Horse's Toes*). I never spoke to Judge Overton personally, and I spent only part of a day in his courtroom. Yet, when I fell ill with cancer the next year, I learned from several sources that Judge Overton had heard and had inquired about my health from mutual acquaintances, asking that his best wishes be conveyed to me. I mourn the passing of this brilliant and compassionate man, and I dedicate this essay to his memory.

# 29 | An Essay on a Pig Roast

ON INDEPENDENCE DAY, 1919, in Toledo, Ohio, Jack Dempsey won the heavyweight crown by knocking out Jess Willard in round three. (Willard, the six-foot six-inch wheat farmer from Kansas, was "the great white hope" who, four years earlier in Havana, had finally KO'd Jack Johnson, the first black heavyweight champ, and primary thorn in the side of racist America.) Dempsey ruled the ring for seven years, until Gene Tunney whipped him in 1926.

Yet during Dempsey's domination of pugilism in its active mode, some mighty impressive fighters were squaring away on other, less physical but equally contentious turfs. One prominent battle occurred entirely within Dempsey's reign, beginning with William Jennings Bryan's decision in 1920 to launch a nationwide legislative campaign against the teaching of evolution and culminating in the Scopes trial of 1925. The main bout may have pitted Bryan against Clarence Darrow at the trial itself, but a preliminary skirmish in 1922, before any state legislature had passed an anti-evolution law, had brought two equally formidable foes together—Bryan again, but this time against Henry Fairfield Osborn, head of the American Museum of Natural History. In some respects, the Bryan-Osborn confrontation was more dramatic than the famous main event three years later. One can hardly imagine two more powerful but more different men: the arrogant, patrician, archconservative Osborn versus the folksy "Great Commoner" from Nebraska. Moreover, while Darrow maintained a certain respect based on genuine affection for

432

Bryan (or at least for his earlier greatness), I detect nothing but pure venom and contempt from Osborn.

The enemy within, as the old saying goes, is always more dangerous than the enemy without. An atheist might have laughed at Bryan or merely felt bewildered. But Osborn was a dedicated theist and a great paleontologist who viewed evolution as the finest expression of God's intent. For Osborn, Bryan was perverting both science and the highest notion of divinity. (Darrow later selected Osborn as one of his potential witnesses in the Scopes trial not only because Osborn was so prominent, socially as well as scientifically, but primarily because trial strategy dictated that religiously devout evolutionists could blunt Bryan's attack on science as intrinsically godless.)

On February 26, 1922, Bryan published an article in the Sunday *New York Times* to further his legislative campaign against the teaching of evolution. Bryan, showing some grasp of the traditional parries against Darwin, but constantly confusing doubts about the mechanism of natural selection with arguments against the fact of evolution itself, rested his case upon a supposed lack of direct evidence for the claims of science:

> The real question is, Did God use evolution as His plan? If it could be shown that man, instead of being made in the image of God, is a development of beasts we would have to accept it, regardless of its effect, for truth is truth and must prevail. But when there is no proof we have a right to consider the effect of the acceptance of an unsupported hypothesis.

The *Times*, having performed its civic duty by granting Bryan a platform, promptly invited Osborn to prepare a reply for the following Sunday. Osborn's answer, published on March 5 and reissued as a slim volume by Charles Scribner's Sons under the title *Evolution and Religion*, integrated two arguments into a single thesis: The direct, primarily geological evidence for evolution is overwhelming, and evolution is not incompatible with religion in any case. As a motto for his approach, and a challenge to Bryan from a source accepted by both men as unimpeachable, Osborn cited a passage from Job (12:8): ". . . speak to the earth, and it shall teach thee." When, on the eve of the Scopes trial, Osborn expanded his essay into a longer attack on Bryan, he dedicated the new book to John Scopes and chose a biting parody of Job for his title—*The Earth Speaks to Bryan* (Charles Scribner's Sons, 1925).

When a man poses such a direct challenge to an adversary, nothing could possibly be more satisfying than a quick confirmation from an unanticipated source. On February 25, 1922, just the day before Bryan's *Times* article, Harold J. Cook, a rancher and consulting geologist, had written to Osborn:

> I have had here, for some little time, a molar tooth from the Upper, or Hipparion phase of the Snake Creek Beds, that very closely approaches the human type. . . . Inasmuch as you are particularly interested in this problem and, in collaboration with Dr. Gregory and others, are in the best position of anyone to accurately determine the relationships of this tooth, if it can be done, I will be glad to send it on to you, should you care to examine and study it.

In those bygone days of an efficient two-penny post, Osborn probably received this letter on the very morning following Bryan's diatribe or, at most, a day or two later. Osborn obtained the tooth itself on March 14, and, with his usual precision (and precisely within the ten-word limit for the basic rate), promptly telegraphed Cook: "Tooth just arrived safely. Looks very promising. Will report immediately." Later that day, Osborn wrote to Cook:

> The instant your package arrived, I sat down with the tooth, in my window, and I said to myself: "It looks one hundred per cent anthropoid." I then took the tooth into Dr. Matthew's room and we have been comparing it with all the books, all the casts and all the drawings, with the conclusion that it is the last right upper molar tooth of some higher Primate. . . . We may cool down tomorrow, but it looks to me as if the first anthropoid ape of America has been found.

But Osborn's enthusiasm only warmed as he studied the tooth and considered the implications. The human fossil record had improved sufficiently to become a source of strength, rather than an embarrassment, to evolutionists, with Cro-Magnon and Neanderthal in Europe (not to mention the fraudulent Piltdown, then considered genuine and strongly supported by Osborn) and *Pithecanthropus* (now called *Homo erectus*) in East Asia. But no fossils of higher apes or human ancestors had ever been found anywhere in the Americas. This absence, in itself, posed no special problem to evolutionists. Humans had evolved in Asia or Africa, and the Americas were an isolated world, accessible primarily by a difficult route of migration over the Bering land bridge. Indeed, to this day, ancient humans are unknown in the New World, and most anthropologists accept a date of 20,000 years or considerably less (probably more like 11,000) for the first peopling of our hemisphere. Moreover, since these first immigrants were members of our stock, *Homo sapiens,* no ancestral species have ever been found—and none probably ever will—in the Americas.

Still, an American anthropoid would certainly be a coup for Osborn's argument that the earth spoke to Bryan in the language of evolution, not to mention the salutary value of a local product for the enduring themes of hoopla, chauvinism, and flag-waving.

Therefore, Osborn's delight—and his confidence—in this highly worn and eroded molar tooth only increased. Within a week or two, he was ready to proclaim the first momentous discovery of a fossil higher primate, perhaps even a direct human ancestor, in America. He honored our hemisphere in choosing the name *Hesperopithecus,* or "ape of the western world." On April 25, less than two months after Bryan's attack, Osborn presented *Hesperopithecus* in two simultaneous papers with the same title and different content: one in the prestigious *Proceedings of the National Academy of Sciences;* the other, containing figures and technical descriptions, in the *Novitates of the American Museum of Natural History*—"*Hesperopithecus,* the First Anthropoid Primate Found in America."

*Hesperopithecus* was good enough news in the abstract, but Osborn particularly exulted in the uncannily happy coincidences of both time and place. Cook had probably written his letter at the very moment that the compositors were setting Bryan's oratory in type. Moreover, for the crowning irony, *Hesperopithecus* had been found in Nebraska—home state of the Great Commoner! If God had permitted a paleontologist to invent a fossil with maximal potential to embarrass Bryan, no one could have bettered *Hesperopithecus* for rhetorical impact. Needless to say, this preciously ironical situation was not lost on Osborn, who inserted the following gloat of triumph into his article for the staid *Proceedings*—about as incongruous in this forum as the erotic poetry of the Song of Songs between Ecclesiastes and Isaiah.

> It has been suggested humorously that the animal should be named *Bryopithecus* after the most distinguished Primate which the State of Nebraska has thus far produced. It is certainly singular that this discovery is announced within six weeks of the day (March 5, 1922) that the author advised William Jennings Bryan to consult a certain passage in the Book of Job, "Speak to the earth and it shall teach thee," and it is a remarkable coincidence that the first earth to speak on this subject is the sandy earth of the Middle Pliocene Snake Creek deposits of western Nebraska.

Old Robert Burns certainly knew his stuff when he lamented the frequent unraveling of the best laid plans of mice and men. Unless you browse in the marginal genre of creationist tracts, you will probably not have encountered *Hesperopithecus* in anything written during the past fifty years (except, perhaps, as a cautionary sentence in a textbook or a paragraph on abandoned hopes in a treatise on the history of science). The reign of *Hesperopithecus* was brief and contentious. In 1927, Osborn's colleague William King Gregory, the man identified in Cook's original letter as the best-qualified expert on primate teeth, threw in the towel with an article in *Science:* "*Hesperopithecus* Apparently not an Ape nor a Man." Expeditions sent out by Osborn in the summers of 1925 and 1926 to collect more material of *Hesperopithecus,* and to test the hypothesis of primate affinity, had amassed a large series to complement the original tooth. But this abundance also doomed Osborn's interpretation—for the worn and eroded *Hesperopithecus* tooth, when compared with others in better and more diagnostic condition, clearly belonged not to a primate but to the extinct peccary *Prosthennops.*

One can hardly blame modern creationists for making hay of this brief but interesting episode in paleontology. After all, they're only getting their fair licks at Osborn, who used the original interpretation to ridicule and lambaste their erstwhile champion Bryan. I don't think I have ever read a modern creationist tract that doesn't feature the tale of "Nebraska Man" in a feint from our remarkable record of genuine human fossils, and an attempted KO of evolution with the one-two punch of Piltdown and *Hesperopithecus*. I write this essay to argue that Nebraska man tells a precisely opposite tale, one that should give creationists pause (though I do admit the purely rhetorical value of a proclaimed primate ancestor later exposed as a fossil pig).

The story of *Hesperopithecus* was certainly embarrassing to Osborn and Gregory in a personal sense, but the sequence of discovery, announcement, testing, and refutation—all done with admirable dispatch, clarity, and honesty—shows science working at its very best. Science is a method for testing claims about the natural world, not an immutable compendium of absolute truths. The fundamentalists, by "knowing" the answers before they start, and then forcing nature into the straitjacket of their discredited preconceptions, lie outside the domain of science—or of any honest intellectual inquiry. The actual story of *Hesperopithecus* could teach creationists a great deal about science as properly practiced if they chose to listen, rather than to scan the surface for cheap shots in the service of debate pursued for immediate advantage, rather than interest in truth.

When we seek a textbook case for the proper operation of science, the correction of certain error offers far more promise than the establishment of probable truth. Confirmed hunches, of course, are more upbeat than discredited hypotheses. Since the worst traditions of "popular" writing falsely equate instruction with sweetness and light, our promotional literature abounds with insipid tales in the heroic mode, although tough stories of disappointment and loss give deeper insight into a methodology that the celebrated philosopher of science Karl Popper once labeled as "conjecture and refutation."

Therefore, I propose that we reexamine the case of Nebraska Man, not as an embarrassment to avoid in polite company, but as an exemplar complete with lessons and ample scope for the primary ingredient of catharsis and popular appeal—the opportunity to laugh at one's self. Consider the story as a chronological sequence of five episodes:

1. *Proposal.* Harold Cook's fossil tooth came from a deposit

about 10 million years old and filled with mammals of Asiatic ancestry. Since paleontologists of Osborn's generation believed that humans and most other higher primates had evolved in Asia, the inclusion of a fossil ape in a fauna filled with Asian migrants seemed entirely reasonable. Osborn wrote to Cook a month before publication:

> The animal is certainly a new genus of anthropoid ape, probably an animal which wandered over here from Asia with the large south Asiatic element which has recently been discovered in our fauna. . . . It is one of the greatest surprises in the history of American paleontology.

Osborn then announced the discovery of *Hesperopithecus* in three publications—technical accounts in the *American Museum Novitates* (April 25, 1922) and in the British journal *Nature* (August 26, 1922), and a shorter notice in the *Proceedings of the National Academy of Sciences* (August 1922, based on an oral report delivered in April).

2. *Proper doubt and statement of alternatives.* Despite all the hoopla and later recrimination, Osborn never identified *Hesperopithecus* as a human ancestor. The tooth had been heavily worn during life, obliterating the distinctive pattern of cusps and crown. Considering both this extensive wear and the further geological erosion of the tooth following the death of its bearer, Osborn knew that he could make no certain identification. He did not cast his net of uncertainty widely enough, however, for he labeled *Hesperopithecus* as an undoubted higher primate. But he remained agnostic about the crucial issue of closer affinity with the various ape branches or the human twig of the primate evolutionary tree.

Osborn described the tooth of *Hesperopithecus* as "a second or third upper molar of the right side of a new genus and species of anthropoid." Osborn did lean toward human affinity, based both on the advice of his colleague Gregory (see point three below) and, no doubt, on personal hope and preference: "On the whole, we think its nearest resemblances are with . . . men rather than with apes." But his formal description left this crucial question entirely open:

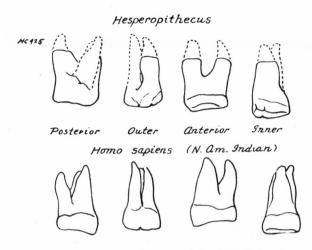

An illustration from Osborn's article of 1922, showing the strong similarity between worn teeth of *Hesperopithecus* and modern humans. NEG. NO. 2A17804. COURTESY DEPARTMENT OF LIBRARY SERVICES, AMERICAN MUSEUM OF NATURAL HISTORY.

The *Hesperopithecus* molar cannot be said to resemble any known type of human molar very closely. It is certainly not closely related to *Pithecanthropus erectus* in the structure of the molar crown. . . . It is therefore a new and independent type of Primate, and we must seek more material before we can determine its relationships.

3. *Encouragement of further study.* If Osborn had been grandstanding with evidence known to be worthless or indecipherable, he would have made his public point and then shut up after locking his useless or incriminating evidence away in a dark drawer in the back room of a large museum collection. Osborn proceeded in exactly the opposite way. He did everything possible to encourage further study and debate, hoping to resolve his own strong uncertainties. (Osborn, by the way, was probably the most pompous, self-assured S.O.B. in the history of American paleontology, a regal patrician secure in his birthright, rather than a scrappy, self-made man. He once published a book devoted en-

tirely to photographs of his medals and awards and to a list of his publications; as an excuse for such vanity, he claimed that he harbored only a selfless desire to inspire young scientists by illustrating the potential rewards of a fine profession. "Osborn stories" are still told by the score wherever vertebrate paleontologists congregate. And when a man's anecdotes outlive him by more than half a century, you know that he was larger than life. Thus, the real news about *Hesperopithecus* must be that, for once, Osborn was expressing genuine puzzlement and uncertainty.)

In any case, Osborn reached out to colleagues throughout the world. He made numerous casts of *Hesperopithecus* and sent them to twenty-six universities and museums in Europe and North America. As a result, he was flooded with alternative interpretations from the world's leading paleoanthropologists. He received sharp criticisms from both sides: from Arthur Smith Woodward, describer of Piltdown, who thought that *Hesperopithecus* was a bear (and I don't mean metaphorically), and from G. Elliot Smith, another "hero" of Piltdown, who became too enthusiastic about the humanity of Osborn's tooth, causing considerable later embarrassment and providing creationists with their "hook." Osborn tried to rein both sides in, beginning his *Nature* article with these words:

> Every discovery directly or indirectly relating to the pre-history of man attracts world-wide attention and is apt to be received either with too great optimism or with too great incredulity. One of my friends, Prof. G. Elliot Smith, has perhaps shown too great optimism in his most interesting newspaper and magazine articles on *Hesperopithecus,* while another of my friends, Dr. A. Smith Woodward, has shown too much incredulity.

Moreover, Osborn immediately enlisted his colleague W. K. Gregory, the acknowledged local expert on primate teeth, to prepare a more extensive study of *Hesperopithecus,* including a formal comparison of the tooth with molars of all great apes and human fossils. Gregory responded with two detailed, technical articles, both published in 1923 with the collaboration of Milo Hellman.

Gregory followed Osborn in caution and legitimate expression of doubt. He began his first article by dividing the characters of the tooth into three categories: those due to wear, to subsequent erosion, and to the genuine taxonomic uniqueness of *Hesperopithecus*. Since the first two categories, representing information lost, tended to overwhelm the last domain of diagnostic biology, Gregory could reach no conclusion beyond a basic placement among the higher primates:

> The type of *Hesperopithecus haroldcookii* represents a hitherto unknown form of higher primates. It combines characters seen in the molars of the chimpanzee, of *Pithecanthropus,* and of man, but, in view of the extremely worn and eroded state of the crown, it is hardly safe to affirm more than that *Hesperopithecus* was structurally related to all three.

In the second and longer article, Gregory and Hellman stuck their necks out a bit more—but in opposite directions. Hellman opted for the human side; Gregory for affinity with "the gorilla-chimpanzee group."

4. *Gathering of additional data.* Osborn knew, of course, that a worn and eroded tooth would never resolve the dilemma of *Hesperopithecus,* no matter how many casts were made or how many paleontologists peered down their microscopes. The answers lay in more data buried in the sands of Nebraska, and Osborn pledged, in his diatribe against Bryan, to make the earth speak further:

> What shall we do with the Nebraska tooth? Shall we destroy it because it jars our long preconceived notion that the family of manlike apes never reached the Western world, or shall we endeavor to interpret it, to discover its real relationship to the apes of Asia and of the more remote Africa. Or shall we continue our excavations, difficult and baffling as they are, in the confident hope, inspired by the admonition of Job, that if we keep on speaking to the earth we shall in time have a more audible and distinct reply [from *The Earth Speaks to Bryan,* p. 43].

To his professional audiences in *Nature,* Osborn made the same pledge with more detail: "We are this season renewing the search with great vigor and expect to run every shovelful of loose river sand which comprises the deposit through a sieve of mesh fine enough to arrest such small objects as these teeth."

Thus, in the summers of 1925 and 1926, Osborn sent a collecting expedition, led by Albert Thomson, to the Snake Creek beds of Nebraska. Several famous paleontologists visited the site and pitched in, including Barnum Brown, the great dinosaur collector; Othenio Abel of Vienna (a dark figure who vitiated the memory of his fine paleontological work by later activity in the Austrian Nazi party); and Osborn himself. They found abundant material to answer their doubts. The earth spoke both audibly and distinctly, but not in the tones that Osborn had anticipated.

5. *Retraction.* After all this buildup and detail, the denouement can only be described as brief, simple, and conclusive. The further expeditions were blessed with success. Abundant new specimens destroyed Osborn's dream for two reasons that could scarcely be challenged. First, the new specimens formed a series from teeth worn as profoundly as *Hesperopithecus* to others of the same species with crown and cusps intact. The diagnostic pattern of the unworn teeth proclaimed pig rather than primate. Second, the unworn teeth could not be distinguished from premolars firmly residing in a peccary's palate found during a previous expedition. Osborn, who was never praised for a charitable nature, simply shut up and never mentioned *Hesperopithecus* again in his numerous succeeding articles on human ancestry. He had enjoyed the glory, but he let Gregory take the heat in a forthright retraction published in *Science* (December 16, 1927):

> Among other material the expedition secured a series of specimens which have led the writer to doubt his former identification of the type as the upper molar of an extinct primate, and to suspect that the type specimen of *Hesperopithecus haroldcookii* may be an upper premolar of a species of *Prosthennops,* an extinct genus related to modern peccaries.

Why should the detractors of science still be drawing such mileage from this simple story of a hypothesis swiftly refuted by science working well? I would divide the reasons into red herrings and a smaller number of allowable points. The red herrings all center on rhetorical peculiarities that anyone skilled in debate could use to advantage. (Debate, remember, is an art form dedicated to the winning of arguments. Truth is one possible weapon, rarely the best, in such an enterprise.) Consider three good lines:

1. "How can you believe those evolutionists if they can make monkeys out of themselves by calling a pig a monkey?" As a trope of rhetoric, given the metaphorical status of pigs in our culture, the true affinity of *Hesperopithecus* became a blessing for creationists. What could possibly sound more foolish than the misidentification of a pig as a primate. My side might have been better off if *Hesperopithecus* had been, say, a deer or an antelope (both members of the order Artiodactyla, along with pigs, and therefore equally far from primates).

Yet anyone who has studied the dental anatomy of mammals knows immediately that this seemingly implausible mix-up of pig for primate is not only easy to understand but represents one of the classic and recurring confusions of the profession. The cheek teeth of pigs and humans are astonishingly and uncannily similar. (I well remember mixing them up more than once in my course on mammalian paleontology, long before I had ever heard the story of *Hesperopithecus*.) Unworn teeth can be told apart by details of the cusps, but isolated and abraded teeth of older animals are very difficult to distinguish. The *Hesperopithecus* tooth, worn so flat and nearly to the roots, was a prime candidate for just such a misidentification.

A wonderfully ironic footnote to this point was unearthed by John Wolf and James S. Mellett in an excellent article on Nebraska Man that served as the basis for my researches (see bibliography). In 1909, the genus *Prosthennops* was described by W. D. Matthew, Osborn's other paleontological colleague at the American Museum of Natural History, and—guess who—the same Harold Cook who would find *Hesperopithecus* ten years later. They explicitly warned their colleagues about the possible confusion of these peccary teeth with the dentition of primates:

> The anterior molars and premolars of this genus of peccaries show a startling resemblance to the teeth of Anthropoidea, and might well be mistaken for them by anyone not familiar with the dentition of Miocene peccaries.

2. "How can you believe those evolutionists if they can base an identification on a single worn tooth?" William Jennings Bryan,

the wily old lawyer, remarked: "These men would destroy the Bible on evidence that would not convict a habitual criminal of a misdemeanor."

My rejoinder may seem like a cavil, but it really isn't. Harold Cook did send but a single tooth to Osborn. (I do not know why he had not heeded his own previous warning of 1909. My guess would be that Cook played no part in writing the manuscript and that Matthew had been sole author of the statement. An old and admirable tradition grants joint authorship to amateur collectors who often find the material that professionals then exploit and describe. Matthew was the pro, Cook the experienced and sharp-eyed local collector.) Osborn sought comparative material in the Museum's collection of fossil mammals and located a very similar tooth found in the same geological strata in 1908. He added this second tooth to the sample and based the genus *Hesperopithecus* on both specimens. (This second tooth had been found by W. D. Matthew, and we must again raise the question of why Matthew didn't heed his own warning of 1909 about mixing up primates and peccaries. For Osborn showed both teeth to Matthew and won his assent for a probable primate identification. In his original description, Osborn wrote of this second tooth: "The specimen belonged to an aged animal and is so water-worn that Doctor Matthew, while inclined to regard it as a primate, did not venture to describe it.")

Thus the old canard about basing a human reconstruction on a single tooth is false. The sample of *Hesperopithecus* included two teeth from the start. You might say that two is only minimally better than one, and still so far from a whole animal that any conclusion must be risible. Not so. One of anything can be a mistake, an oddball, an isolated peculiarity; two, on the other hand, is the beginning of a pattern. Second specimens always provide a great increment of respect. The Piltdown fraud, for example, did not take hold until the forgers concocted a second specimen.

3. "How can you believe those evolutionists if they reconstruct an entire man—hair, skin, and all—from a single tooth?" On this issue, Osborn and Gregory were unjustly sandbagged by an over-zealous colleague. In England, G. Elliot Smith collaborated with the well-known scientific artist Amedee Forestier to produce a graphic reconstruction of a *Hesperopithecus* couple in a forest sur-

The infamous restoration of *Hesperopithecus* published in the *Illustrated London News* in 1922. NEG. NO. 2A17487. COURTESY DEPARTMENT OF LIBRARY SERVICES, AMERICAN MUSEUM OF NATURAL HISTORY.

rounded by other members of the Snake Creek fauna. Forestier, of course, could learn nothing from the tooth and actually based his reconstruction on the conventional rendering of *Pithecanthropus*, or Java man.

Forestier's figure is the one ridiculed and reproduced by creationists, and who can blame them? The attempt to reconstruct an entire creature from a single tooth is absolute folly—especially in this case when the authors of *Hesperopithecus* had declined to decide whether their creature was ape or human. Osborn had explicitly warned against such an attempt by pointing out how organs evolve at different rates, and how teeth of one type can be found in bodies of a different form. (Ironically, he cited Piltdown as an example of this phenomenon, arguing that, by teeth alone, the "man" would have been called an ape. How prescient in retrospect, since Piltdown is a fraud made of orangutan teeth and a human skull.)

Thus, Osborn explicitly repudiated the major debating point continually raised by modern creationists—the nonsense of reconstructing an entire creature from a single tooth. He said so obliquely and with gentle satire in his technical article for *Nature,* complaining that G. E. Smith had shown "too great optimism in his most interesting newspaper and magazine articles on *Hesperopithecus.*" The *New York Times* reported a more direct quotation: "Such a drawing or 'reconstruction' would doubtless be only a figment of the imagination of no scientific value, and undoubtedly inaccurate."

Among the smaller number of allowable points, I can hardly blame creationists for gloating over the propaganda value of this story, especially since Osborn had so shamelessly used the original report to tweak Bryan. Tit for tat.

I can specify only one possibly legitimate point of criticism against Osborn and Gregory. Perhaps they were hasty. Perhaps they should have waited and not published so quickly. Perhaps they should have sent out their later expedition before committing anything to writing, for then the teeth would have been officially identified as peccaries first, last, and always. Perhaps they proceeded too rapidly because they couldn't resist such a nifty opportunity to score a rhetorical point at Bryan's expense. I am not bothered by the small sample of only two teeth. Single teeth, when well preserved, can be absolutely diagnostic of a broad taxonomic group. The argument for caution lay in the worn and eroded character of both premolars. Matthew had left the second tooth of 1908 in a museum drawer; why hadn't Osborn shown similar restraint?

But look at the case from a different angle. The resolution of *Hesperopithecus* may have been personally embarrassing for Osborn and Gregory, but the denouement was only invigorating and positive for the institution of science. A puzzle had been noted and swiftly solved, though not in the manner anticipated by the original authors. In fact, I would argue that Osborn's decision to publish, however poor his evidence and tentative his conclusions, was the most positive step he could have taken to secure a resolution. The published descriptions were properly cautious and noncommittal. They focused attention on the specimens, provided a series of good illustrations and measurements, provoked a rash of hypotheses for interpretation, and inspired the subsequent study and collection that soon resolved the issue. If Osborn had left the molar in a museum drawer, as Matthew had for the second tooth found in 1908, persistent anomaly would have been the only outcome. Conjecture and refutation is a chancy game with more losers than winners.

I have used the word *irony* too may times in this essay, for the story of *Hesperopithecus* is awash in this quintessential consequence of human foibles. But I must beg your indulgence for one last round. As their major pitch, modern fundamentalists argue that their brand of biblical literalism represents a genuine discipline called "scientific creationism." They use the case of Nebraska Man, in their rhetorical version, to bolster this claim, by arguing that conventional science is too foolish to merit the name and that the torch should pass to them.

As the greatest irony of all, they could use the story of *Hesperopithecus,* if they understood it properly, to advance their general argument. Instead, they focus on their usual ridicule and rhetoric, thereby showing their true stripes even more clearly. The real message of *Hesperopithecus* proclaims that science moves forward by admitting and correcting its errors. If creationists really wanted to ape the procedures of science, they would take this theme to heart. They would hold up their most ballyhooed, and now most thoroughly discredited, empirical claim—the coexistence of dinosaur and human footprints in the Paluxy Creek beds near Dallas—and publicly announce their error and its welcome correction. (The supposed human footprints turn out to be either random depressions in the hummocky limestone surface or partial dinosaur heel prints that vaguely resemble a human foot when the dinosaur toe strikes are not preserved.) But the world of creationists is too imbued with irrefutable dogma, and they don't seem able even to grasp enough about science to put up a good show in imitation.

I can hardly expect them to seek advice from me. May they, therefore, learn the virtue of admitting error from their favorite source of authority, a work so full of moral wisdom and intellectual value that such a theme of basic honesty must win special prominence. I remind my adversaries, in the wonderful mixed metaphors of Proverbs (25:11,14), that "a word fitly spoken is like apples of gold in pictures of silver. . . . Whoso boasteth himself of a false gift is like clouds and wind without rain."

# 30 | Justice Scalia's Misunderstanding

CHARLES LYELL, defending both his version of geology and his designation of James Hutton as its intellectual father, described Richard Kirwan as a man "who possessed much greater authority in the scientific world than he was entitled by his talents to enjoy."

Kirwan, chemist, mineralogist, and president of the Royal Academy of Dublin, did not incur Lyell's wrath for a mere scientific disagreement, but for saddling Hutton with the most serious indictment of all—atheism and impiety. Kirwan based his accusations on the unlikely charge that Hutton had placed the earth's origin beyond the domain of what science could consider or (in a stronger claim) had even denied that a point of origin could be inferred at all. Kirwan wrote in 1799:

> Recent experience has shown that the obscurity in which the philosophical knowledge of this [original] state has hitherto been involved, has proved too favorable to the structure of various systems of atheism or infidelity, as these have been in their turn to turbulence and immorality, not to endeavor to dispel it by all the lights which modern geological researches have struck out. Thus it will be found that geology naturally ripens . . . into religion, as this does into morality.

In our more secular age, we may fail to grasp the incendiary character of such a charge at the end of the eighteenth century, when intellectual respectability in Britain absolutely demanded an affirmation of religious fealty, and when fear of spreading revolution from France and America equated any departure from orthodoxy with encouragement of social anarchy. Calling someone an atheist in those best and worst of all times invited the same predictable reaction as asking Cyrano how many sparrows had perched up there or standing up in a Boston bar and announcing that DiMaggio was a better hitter than Williams.

Thus, Hutton's champions leaped to his defense, first his contemporary and Boswell, John Playfair, who wrote (in 1802) that

> such poisoned weapons as he [Kirwan] was preparing to use, are hardly ever allowable in scientific contest, as having a less direct tendency to overthrow the system, than to hurt the person of an adversary, and to wound, perhaps incurably, his mind, his reputation, or his peace.

Thirty years later, Charles Lyell was still fuming:

> We cannot estimate the malevolence of such a persecution, by the pain which similar insinuations might now inflict; for although charges of infidelity and atheism must always be odious, they were injurious in the extreme at that moment of political excitement [*Principles of Geology*, 1830].

(Indeed, Kirwan noted that his book had been ready for the printers in 1798 but had been delayed for a year by "the confusion arising from the rebellion then raging in Ireland"—the great Irish peasant revolt of 1798, squelched by Viscount Castlereagh, uncle of Darwin's Captain FitzRoy [see Essay 1 for much more on Castlereagh].)

Kirwan's accusation centered upon the last sentence of Hutton's *Theory of the Earth* (original version of 1788)—the most famous words ever written by a geologist (quoted in all textbooks, and often emblazoned on the coffee mugs and T-shirts of my colleagues):

> The result, therefore, of our present enquiry is, that we find no vestige of a beginning—no prospect of an end.

Kirwan interpreted both this motto, and Hutton's entire argument, as a claim for the earth's eternity (or at least as a statement of necessary agnosticism about the nature of its origin). But if the earth be eternal, then God did not make it. And if we need no God to fashion our planet, then do we need him at all? Even the weaker version of Hutton as agnostic about the earth's origin supported a charge of atheism in Kirwan's view—for if we cannot know that God made the earth at a certain time, then biblical authority is dethroned, and we must wallow in uncertainty about the one matter that demands our total confidence.

It is, I suppose, a testimony to human carelessness and to our tendency to substitute quips for analysis that so many key phrases, the mottoes of our social mythology, have standard interpretations quite contrary to their intended meanings. Kirwan's reading has prevailed. Most geologists still think that Hutton was advocating an earth of unlimited duration—though we now view such a claim as heroic rather than impious.

Yet Kirwan's charge was more than merely vicious—it was dead wrong. Moreover, in understanding why Kirwan erred (and why we still do), and in recovering what Hutton really meant, we illustrate perhaps the most important principle that we can state about science as a way of knowing. Our failure to grasp the principle underlies much public misperception about science. In particular, Justice Scalia's recent dissent in the Louisiana "creation science" case rests upon this error in discussing the character of evolutionary arguments. We all rejoiced when the Supreme Court ended a long episode in American history and voided the last law that would have forced teachers to "balance" instruction in evolution with fundamentalist biblical literalism masquerading under the oxymoron "creation science." I now add a tiny hurrah in postscript by pointing out that the dissenting argument rests, in large part, upon a misunderstanding of science.

Hutton replied to Kirwan's original attack by expanding his 1788 treatise into a cumbersome work, *The Theory of the Earth* (1795). With forty-page quotations in French and repetitive, involuted justifications, Hutton's new work condemned his theory to unreadability. Fortunately, his friend John Playfair, a mathematician and outstanding prose stylist, composed the most elegant pony ever written and published his *Illustrations of the Huttonian Theory of the Earth* in 1802. Playfair presents a two-part refutation for Kirwan's charge of atheism.

1. Hutton neither argued for the earth's eternity nor claimed

that we could say nothing about its origin. In his greatest contribution, Hutton tried to develop a cyclical theory for the history of the earth's surface, a notion to match the Newtonian vision of continuous planetary revolution about the sun. The materials of the earth's surface, he argued, passed through a cycle of perfect repetition in broad scale. Consider the three major stages. First, mountains erode and their products are accumulated as thick sequences of layered sediments in the ocean. Second, sediments consolidate and their weight melts the lower layers, forming magmas. Third, the pressure of these magmas forces the sediments up to form new mountains (with solidified magmas at their core), while the old, eroded continents become new ocean basins. The cycle then starts again as mountains (at the site of old oceans) shed their sediments into ocean basins (at the site of old continents). Land and sea change positions in an endless dance, but the earth itself remains fundamentally the same. Playfair writes:

> It is the peculiar excellence of this theory . . . that it makes the decay of one part subservient to the restoration of another, and gives stability to the whole, not by perpetuating individuals, but by reproducing them in succession.

We can easily grasp the revolutionary nature of this theory for concepts of time. Most previous geologies had envisioned an earth of short duration, moving in a single irreversible direction, as its original mountains eroded into the sea. By supplying a "concept of repair" in his view of magmas as uplifting forces, Hutton burst the strictures of time. No more did continents erode once into oblivion; they could form anew from the products of their own decay and the earth could cycle on and on.

This cyclical theory has engendered the false view that Hutton considered the earth eternal. True, the mechanics of the cycle provide no insight into beginnings or endings, for laws of the cycle can only produce a continuous repetition and therefore contain no notion of birth, death, or even of aging. But this conclusion only specifies that laws of nature's *present order* cannot specify beginnings or ends. Beginnings and ends may exist—in fact, Hutton considered a concept of starts and stops absolutely essential for any rational understanding—but we cannot learn anything about this vital subject from nature's present laws. Hutton, who was a devoted theist despite Kirwan's charge, argued that God had made a beginning, and would ordain an end, by summoning forces outside the current order of nature. For the stable period between, he had ordained laws that impart no directionality and therefore permit no insight into beginnings and ends.

Note how carefully Hutton chose the words of his celebrated motto. "No *vestige* of a beginning" because the earth has been through so many cycles since then that all traces of an original state have vanished. But the earth certainly had an original state. "No *prospect* of an end" because the current laws of nature provide no insight into a termination that must surely occur. Playfair describes Hutton's view of God:

> He may put an end, as he no doubt gave a beginning, to the present system, at some determinate period; but we may safely conclude, that this great catastrophe will not be brought about by any of the laws now existing, and that it is not indicated by any thing which we perceive.

2. Hutton did not view our inability to specify beginnings and ends as a baleful limitation of science but as a powerful affirmation of proper scientific methodology. Let theology deal with ultimate origins, and let science be the art of the empirically soluble.

The British tradition of speculative geology—from Burnet, Whiston, and Woodward in the late seventeenth century to Kirwan himself at the tail end of the eighteenth—had focused upon reconstructions of the earth's origin, primarily to justify the Mosaic narrative as scientifically plausible. Hutton argued that such attempts could not qualify as proper science, for they could only produce speculations about a distant past devoid of evidence to test any assertion (no vestige of a beginning). The subject of origins may be vital and fascinating, far more compelling than the humdrum of quotidian forces that drive the present cycle of uplift, erosion, deposition, and consolidation. But science is not speculation about unattainable ultimates; it is a way of knowing based upon laws now in operation and results subject to observation and inference. We acknowledge limits in order to proceed with power and confidence.

Hutton therefore attacked the old tradition of speculation about the earth's origin as an exercise in futile unprovability. Better to focus upon what we can know and test, leaving aside what the methods of science cannot touch, however fascinating the subject. Playfair stresses this theme more forcefully (and more often) than any other in his exposition of Hutton's theory. He regards Hutton's treatise as, above all, an elegant statement of proper scientific methodology—and he locates Hutton's wisdom primarily in his friend's decision to eschew the subject of ultimate origins and to focus on the earth's present operation. Playfair begins by criticizing the old manner of theorizing:

> The sole object of such theories has hitherto been, to explain the manner in which the present laws of the mineral kingdom were first established, or began to exist, without treating of the manner in which they now proceed.

He then evaluates this puerile strategy in one of his best prose flourishes:

> The absurdity of such an undertaking admits of no apology; and the smile which it might excite, if addressed merely to the fancy, gives place to indignation when it assumes the air of philosophic investigation.

Hutton, on the other hand, established the basis of a proper geological science by avoiding subjects "altogether beyond the limits of philosophical investigation." Hutton's explorations "never extended to the first origin of substances, but were confined entirely to their changes." Playfair elaborated:

> He has indeed no where treated of the first origin of any of the earths, or of any substance whatsoever, but only of the transformations which bodies have undergone since the present laws of nature were established. He considered this last as all that a science, built on experiment and observation, can possibly extend to; and willingly left, to more presumptuous inquirers, the task of carrying their reasonings beyond the boundaries of nature.

Finally, to Kirwan's charge that Hutton had limited science by his "evasion" of origins, Playfair responded that his friend had strengthened science by his positive program of studying what could be resolved:

> Instead of an *evasion,* therefore, any one who considers the subject fairly, will see, in Dr. Hutton's reasoning, nothing but the caution of a philosopher, who wisely confines his theory within the same limits by which nature has confined his experience and observation.

This all happened a long time ago, and in a context foreign to our concerns. But Hutton's methodological wisdom, and Playfair's eloquent warning, could not be more relevant today—for basic principles of empirical science have an underlying generality that transcends time. Practicing scientists have largely (but not always) imbibed Hutton's wisdom about restricting inquiry to questions that can be answered. But Kirwan's error of equating the best in science with the biggest questions about ultimate meanings continues to be the most common of popular misunderstandings.

I have written these monthly essays for nearly twenty years, and they have brought me an enormous correspondence from nonprofessionals about all aspects of science. From sheer volume, I obtain a pretty good sense of strengths and weaknesses in public perceptions. I have found that one common misconception surpasses all others. People will write, telling me that they have developed a revolutionary theory, one that will expand the boundaries of science. These theories, usually described in several pages of single-spaced typescript, are speculations about the deepest ultimate questions we can ask—what is the nature of life? the origin of the universe? the beginning of time?

But thoughts are cheap. Any person of intelligence can devise his half dozen before breakfast. Scientists can also spin out ideas about ultimates. We don't (or, rather, we confine them to our private thoughts) because we cannot devise ways to test them, to decide whether they are right or wrong. What good to science is a lovely idea that cannot, as a matter of principle, ever be affirmed or denied?

The following homily may seem paradoxical, but it embodies

Hutton's wisdom: The best science often proceeds by putting aside the overarching generality and focusing instead on a smaller question that can be reliably answered. In so doing, scientists show their intuitive feel for the fruitful, not their narrowness or paltriness of spirit. In this way we sneak up on big questions that only repel us if we try to engulf them in one fell speculation. Newton could not discover the nature of gravity, but he could devise a mathematics that unified the motion of a carriage with the revolution of the moon (and the drop of an apple). Darwin never tried to grasp the meaning of life (or even the manner of its origin on our planet), but he did develop a powerful theory to explain its manner of change through time. Hutton did not discover how our earth originated, but he developed some powerful and testable ideas about how it ticked. You might almost define a good scientist as a person with the horse sense to discern the largest answerable question—and to shun useless issues that sound grander.

Hutton's positive principle of restriction to the doable also defines the domain and procedures of evolutionary biology, my own discipline. Evolution is not the study of life's ultimate origin as a path toward discerning its deepest meaning. Evolution, in fact, is not the study of origins at all. Even the more restricted (and scientifically permissible) question of life's origin on our earth lies outside its domain. (This interesting problem, I suspect, falls primarily within the purview of chemistry and the physics of self-organizing systems.) Evolution studies the pathways and mechanisms of organic change following the origin of life. Not exactly a shabby subject either—what with such resolvable questions as "How, when, and where did humans evolve?"; "How do mass extinction, continental drift, competition among species, climatic change, and inherited constraints of form and development interact to influence the manner and rate of evolutionary change?"; and "How do the branches of life's tree fit together?" to mention just a few among thousands equally exciting.

In their recently aborted struggle to inject Genesis literalism into science classrooms, fundamentalist groups followed their usual opportunistic strategy of arguing two contradictory sides of a question when a supposed rhetorical advantage could be extracted from each. Their main pseudoargument held that Genesis literalism is not religion at all, but really an alternative form of science not acknowledged by professional biologists too hidebound and dogmatic to appreciate the cutting edge of their own discipline. When we successfully pointed out that "creation science"—as an untestable set of dogmatic proposals—could not qualify as science by any standard definition, they turned around and shamelessly argued the other side. (They actually pulled off the neater trick of holding both positions simultaneously.) Now they argued that, yes indeed, creation science is religion, but evolution is equally religious.

To support this dubious claim, they tumbled (as a conscious trick of rhetoric, I suspect) right into Kirwan's error. They ignored what evolutionists actually do and misrepresented our science as the study of life's ultimate origin. They then pointed out, as Hutton had, that questions of ultimate origins are not resolvable by science. Thus, they claimed, creation science and evolution science are symmetrical—that is, equally religious. Creation science isn't science because it rests upon the untestable fashioning of life *ex nihilo* by God. Evolution science isn't science because it tries, as its major aim, to resolve the unresolvable and ultimate origin of life. But we do no such thing. We understand Hutton's wisdom—"he has nowhere treated of the first origin . . . of any substance . . . but only of the transformations which bodies have undergone. . . ."

Our legal battle with creationists started in the 1920s and reached an early climax with the conviction of John Scopes in 1925. After some quiescence, the conflict began in earnest again during the 1970s and has haunted us ever since. Finally, in June 1987, the Supreme Court ended this major chapter in American history with a decisive 7–2 vote, striking down the last creationist statute, the Louisiana equal time act, as a ruse to inject religion into science classrooms in violation of First Amendment guarantees for separation of church and state.

I don't mean to appear ungrateful, but we fallible humans are always seeking perfection in others. I couldn't help wondering how two justices could have ruled the other way. I may not be politically astute, but I am not totally naive either. I have read Justice Scalia's long dissent carefully, and I recognize that its main thrust lies in legal issues supporting the extreme judicial conservatism espoused by Scalia and the other dissenter, Chief

Justice Rehnquist. Nonetheless, though forming only part of his rationale, Scalia's argument relies crucially upon a false concept of science—Kirwan's error again. I regret to say that Justice Scalia does not understand the subject matter of evolutionary biology. He has simply adopted the creationists' definition and thereby repeated their willful mistake.

Justice Scalia writes, in his key statement on scientific evidence:

> The people of Louisiana, including those who are Christian fundamentalists, are quite entitled, as a secular matter, to have whatever scientific evidence there may be against evolution presented in their schools.

I simply don't see the point of this statement. Of course they are so entitled, and absolutely nothing prevents such a presentation, if evidence there be. The equal time law forces teaching of creation science, but nothing prevented it before, and nothing prevents it now. Teachers were, and still are, free to teach creation science. They don't because they recognize it as a ruse and a sham.

Scalia does acknowledge that the law would be unconstitutional if creation science is free of evidence—as it is—and if it merely restates the Book of Genesis—as it does:

> Perhaps what the Louisiana Legislature has done is unconstitutional because there is no such evidence, and the scheme they have established will amount to no more than a presentation of the Book of Genesis.

Scalia therefore admits that the issue is not merely legal and does hinge on a question of scientific fact. He then buys the creationist argument and denies that we have sufficient evidence to render this judgment of unconstitutionality. Continuing directly from the last statement, he writes:

> But we cannot say that on the evidence before us. . . . Infinitely less can we say (or should we say) that the scientific evidence for evolution is so conclusive that no one would be gullible enough to believe that there is any real scientific evidence to the contrary.

But this is exactly what I, and all scientists, do say. We are not blessed with absolute certainty about any fact of nature, but evolution is as well confirmed as anything we know—surely as well as the earth's shape and position (and we don't require equal time for flat-earthers and those who believe that our planet resides at the center of the universe). We have oodles to learn about how evolution happened, but we have adequate proof that living forms are connected by bonds of genealogical descent.

So I asked myself, how could Justice Scalia be so uninformed about the state of our basic knowledge? And then I remembered something peculiar that bothered me, but did not quite register, when I first read his dissent. I went back to his characterization of evolution and what did I find (repeated, by the way, more than a dozen times, so we know that the argument represents no one-time slip of his pen, but a consistent definition)?

Justice Scalia has defined evolution as the search for life's origin—and nothing more. He keeps speaking about "the current state of scientific evidence about the origin of life" when he means to designate evolution. He writes that "the legislature wanted to ensure that students would be free to decide for themselves how life began based upon a fair and balanced presentation of the scientific evidence." Never does he even hint that evolution might be the study of how life changes after it originates—the entire panoply of transformation from simple molecules to all modern, multicellular complexity.

Moreover, to make matters worse, Scalia doesn't even acknowledge the scientific side of the origin of life on earth. He argues that a creationist law might have a secular purpose so long as we can envisage a concept of creation not involving a personal God "who is the object of religious veneration." He then points out that many such concepts exist, stretching back to Aristotle's notion of an unmoved mover. In the oral argument before the Court, which I attended on December 10, 1986, Scalia pressed this point even more forcefully with counsel for our side. He sparred:

> What about Aristotle's view of a first cause, an unmoved mover? Would that be a creationist view? I don't think Aristotle considered himself as a theologian as opposed to a philosopher.

In fact, he probably considered himself a scientist. . . . Well, then, you could believe in a first cause, an unmoved mover, that may be impersonal, and has no obligation of obedience or veneration from men, and in fact, doesn't care what's happening to mankind. And believe in creation. [From the official transcript, and omitting the responses of our lawyer.]

Following this theme, Scalia presents his most confused statement in the written dissent:

Creation science, its proponents insist, no more must explain whence life came than evolution must explain whence came the inanimate materials from which it says life evolved. But even if that were not so, to posit a past creator is not to posit the eternal and personal God who is the object of religious veneration.

True indeed; one might be a creationist in some vernacular sense by maintaining a highly abstract and impersonal view of a creator. But Aristotle's unmoved mover is no more part of science than the Lord of Genesis. Science does not deal with questions of ultimate origins. We would object just as strongly if the Aristotelophiles of Delaware forced a law through the state legislature requiring that creation of each species *ex nihilo* by an unmoved mover be presented every time evolution is discussed in class. The difference lies only in historical circumstance, not the logic of argument. The unmoved mover doesn't pack much political punch; fundamentalism ranks among our most potent irrationalisms.

Consider also, indeed especially, Scalia's false concept of science. He equates creation and evolution because creationists can't explain life's beginning, while evolutionists can't resolve the ultimate origin of the inorganic components that later aggregated to life. But this inability is the very heart of creationist logic and the central reason why their doctrine is not science, while science's inability to specify the ultimate origin of matter is irrelevant because we are not trying to do any such thing. We know that we can't, and we don't even consider such a question as part of science.

We understand Hutton's wisdom. We do not search for unattainable ultimates. We define evolution, using Darwin's phrase, as "descent with modification" from prior living things. Our documentation of life's evolutionary tree records one of science's greatest triumphs, a profoundly liberating discovery on the oldest maxim that truth can make us free. We have made this discovery by recognizing what can be answered and what must be left alone. If Justice Scalia heeded our definitions and our practices, he would understand why creationism cannot qualify as science. He would also, by the way, sense the excitement of evolution and its evidence; no person of substance could be unmoved by something so interesting. Only Aristotle's creator may be so impassive.

Don Quixote recognized "no limits but the sky," but became thereby the literary embodiment of unattainable reverie. G. K. Chesterton understood that any discipline must define its borders of fruitfulness. He spoke for painting, but you may substitute any creative enterprise: "Art is limitation: the essence of every picture is the frame."

# 9 | Numbers and Probability

# 31 | The Streak of Streaks*

MY FATHER was a court stenographer. At his less than princely salary, we watched Yankee games from the bleachers or high in the third deck. But one of the judges had season tickets, so we occasionally sat in the lower boxes when hizzoner couldn't attend. One afternoon, while DiMaggio was going 0 for 4 against, of all people, the lowly St. Louis Browns, the great man fouled one in our direction. "Catch it, Dad," I screamed. "You never get them," he replied, but stuck up his hand like the Statue of Liberty—and the ball fell right in. I mailed it to DiMaggio, and, bless him, he actually sent the ball back, signed and in a box marked "insured." Insured, that is, to make me the envy of the neighborhood, and DiMaggio the model and hero of my life.

I met DiMaggio a few years ago on a small playing field at the Presidio of San Francisco. My son, wearing DiMaggio's old number 5 on his Little League jersey, accompanied me, exactly one generation after my father caught that ball. DiMaggio gave him a pointer or two on batting and then signed a baseball for him. One generation passeth away, and another generation cometh: But the earth abideth forever.

My son, uncoached by Dad, and given the chance that comes but once in a lifetime, asked DiMaggio as his only query about life

---

*This essay originally appeared in the *New York Review of Books* as a review of Michael Seidel's *Streak: Joe DiMaggio and the Summer of 1941* (New York: McGraw-Hill, 1988). I have excised the references to Seidel's book in order to forge a more general essay, but I thank him both for the impetus and for writing such a fine book.

and career: "Suppose you had walked every time up during one game of your 56-game hitting streak? Would the streak have been over?" DiMaggio replied that, under 1941 rules, the streak would have ended, but that this unfair statute has since been revised, and such a game would not count today.

My son's choice for a single question tells us something vital about the nature of legend. A man may labor for a professional lifetime, especially in sport or in battle, but posterity needs a single transcendant event to fix him in permanent memory. Every hero must be a Wellington on the right side of his personal Waterloo; generality of excellence is too diffuse. The unambiguous factuality of a single achievement is adamantine. Detractors can argue forever about the general tenor of your life and works, but they can never erase a great event.

In 1941, as I gestated in my mother's womb, Joe DiMaggio got at least one hit in each of 56 successive games. Most records are only incrementally superior to runners-up; Roger Maris hit 61 homers in 1961, but Babe Ruth hit 60 in 1927 and 59 in 1921, while Hank Greenberg (1938) and Jimmy Foxx (1932) both hit 58. But DiMaggio's 56-game hitting streak is ridiculously, almost unreachably far from all challengers (Wee Willie Keeler and Pete Rose, both with 44, come second). Among sabermetricians (a happy neologism based on an acronym for members of the Society for American Baseball Research, and referring to the statistical mavens of the sport)—a contentious lot not known for agreement about anything—we find virtual consensus that DiMaggio's 56-game hitting streak is the greatest accomplishment in the history of baseball, if not all modern sport.

The reasons for this respect are not far to seek. Single moments of unexpected supremacy—Johnny Vander Meer's back-to-back no-hitters in 1938, Don Larsen's perfect game in the 1956 World Series—can occur at any time to almost anybody, and have an irreducibly capricious character. Achievements of a full season—such as Maris's 61 homers in 1961 and Ted Williams's batting average of .406, also posted in 1941 and not equaled since—have a certain overall majesty, but they don't demand unfailing consistency every single day; you can slump for a while, so long as your average holds. But a streak must be absolutely exceptionless; you are not allowed a single day of subpar play, or even bad luck. You bat only four or five times in an average game. Sometimes two or three of these efforts yield walks, and you get only one or two shots at a hit. Moreover, as tension mounts and notice increases, your life becomes unbearable. Reporters dog your every step; fans are even more intrusive than usual (one stole DiMaggio's favorite bat right in the middle of his streak). You cannot make a single mistake.

Thus Joe DiMaggio's 56-game hitting streak is both the greatest factual achievement in the history of baseball and a principal icon of American mythology. What shall we do with such a central item of our cultural history?

Statistics and mythology may strike us as the most unlikely of bedfellows. How can we quantify Caruso or measure *Middlemarch*? *But if God could mete out heaven with the span (Isaiah 40:12), perhaps we can say something useful about hitting streaks. The statistics of "runs," defined as continuous series of good or bad results (including baseball's streaks and slumps), is a well-developed branch of the profession, and can yield clear— but wildly counterintuitive—results. (The fact that we find these conclusions so surprising is the key to appreciating DiMaggio's achievement, the point of this article, and the gateway to an important insight about the human mind.)*

*Start with a phenomenon that nearly everyone both accepts and considers well understood—"hot hands" in basketball. Now and then, someone just gets hot, and can't be stopped. Basket after basket falls in—or out as with "cold hands," when a man can't buy a bucket for love or money (choose your cliché). The reason for this phenomenon is clear enough: It lies embodied in the maxim, "When you're hot, you're hot; and when you're not, you're not." You get that touch, build confidence; all nervousness fades, you find your rhythm; swish, swish, swish. Or you miss a* few, get rattled, endure the booing, experience despair; hands start shaking and you realize that you shoulda stood in bed.

Everybody knows about hot hands. The only problem is that no such phenomenon exists. Stanford psychologist Amos Tversky studied every basket made by the Philadelphia 76ers for more than a season. He found, first of all, that the probability of making a second basket did not rise following a successful shot. Moreover, the number of "runs," or baskets in succession, was no greater than what a standard random, or coin-tossing, model would predict. (If the chance of making each basket is 0.5, for example, a reasonable value for good shooters, five hits in a row will occur, on average, once in 32 sequences—just as you can expect to toss five successive heads about once in 32 times, or $0.5^5$.)

Of course Larry Bird, the great forward of the Boston Celtics, will have more sequences of five than Joe Airball—but not because he has greater will or gets in that magic rhythm more often. Larry has longer runs because his average success rate is so much higher, and random models predict more frequent and longer sequences. If Larry shoots field goals at 0.6 probability of success, he will get five in a row about once every 13 sequences $(0.6^5)$. If Joe, by contrast, shoots only 0.3, he will get his five straight only about once in 412 times. In other words, we need no special explanation for the apparent pattern of long runs. There is no ineffable "causality of circumstance" (to coin a phrase), no definite reason born of the particulars that make for heroic myths—courage in the clinch, strength in adversity, etc. You only have to know a person's ordinary play in order to predict his sequences. (I rather suspect that we are convinced of the contrary not only because we need myths so badly, but also because we remember the successes and simply allow the failures to fade from memory. More on this later.) But how does this revisionist pessimism work for baseball?

My colleague Ed Purcell, Nobel laureate in physics but, for purposes of this subject, just another baseball fan, has done a comprehensive study of all baseball streak and slump records. His firm conclusion is easily and swiftly summarized. Nothing ever happened in baseball above and beyond the frequency predicted by coin-tossing models. The longest runs of wins or losses are as long as they should be, and occur about as often as they ought to. Even the hapless Orioles, at 0 and 21 to start the 1988 season, only fell victim to the laws of probability (and not to the vengeful God of racism, out to punish major league baseball's only black manager).*

But "treasure your exceptions," as the old motto goes. Pur-

---

*When I wrote this essay, Frank Robinson, the Baltimore skipper, was the only black man at the helm of a major league team. For more on the stats of Baltimore's slump, see my article "Winning and Losing: It's All in the Game," *Rotunda*, Spring 1989.

cell's rule has but one major exception, one sequence so many standard deviations above the expected distribution that it should never have occurred at all: Joe DiMaggio's 56-game hitting streak in 1941. The intuition of baseball aficionados has been vindicated. Purcell calculated that to make it likely (probability greater than 50 percent) that a run of even 50 games will occur once in the history of baseball up to now (and 56 is a lot more than 50 in this kind of league), baseball's rosters would have to include either four lifetime .400 batters or 52 lifetime .350 batters over careers of 1,000 games. In actuality, only three men have lifetime batting averages in excess of .350, and no one is anywhere near .400 (Ty Cobb at .367, Rogers Hornsby at .358, and Shoeless Joe Jackson at .356). DiMaggio's streak is the most extraordinary thing that ever happened in American sports. He sits on the shoulders of two bearers—mythology and science. For Joe DiMaggio accomplished what no other ballplayer has done. He beat the hardest taskmaster of all, a woman who makes Nolan Ryan's fastball look like a cantaloupe in slow motion—Lady Luck.

A larger issue lies behind basic documentation and simple appreciation. For we don't understand the truly special character of DiMaggio's record because we are so poorly equipped, whether by habits of culture or by our modes of cognition, to grasp the workings of random processes and patterning in nature.

Omar Khayyám, the old Persian tentmaker, understood the quandary of our lives (*Rubaiyat of Omar Khayyám*, Edward Fitzgerald, trans.):

> Into this Universe, and Why not knowing,
> Nor Whence, like Water willy-nilly flowing;
> And out of it, as Wind along the Waste,
> I know not Whither, willy-nilly blowing.

But we cannot bear it. We must have comforting answers. We see pattern, for pattern surely exists, even in a purely random world. (Only a highly nonrandom universe could possibly cancel out the clumping that we perceive as pattern. We think we see constellations because stars are dispersed at random in the heavens, and therefore clump in our sight—see Essay 17.) Our error lies not in the perception of pattern but in automatically imbuing pattern with meaning, especially with meaning that can bring us comfort, or dispel confusion. Again, Omar took the more honest approach:

> Ah, love! could you and I with Fate conspire
> To grasp this sorry Scheme of Things entire,
> Would not we shatter it to bits—and then
> Re-mould it nearer to the Heart's Desire!

We, instead, have tried to impose that "heart's desire" upon the actual earth and its largely random patterns (Alexander Pope, *Essay on Man,* end of Epistle 1):

> All Nature is but Art, unknown to thee;
> All Chance, Direction, which thou canst not see;
> All Discord, Harmony not understood:
> All partial Evil, universal Good.

Sorry to wax so poetic and tendentious about something that leads back to DiMaggio's hitting streak, but this broader setting forms the source of our misinterpretation. We believe in "hot hands" because we must impart meaning to a pattern—and we like meanings that tell stories about heroism, valor, and excellence. We believe that long streaks and slumps must have direct causes internal to the sequence itself, and we have no feel for the frequency and length of sequences in random data. Thus, while we understand that DiMaggio's hitting streak was the longest ever, we don't appreciate its truly special character because we view all the others as equally patterned by cause, only a little shorter. We distinguish DiMaggio's feat merely by quantity along a continuum of courage; we should, instead, view his 56-game hitting streak as a unique assault upon the otherwise unblemished record of Dame Probability.

Amos Tversky, who studied "hot hands," has performed, with Daniel Kahneman, a series of elegant psychological experiments. These long-term studies have provided our finest insight into "natural reasoning" and its curious departure from logical truth. To cite an example, they construct a fictional description of a young woman: "Linda is 31 years old, single, outspoken, and very bright. She majored in philosophy. As a student, she was deeply concerned with issues of discrimination and social justice, and also participated in anti-nuclear demonstrations." Subjects are then given a list of hypothetical statements about Linda: They must rank these in order of presumed likelihood, most to least probable. Tversky and Kahneman list eight statements, but five are a blind, and only three make up the true experiment:

Linda is active in the feminist movement;
Linda is a bank teller;
Linda is a bank teller and is active in the feminist movement.

Now it simply must be true that the third statement is least likely, since any conjunction has to be less probable than either of its parts considered separately. Everybody can understand this when the principle is explained explicitly and patiently. But all groups of subjects, sophisticated students who have pondered logic and probability as well as folks off the street corner, rank the last statement as more probable than the second. (I am particularly fond of this example because I know that the third statement is least probable, yet a little homunculus in my head continues to jump up and down, shouting at me—"but she can't just be a bank teller; read the description.")

Why do we so consistently make this simple logical error? Tversky and Kahneman argue, correctly I think, that our minds are not built (for whatever reason) to work by the rules of probability, though these rules clearly govern our universe. We do something else that usually serves us well, but fails in crucial instances: We "match to type." We abstract what we consider the "essence" of an entity, and then arrange our judgments by their degree of similarity to this assumed type. Since we are given a "type" for Linda that implies feminism, but definitely not a bank job, we rank any statement matching the type as more probable than another that only contains material contrary to the type. This propensity may help us to understand an entire range of human preferences, from Plato's theory of form to modern stereotyping of race or gender.

We might also understand the world better, and free ourselves of unseemly prejudice, if we properly grasped the workings of probability and its inexorable hold, through laws of logic, upon much of nature's pattern. "Matching to type" is one common error; failure to understand random patterning in streaks and slumps is another—hence Tversky's study of both the fictional Linda and the 76ers' baskets. Our failure to appreciate the uniqueness of DiMaggio's streak derives from the same unnatural and uncomfortable relationship that we maintain with probability. (If we knew Lady Luck better, Las Vegas might still be a roadstop in the desert.)

My favorite illustration of this basic misunderstanding, as applied to DiMaggio's hitting streak, appeared in a recent article by baseball writer John Holway, "A Little Help from His Friends," and subtitled "Hits or Hype in '41" (*Sports Heritage*, 1987). Holway points out that five of DiMaggio's successes were narrow escapes and lucky breaks. He received two benefits-of-the-doubt from official scorers on plays that might have been judged as errors. In each of two games, his only hit was a cheapie. In game 16, a ball dropped untouched in the outfield and had to be called a hit, even though the ball had been misjudged and could have been caught; in game 54, DiMaggio dribbled one down the third-base line, easily beating the throw because the third baseman, expecting the usual, was playing far back. The fifth incident is an oft-told tale, perhaps the most interesting story of the streak. In game 38, DiMaggio was 0 for 3 going into the last inning. Scheduled to bat fourth, he might have been denied a chance to hit at all. Johnny Sturm popped up to begin the inning, but Red Rolfe then walked. Slugger Tommy Henrich, up next, was suddenly swept with a premonitory fear: Suppose I ground into a double play and end the inning? An elegant solution immediately occurred to him: Why not bunt (an odd strategy for a power hitter). Henrich laid down a beauty; DiMaggio, up next, promptly drilled a double to left.

I enjoyed Holway's account, but his premise is entirely, almost preciously, wrong. First of all, none of the five incidents represents an egregious miscall. The two hits were less than elegant, but undoubtedly legitimate; the two boosts from official scorers were close calls on judgment plays, not gifts. As for Henrich, I can only repeat manager Joe McCarthy's comment when Tommy asked him for permission to bunt: "Yeah, that's a good idea." Not a terrible strategy either—to put a man into scoring position for an insurance run when you're up 3–1.

But these details do not touch the main point: Holway's premise is false because he accepts the conventional mythology about long sequences. He believes that streaks are unbroken runs of causal courage—so that any prolongation by hook-or-crook becomes an outrage against the deep meaning of the phenomenon. But extended sequences are not pure exercises in valor. Long streaks always are, and must be, a matter of extraordinary luck imposed upon great skill. Please don't make the vulgar mistake of thinking that Purcell or Tversky or I or anyone else would attribute a long streak to "just luck"—as though everyone's chances are exactly the same, and streaks represent nothing more than the lucky atom that kept moving in one direction. Long hitting streaks happen to the greatest players—Sisler, Keeler, DiMaggio, Rose—because their general chance of getting a hit is so much higher than average. Just as Joe Airball cannot match Larry Bird for runs of baskets, Joe's cousin Bill Ofer, with a lifetime batting average of .184, will never have a streak to match DiMaggio's with a lifetime average of .325. The statistics show something else, and something fascinating: There is no "causality of circumstance," no "extra" that the great can draw from the soul of their valor to extend a streak beyond the ordinary expectation of coin-tossing models for a series of unconnected events, each occurring with a characteristic probability for that particular player. Good players have higher characteristic probabilities, hence longer streaks.

Of course DiMaggio had a little luck during his streak. That's what streaks are all about. No long sequence has ever been entirely sustained in any other way (the Orioles almost won several of those 21 games). DiMaggio's remarkable achievement—its uniqueness, in the unvarnished literal sense of that word—lies in whatever he did to extend his success well beyond the reasonable expectations of random models that have governed every other streak or slump in the history of baseball.

Probability does pervade the universe—and in this sense, the old chestnut about baseball imitating life really has validity. The statistics of streaks and slumps, properly understood, do teach an important lesson about epistemology, and life in general. The history of a species, or any natural phenomenon that requires unbroken continuity in a world of trouble, works like a batting streak. All are games of a gambler playing with a limited stake against a house with infinite resources. The gambler must eventually go bust. His aim can only be to stick around as long as possible, to have some fun while he's at it, and, if he happens to be a moral agent as well, to worry about staying the course with honor. The best of us will try to live by a few simple rules: Do justly, love mercy, walk humbly with thy God, and never draw to an inside straight.

DiMaggio's hitting streak is the finest of legitimate legends because it embodies the essence of the battle that truly defines our lives. DiMaggio activated the greatest and most unattainable dream of all humanity, the hope and chimera of all sages and shamans: He cheated death, at least for a while.

# 32 | The Median Isn't the Message

MY LIFE HAS RECENTLY intersected, in a most personal way, two of Mark Twain's famous quips. One I shall defer to the end of this essay. The other (sometimes attributed to Disraeli) identifies three species of mendacity, each worse than the one before—lies, damned lies, and statistics.

Consider the standard example of stretching truth with numbers—a case quite relevant to my story. Statistics recognizes different measures of an "average," or central tendency. The *mean* represents our usual concept of an overall average—add up the items and divide them by the number of sharers (100 candy bars collected for five kids next Halloween will yield 20 for each in a fair world). The *median,* a different measure of central tendency, is the halfway point. If I line up five kids by height, the median child is shorter than two and taller than the other two (who might have trouble getting their mean share of the candy). A politician in power might say with pride, "The mean income of our citizens is $15,000 per year." The leader of the opposition might retort, "But half our citizens make less than $10,000 per year." Both are right, but neither cites a statistic with impassive objectivity. The first invokes a mean, the second a median. (Means are higher than medians in such cases because one millionaire may outweigh hundreds of poor people in setting a mean, but can balance only one mendicant in calculating a median.)

The larger issue that creates a common distrust or contempt for statistics is more troubling. Many people make an unfortunate and invalid separation between heart and mind, or feeling and intellect. In some contemporary traditions, abetted by attitudes

473

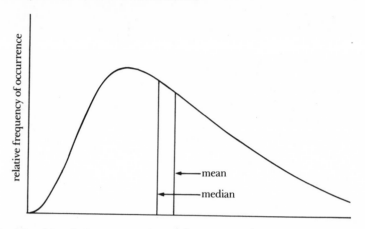

A right-skewed distribution showing that means must be higher than medians, and that the right side of the distribution extends out into a long tail. BEN GAMIT.

sterotypically centered upon Southern California, feelings are exalted as more "real" and the only proper basis for action, while intellect gets short shrift as a hang-up of outmoded elitism. Statistics, in this absurd dichotomy, often becomes the symbol of the enemy. As Hilaire Belloc wrote, "Statistics are the triumph of the quantitative method, and the quantitative method is the victory of sterility and death."

This is a personal story of statistics, properly interpreted, as profoundly nurturant and life-giving. It declares holy war on the downgrading of intellect by telling a small story to illustrate the utility of dry, academic knowledge about science. Heart and head are focal points of one body, one personality.

In July 1982, I learned that I was suffering from abdominal mesothelioma, a rare and serious cancer usually associated with exposure to asbestos. When I revived after surgery, I asked my first question of my doctor and chemotherapist: "What is the best technical literature about mesothelioma?" She replied, with a touch of diplomacy (the only departure she has ever made from direct frankness), that the medical literature contained nothing really worth reading.

Of course, trying to keep an intellectual away from literature works about as well as recommending chastity to *Homo sapiens,* the sexiest primate of all. As soon as I could walk, I made a bee-line for Harvard's Countway medical library and punched meso-thelioma into the computer's bibliographic search program. An hour later, surrounded by the latest literature on abdominal mesothelioma, I realized with a gulp why my doctor had offered that humane advice. The literature couldn't have been more bru-tally clear: Mesothelioma is incurable, with a median mortality of only eight months after discovery. I sat stunned for about fifteen minutes, then smiled and said to myself: So that's why they didn't give me anything to read. Then my mind started to work again, thank goodness.

If a little learning could ever be a dangerous thing, I had en-countered a classic example. Attitude clearly matters in fighting cancer. We don't know why (from my old-style materialistic per-spective, I suspect that mental states feed back upon the immune system). But match people with the same cancer for age, class, health, and socioeconomic status, and, in general, those with pos-itive attitudes, with a strong will and purpose for living, with com-mitment to struggle, and with an active response to aiding their own treatment and not just a passive acceptance of anything doc-tors say tend to live longer. A few months later I asked Sir Peter Medawar, my personal scientific guru and a Nobelist in immunol-ogy, what the best prescription for success against cancer might be. "A sanguine personality," he replied. Fortunately (since one can't reconstruct oneself at short notice and for a definite pur-pose), I am, if anything, even-tempered and confident in just this manner.

Hence the dilemma for humane doctors: Since attitude matters so critically, should such a somber conclusion be advertised, es-pecially since few people have sufficient understanding of statis-tics to evaluate what the statements really mean? From years of experience with the small-scale evolution of Bahamian land snails treated quantitatively, I have developed this technical knowl-edge—and I am convinced that it played a major role in saving my life. Knowledge is indeed power, as Francis Bacon proclaimed.

The problem may be briefly stated: What does "median mor-tality of eight months" signify in our vernacular? I suspect that most people, without training in statistics, would read such a statement as "I will probably be dead in eight months"—the very conclusion that must be avoided, both because this formulation is false, and because attitude matters so much.

I was not, of course, overjoyed, but I didn't read the statement in this vernacular way either. My technical training enjoined a different perspective on "eight months median mortality." The point may seem subtle, but the consequences can be profound. Moreover, this perspective embodies the distinctive way of thinking in my own field of evolutionary biology and natural history.

We still carry the historical baggage of a Platonic heritage that seeks sharp essences and definite boundaries. (Thus we hope to find an unambiguous "beginning of life" or "definition of death," although nature often comes to us as irreducible continua.) This Platonic heritage, with its emphasis on clear distinctions and separated immutable entities, leads us to view statistical measures of central tendency wrongly, indeed opposite to the appropriate interpretation in our actual world of variation, shadings, and continua. In short, we view means and medians as hard "realities," and the variation that permits their calculation as a set of transient and imperfect measurements of this hidden essence. If the median is the reality and variation around the median just a device for calculation, then "I will probably be dead in eight months" may pass as a reasonable interpretation.

But all evolutionary biologists know that variation itself is nature's only irreducible essence. Variation is the hard reality, not a set of imperfect measures for a central tendency. Means and medians are the abstractions. Therefore, I looked at the mesothelioma statistics quite differently—and not only because I am an optimist who tends to see the doughnut instead of the hole, but primarily because I know that variation itself is the reality. I had to place myself amidst the variation.

When I learned about the eight-month median, my first intellectual reaction was: Fine, half the people will live longer; now what are my chances of being in that half. I read for a furious and nervous hour and concluded, with relief: damned good. I possessed every one of the characteristics conferring a probability of longer life: I was young; my disease had been recognized in a relatively early stage; I would receive the nation's best medical treatment; I had the world to live for; I knew how to read the data properly and not despair.

Another technical point then added even more solace. I immediately recognized that the distribution of variation about the eight-month median would almost surely be what statisticians call "right skewed." (In a symmetrical distribution, the profile of variation to the left of the central tendency is a mirror image of variation to the right. Skewed distributions are asymmetrical, with variation stretching out more in one direction than the other— left skewed if extended to the left, right skewed if stretched out to the right.) The distribution of variation had to be right skewed, I reasoned. After all, the left of the distribution contains an irrevocable lower boundary of zero (since mesothelioma can only be identified at death or before). Thus, little space exists for the distribution's lower (or left) half—it must be scrunched up between zero and eight months. But the upper (or right) half can extend out for years and years, even if nobody ultimately survives. The distribution must be right skewed, and I needed to know how long the extended tail ran—for I had already concluded that my favorable profile made me a good candidate for the right half of the curve.

The distribution was, indeed, strongly right skewed, with a long tail (however small) that extended for several years above the eight-month median. I saw no reason why I shouldn't be in that small tail, and I breathed a very long sigh of relief. My technical knowledge had helped. I had read the graph correctly. I had asked the right question and found the answers. I had obtained, in all probability, that most precious of all possible gifts in the circumstances—substantial time. I didn't have to stop and immediately follow Isaiah's injunction to Hezekiah—set thine house in order: for thou shalt die, and not live. I would have time to think, to plan, and to fight.

One final point about statistical distributions. They apply only to a prescribed set of circumstances—in this case to survival with mesothelioma under conventional modes of treatment. If circumstances change, the distribution may alter. I was placed on an experimental protocol of treatment and, if fortune holds, will be in the first cohort of a new distribution with high median and a right tail extending to death by natural causes at advanced old age.*

---

*So far so good.

It has become, in my view, a bit too trendy to regard the acceptance of death as something tantamount to intrinsic dignity. Of course I agree with the preacher of Ecclesiastes that there is a time to love and a time to die—and when my skein runs out I hope to face the end calmly and in my own way. For most situations, however, I prefer the more martial view that death is the ultimate enemy—and I find nothing reproachable in those who rage mightily against the dying of the light.

The swords of battle are numerous, and none more effective than humor. My death was announced at a meeting of my colleagues in Scotland, and I almost experienced the delicious pleasure of reading my obituary penned by one of my best friends (the so-and-so got suspicious and checked; he too is a statistician, and didn't expect to find me so far out on the left tail). Still, the incident provided my first good laugh after the diagnosis. Just think, I almost got to repeat Mark Twain's most famous line of all: The reports of my death are greatly exaggerated.*

---

*Since writing this, my death has actually been reported in two European magazines, five years apart. *Fama volat* (and lasts a long time). I squawked very loudly both times and demanded a retraction; guess I just don't have Mr. Clemens's *savoir faire.*

# 33 | The Ant and the Plant

WHY DO WE CARE so much about size and number? My friend Ralph Keyes, who tips the charts with me at five feet seven and a half inches, wrote an entire book about our obsession with this supposedly irrelevant subject—*The Height of Your Life*. He documented the extraordinary steps that short politicians and film stars often take to avoid discovery of their secret. (Ralph couldn't penetrate the subterfuges of Jimmy Carter's staff to discover the height of our shortest recent president, who is at least an inch or two taller than Ralph and me, and therefore at, or not far from, the American average.) The most amusing item in Ralph's book is an old publicity shot of a short Humphrey Bogart with two of his leading ladies, Lauren Bacall and Katharine Hepburn. They have just emerged from an airplane. Bogie is on the first step of the gangplank; the two women stand on the ground.

Why do we so stupidly equate more with better? Penises and automobiles, two objects frequently graded for size by foolish men, work just as well, and often more efficiently, at smaller than average lengths. Extremes in body size almost always entail tragic consequences (at least off the basketball court). Robert Wadlow, just shy of nine feet and the tallest human ever recorded, died at age twenty-two from infection caused by a faulty ankle brace needed for supplementary support since his legs could not adequately carry his body. Moving beyond the pathology of extreme individuals, entire species of unusually large body size generally have short geological lifetimes. I doubt that their problem lies in biomechanical inefficiency, as earlier theories of lumbering dino-

479

saurs held. Rather, large creatures tend to be anatomically specialized and form relatively small populations (fewer brontosauruses than boll weevils)—perhaps the two strongest detriments to extended survival in a world of large and capricious environmental fluctuation over time.

While most people do understand that large size does not guarantee long-term success, the myth of "more is better" still pervades our interpretations. I have, for example, noted with surprise, as I have monitored the impressions of students and correspondents for more than twenty years, how many people assume, as almost logically necessary a priori, that evolutionary "progress" and complexity should correlate with the amount of DNA in an organism's cell—the ultimate baseline for more is better. Not so. The very simplest creatures, including viruses at the low end, followed by bacteria and other prokaryotic organisms, do have relatively little DNA. But as soon as we reach multicellular life, based on eukaryotic cells with nuclei and chromosomes, the correlation breaks down completely. Mammals stand squarely in the middle of the pack, with $10^9$–$10^{10}$ nucleotides per haploid cell. The largest values, ranging to nearly 100 times more DNA than the most richly endowed mammals, belong to salamanders and to some flowering plants.

Many species of plants arise by polyploidy, or doubling of chromosome number. These doublings often run through several cycles among a group of closely related species, so the amount of DNA can increase greatly—and the high DNA content of some polyploid plants has never been much of a mystery. On the other hand, the extreme values for amphibians once puzzled zoologists sufficiently that they gave the phenomenon a name—"the C-value paradox." However, since the discovery that so little of the total DNA codes actively for enzymes and proteins, this hundredfold difference between some mammals and salamanders seems less troubling. Most DNA consists of repeated copies; much of it codes for nothing and may represent "junk" in terms of an organism's morphology. The hundredfold difference does not mean that salamanders have 100 times more active genes than mammals, for the disparity occurs chiefly in nonessential, or noncoding, regions. (We would still, of course, like to know why some groups accumulate more junk and more repetitions, but such differences do not merit special recognition as a formal paradox.)

This essay considers another expression of maximum and minimum, and another test of correlation between quantity and quality—numbers of chromosomes. We have voluminous data on average differences in number of chromosomes among groups of organisms, and some patterns surely emerge. Diptera (flies and their allies) tend to have few per cell; *Drosophila,* the great laboratory stalwart (and largely for this reason), harbors four pairs per diploid cell. Birds tend to have many. Instead of providing a compendium for these well-chronicled differences, I shall focus on the extreme cases of more and less among organisms. Extreme values may titillate our fancy, but they are also unusually instructive for recognizing and specifying generalities. Exceptions do prove rules. (The etymology of that cliché, usually mistaken for a reversed meaning, is not "prove" in the sense of verify, but "probe" in the sense of test or challenge. This definition, from the Latin *probare,* is not entirely archaic in English—consider printer's proof, or a proving ground for testing weapons.)

Until two years ago, the lowest number of chromosomes, a commandingly minimal one pair, had been found for only a single organism—a nematode worm, appropriately honored in its subspecific name as *Parascaris equorum univalens.* This minimal complement had been discovered long ago, in 1887, by Theodor Boveri, the greatest cytologist (student of cellular architecture) of the late nineteenth century. Boveri (1862–1915) was a great intellectual in the European tradition—a complex and fascinating man who lived for the laboratory, but who also played the piano and painted with professional competence. His short life was scarred by fits of depression, and he died in despondency as the First World War enveloped Europe. Of Boveri's many scientific discoveries, the two greatest centered on chromosomes. First, he established their individuality and shifted attention from the nucleus as a whole to chromosomes as the agent of inheritance (in years before the rediscovery of Mendel's laws). Second, he demonstrated the differential value of chromosomes. Before Boveri's experiments, many scientists had conjectured that each chromosome carried all the hereditary information, and that organisms with many chromosomes carried more copies of this totality. Boveri proved that each chromosome carried only part of the hereditary information (some of the genes, as we might say today), and that the full complement built the organism through a complex orchestration of development.

Boveri took great interest in his discovery of an organism that carried but one pair of chromosomes per cell—and therefore did place all its hereditary information into one package. But Boveri quickly discovered that *P. equorum univalens,* though no imposter in its claims to minimalism, was not entirely consistent either. Only the cells of the germ line, those destined to produce eggs and sperm by meiosis, kept all the hereditary material together in a single pair of chromosomes. In cells destined to form body tissues, this chromosome fractured several times during the first cleavage divisions of early embryology, leading to adult cells with up to seventy chromosomes!

Finally, in 1986, Australian zoologists Michael W. J. Crosland and Ross H. Crozier reported a remarkable new species within a closely related group of ants, previously united into the overextended species *Myrmecia pilosula* (see their article of 1988 cited in the bibliography). This name falsely amalgamates several distinct species sharing a similar body form, but carrying different numbers of chromosomes in their cells. Species with nine, ten, sixteen, twenty-four, thirty, thirty-one, and thirty-two pairs of chromosomes have been described. Obviously, this complex of forms has evolved some way of speciating in concert with substantial changes in chromosome number.

On February 24, 1985, on the Tidbinbilla Nature Reserve near Canberra, Crosland and Crozier collected a colony of winged males and females, plus a mated queen with pupae and more than 100 workers. All workers tested from this colony carried but a single pair of chromosomes in their cells—all of them, not just cells of a particular type. An unambiguous example of chromosomal minimalism had finally been discovered, almost exactly 100 years after Boveri found only one pair of chromosomes in the germ line cells of *Parascaris.*

But the story of *M. pilosula* is even better, deliciously so. If you were out searching for absolute minimalism, you would have to root for finding your single pair of chromosomes in an ant, bee, or wasp—for the following interesting reason: The Hymenoptera, and just a few other creatures, reproduce by an unusual genetic system called haplodiploidy. In most animals, all body cells contain chromosomes in pairs, and sex is determined by maternal and paternal contributions (or noncontributions in some cases) to a single pair. But haplodiploid organisms specify sex by a different route. Reproductive females usually store sperm, often for long periods. Genetic females (including the functionally neuter workers) arise from fertilized eggs, and therefore contain chromosomes in pairs. But males are produced when the queen fails to fertilize a developing egg with stored sperm (in most other animal groups unfertilized eggs are inviable). Thus, the cells of male ants, bees, and wasps do not carry chromosomes in pairs and bear only the single set inherited from their mother. These males have no father, and their cells contain only half the chromosomes of females—a condition called haploid, as opposed to the diploid, or paired, complement of their sisters. (The entire system therefore receives the name haplodiploid, or male-female in this case.)

Haplodiploidy implies, of course, that males of the Tidbinbilla colony of *M. pilosula* have a truly and absolutely minimal number of one chromosome per cell. Not even a single pair—just one. The only lower possibility is disappearance. Crosland and Crozier checked just to be sure. The males of their colony contained a single chromosome per cell.

If we have reached a limit in the search for less, the other extreme seems more open-ended. How many chromosomes can a cell contain and still undergo the orderly divisions of mitosis and meiosis? Can hundreds of chromosomes line up neatly along a mitotic spindle and divide precisely to place an equal complement into each daughter cell? At what point do things become so crowded that this most elegant of biological mechanisms breaks down?

Maximal numbers are most easily reached by polyploidy, or doubling of chromosomes. This process occurs in two basic modes with differing evolutionary significances. In autoploidy, a cell doubles its own complement, forming, initially at least, a cell with two sets of identical pairs. Thus, the new autoploid usually looks like its parent. Autoploidy is not a mechanism for rapid evolution of form, though the redundancy introduced by doubling does permit considerable evolutionary divergence afterward—as one member of the duplicated pair becomes free to change. On the other hand, alloploidy, the second mode of doubling, can produce viable hybrids between distant species and can serve as a mechanism for sudden and substantial changes in form. Hybrids, with different forms and numbers of maternal and paternal chromosomes, will usually be sterile because chromosomes have no partners for pairing before meiosis—the "reduction division" that produces sex cells with half the genetic information of body cells. But if the precursors of sex cells undergo polyploidy, then each chromosome will find a partner in the duplicated version of its own form.

Since polyploidy is so much more common in plants than animals, we should search for maximalism in our gardens, not our zoos. The numerical importance of polyploidy in plants can best be appreciated in a wonderful graph that I first encountered, when a graduate student, in Verne Grant's *The Origin of Adaptations*. This graph is a frequency distribution for chromosome pairs in monocot plants. For ten pairs of chromosomes and higher, without exception, all peaks are for even numbers of chromosome pairs.

At first inadequate sight, this pattern doesn't make sense in the deepest possible way. Biology is not numerology; its regularities do not take the form of such abstractions as "cleave to even numbers." Such a graph will not be satisfying until we figure out a biological mechanism that, as a side consequence and not because evens are better than odds per se, produces an imbalance of species with chromosomes in pairs of even numbers. The resolution is elegantly simple in this case. Polyploidy is very common in plants, and every number, odd or even, when doubled, yields an even number. The peaks therefore indicate the prevalence of polyploidy in plants. Estimates range as high as 50 percent for the number of angiosperm species produced by polyploidy.

Since polyploidy can continue in cycles—doubling followed by redoubling—chromosome numbers, like the pot in a poker game with table stakes, can rise alarmingly from small beginnings. The champions among all organisms are ferns in the family Ophioglossaceae. The genus *Ophioglossum* exhibits a basic number of 120 chromosome pairs, the lowest value among living species. (Such a high number must, itself, be derived from earlier incidents of polyploidy among species now extinct. The basic number for the entire family, 15 pairs, may have been the starting

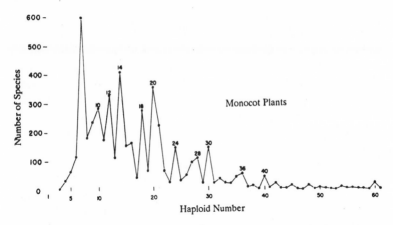

Frequency distribution for the number of chromosome pairs in monocot plants. Note that all peaks are for even numbers of chromosomes. This occurs because so many plant species are produced by polyploidy, or doubling of chromosome number, and a doubling of any number, odd or even, produces an even number. FROM VERNE GRANT, *THE ORIGIN OF ADAPTATIONS,* 1963.

point.) In any case, cycles of polyploidy have proceeded onward from this already large beginning of 120 pairs. The all-time champion, not only in *Ophioglossum,* but among all organisms, is *Ophioglossum reticulatum,* with about 630 pairs of chromosomes, or 1,260 per cell! (The total need not be an exact multiple of 120, because doubling may be imperfect, and secondary gains or losses for individual chromosomes are common.)

The very idea of a nucleus with 1,260 chromosomes, all obeying the rules of precise alignment and division as cells proliferate, inspired G. Ledyard Stebbins, our greatest living evolutionary botanist, to a rare emotion for a scientific paper—rapture (since Ledyard and I share a passion for Gilbert and Sullivan, I will write, for his sake, "modified rapture"—and he will know the reference and meaning): "At meiosis, these chromosomes pair regularly to form about 630 bivalents, a feat which to cytologists is as remarkable a wonder of nature as are the fantastic elaborations of form exhibited by orchids, insectivorous plants, and many animals" (see Stebbins, 1966, in the bibliography).

In fifteen years of writing these monthly essays, I have specialized in trying to draw general messages from particulars. But this time, I am stumped. I don't know what deep truth of nature emerges from the documentation of minimal and maximal chromosome numbers. Oh, I can cite some clichés and platitudes: Quantity is not quality; good things come in small packages. I can also state the obvious conclusion that inheritance and development do not depend primarily upon the number of distinct rods holding hereditary information—but this fact has been featured in textbooks of genetics for more than seventy years.

No, I think that every once in a while, we must simply let a fact stand by itself, for its own absolutely unvarnished fascination. Has your day not been brightened just a bit by learning that a plant can orchestrate the division of its cells by splitting 630 pairs of chromosomes with unerring accuracy—or that an ant, looking much like others, can gallivant about with an absolute minimum of one chromosome per cell? If so, I have earned my keep, and can go cultivate my garden. I think I'll try growing some ferns. Then I might take some colchicine, which often induces polyploidy, and maybe, just maybe. . . .

# 10 | Planets as Persons

## *Prologue*

The *Voyager* expedition represents the greatest technological and intellectual triumph of our century. The fact that this tiny, relatively inexpensive machine could explore and photograph every outer planet except Pluto (but including Neptune, now the most distant planet, as Pluto temporarily moves closer to the sun on its eccentric orbit) is not only, as the cliché goes, a triumph of the human spirit (not to mention good old American tinkering and know-how), but also a living proof that billions of bucks, bureaucratic immuring, and hush-hush military spin-offs need not power our space program—and that knowledge and wonder really could be the main motivation and reward.

Such a triumph must be celebrated by any writer in natural history. I have chosen my own idiosyncratic mode. *Voyager*'s results convey many messages. These two essays, with their common theme, embody my reading of the main lesson from the standpoint of an evolutionary biologist: Planets are like organisms in that they have irreducible individuality and must therefore be explained by methods of historical analysis; they are not like molecules in a chemical equation. Planets therefore affirm the larger goal of unity among sciences by showing that methods of one approach (biological-historical) apply to cardinal objects of another mode often viewed as disparate or even opposed (physical-experimental).

# 34 | The Face of Miranda

WHEN MIRANDA, confined for all her conscious life on Prospero's magic island, saw a group of men for the first time, she exclaimed, "O, wonder! How many goodly creatures are there here! How beauteous mankind is! O brave new world, that has such people in't" (the source, of course, for Aldous Huxley's more sardonic citation). Now, almost 400 years after Miranda spoke through Shakespeare, we have returned the favor, gazed for the first time upon Miranda and found her every bit as wonderful—"so perfect and so peerless . . . she will outstrip all praise and make it halt behind her."

Prospero used all his magic to import his visitors by tempest. We have seen Miranda, the innermost large moon of Uranus, through the most stunning feat of technical precision in all our history. Ariel himself, Prospero's agent of magic (and also another moon of Uranus), would have been astounded. For we have sent a small probe hurtling though space for nine years, boosting it with the gravitational slings of both Jupiter and Saturn toward distant Uranus, there to transmit a signal across 2 billion miles and three light-hours, showing the face of Miranda with the same clarity that Prospero beheld when he gazed upon his daughter's beauty and exclaimed, "Thou didst smile, infused with a fortitude from heaven."

It is easy to wax poetic about this feat (especially with a little help from the Bard himself). *Voyager*'s data from Jupiter, Saturn, and now Uranus have supplied more scientific return for expended output than anything else that space exploration ever dared or dreamed. In the chorus of praise, however, we have not always recognized how much this new information has transcended the visually dazzling—how deeply our *ideas* about the formation and history of the solar system have been changed. This confluence of aesthetics and intellect must be celebrated above all—and I should like to record my delight by thoroughly repudiating an early essay in this series (March 1977) as an illustration both of our new understanding and of the vital generalization so obtained.

My story is the tale of an old and eminently reasonable hypothesis, proposed long ago and beautifully affirmed by the first explorations of other worlds—our moon, then Mercury, and finally Mars. Then, at the height of its triumph, the theory begins to unravel, first at the moons of Jupiter, then at the surface of Venus, and finally and irretrievably, in the face of Miranda.

The initial hypothesis sought to explain the surfaces (and inferred histories) of rocky planets and moons as simple consequences of their differences in size. Why, in particular, is the earth so different from the moon? Our moon is a dead world, covered with impact craters that have not eroded away since their formation, often billions of years ago. The earth, by contrast, is a dynamic world of relative smoothness.

This difference, we assume, is a result of historical divergence, not initial disparity. Billions of years ago, when the planets were young and our portion of space still abounded with debris not yet swept up in planets and moons, the earth must have been as intensely cratered as the moon. The current difference must therefore be a result of the moon's retention, and the earth's obliteration, of their early histories. Why the difference?

On earth, both internal and external "machines" recycle the landscape on a scale of millions of years. The atmosphere (external machine) generates agents of erosion—running water, wind, and ice—that quickly obliterate the topography of any crater. Yet even without rain and wind, the earth's internal activity of volcanism, earthquakes, and ultimately of plate tectonics itself would eventually disaggregate and erase any old topography. Surfaces do not last for billions of years on an active planet. But neither machine works on the moon. With no atmosphere, erosion proceeds (even in geological time) at a snail's pace. Likewise, the moon is a rigid body with a crust 600 miles thick. Moonquakes do not fracture the lunar surface and volcanoes do not rise from the tiny molten core.

The earth's activity and moon's silence are consequences of a single factor—size. Large bodies have much lower ratios of surface to volume than small bodies of the same shape, since surfaces (length × length) grow so much more slowly than volumes (length × length × length) as size increases. Our planet powers its two machines by low surface-to-volume ratios. The earth generates heat (by radioactivity) over its relatively large volume and then loses this heat through its relatively small surface—thus remaining hot and active enough to propel plate tectonics. The moon, by contrast, and by virtue of its higher surface-to-volume ratio, lost most of its internal heat long ago, and solidified nearly throughout. Likewise, the earth's large mass generates enough gravity to hold an atmosphere and power its external machine, while the moon lost any gases once produced.

As planetary exploration began, this "size-dependent" theory of planetary surfaces and their histories received its first tests and passed elegantly. The first photos of Mercury showed nothing but craters—as expected for a body about the same size as our moon.

Mars posed a clearer and more crucial test. As a planet about midway in size between the earth and moon, it should preserve some of its early topography, but also display the action of weak internal and external machines. The *Surveyor* flyby and *Viking* landings affirmed this prediction. The surface of Mars is about 50 percent cratered. The remaining areas show abundant signs of erosion, primarily by winds today (dune fields and etched boulders) and by running water in the past (now frozen), and internal churning more limited than on the earth. Most intriguing are signs of incipient (but unrealized) plate tectonics—as though the Martian crust remains pliant enough to fracture, but too rigid to move.

At this point in space exploration, I felt confident enough to write an essay extolling the size hypothesis as a sufficient and elegantly simple explanation of planetary surfaces and their histories. Contrasting the earth with the smaller bodies then known, I wrote (in March 1977) that "the difference arises from a disarmingly simple fact—*size itself, and nothing else:* the earth is a good deal larger than its neighbors."

The first test after my essay appeared would be *Voyager's* photographic survey of the Galilean satellites of Jupiter—the four moon-sized rocky bodies that, by the size hypothesis, would surely be intensely cratered worlds, cold and dead. Thus, I waited with confidence as *Voyager* approached Io, the innermost moon of Jupiter. The first photos, distant and fuzzy, revealed some circular structures initially read as craters. Well and good. But the next day brought sharp photos, and evoked both wonder and surprise. The circles were not craters, but giant volcanoes, spewing forth lakes of sulfur. In fact, not a single crater could be found on Io, the most active satellite in the solar system. Yet, as a body smaller than our moon, Io should have been cold and cratered.

The explanation now offered for Io's intense activity had been predicted just a few days before the photos arrived. Io is so close to giant Jupiter that the interplay between Jupiter's gravitational tug and the reverse pull of the three other large satellites from the opposite side keeps the interior of Io fluid enough to resist rigidification.

As this information arrived, I could only stand by in awe and reflect that Io had been misnamed. The four Galilean satellites honor some of Jove's many lovers—an ecumenical assortment including his homosexual partner Ganymede; the nymph Callisto; and Europa, the mother of King Minos. Io, the fallen priestess, was changed to a heifer by jealous Hera, afflicted with a gadfly, and sent to roam Europe, where she forded (and indirectly named) the Bosporus (literally, the cow crossing) and finally emerged, human again, in Egypt. I thought that this innermost moon should be renamed Semele, to honor another lover who made the mistake of demanding that Jove appear to her in his true, rather than his disguised (and muted) human, form—and was immediately burned to a crisp!

Is the size hypothesis therefore wrong because Io violated its prediction? The principle of surfaces and volumes, as a basic law of physics and the geometry of space, is surely correct. Io does not challenge its validity, but only its scope. The size hypothesis does not merely claim that the surface-to-volume principle operates—for this we can scarcely doubt. The hypothesis insists, rather, that the surface-to-volume principle so dominates all other potential forces that we need invoke nothing further to understand the history and topography of rocky planetary surfaces. Io does not refute the principle. But Io does prove dramatically that other circumstances—in this case proximity to opposing gravitational sources—can so override the surface-to-volume rule that its predictions fail or, in this case (even worse), are diametrically refuted.

Planetary surfaces lie in the domain of complex historical sciences, where modes of explanation differ from the stereotypes of simple and well-controlled laboratory experiments. We are not trying to demonstrate the validity of physical laws. Rather, we must try to assess the relative importance of several complex and interacting forces. The validity of the surface-to-volume principle was never at issue, only its relative importance—and Io has challenged its domination.

We must therefore know, in order to judge the size hypothesis, whether Io is a lone exception in a singular circumstance or a general reminder that the surface-to-volume principle ranks as only one among many competing influences—and therefore not as *the* determinant of planetary surfaces and their history. The test will not center upon arguments about the laws of physics, but upon observations of other bodies, for we must establish, empirically, the relative importance of a hypothesis that worked until *Voyager* photographed Io.

Venus was the next candidate. Our sister planet, although closer to us than any other, had remained shrouded (literally) in mystery by its dense cover of clouds. But Russian and American probes have now mapped the Venusian surface with radio waves that can penetrate the clouds, as wavelengths in the visible spectrum cannot. Results are ambiguous and still under analysis, but proponents of the size hypothesis can scarcely react with unalloyed pleasure. Venus and Earth are just about the same size and Venus should, by the surface-to-volume hypothesis, be as active as our planet. Our sister world is, to be sure, no dead body. We have seen high mountains, giant rifts, and other signs of extensive tectonic activity. But Venus also seems to maintain too much old and cratered terrain for a body of its size, according to the principle of surfaces and volumes alone.

Scientists have advanced many explanations for the difference between Venus and Earth. Perhaps tidal forces generated by the moon's gravity keep Earth in its high state of geological flux. Venus has no satellite. Perhaps the high surface temperature of Venus, generated by a greenhouse effect under its dense cover of clouds, keeps the surface too pliable to form the thin and rigid plates that, in their constant motion, keep Earth's surface so active.

*Voyager* then moved toward Uranus and a final test. By this time, buffeted by Io and Venus, I was holding out little hope for the size hypothesis (and also wishing that the *Rubáiyát* had not spoken so truly about the moving finger, and that my publishers might deep-six all unsold copies of *The Panda's Thumb,* with its reprint of my original 1977 essay). In fact, anticipating final defeat for that elegantly simple proposal of earlier years, I actually managed to turn disappointment into a modest professorial coup. I have long believed that examinations have little intellectual value, existing only to fulfill, and ever so imperfectly at that, any large institution's need for assessment by number. Yet, for the first time, the moons of Uranus allowed me to ask an examination question with some intellectual interest and integrity.

I realized that the final examination for my large undergraduate course had been set for the morning of January 24, at the very hour that *Voyager* would be relaying photographs of Uranian moons to earth. I therefore predicted that Miranda, although the smallest of five major moons, would be most active among them, and asked the students to justify (or reject) such a speculation—though the conjecture itself is absurd under the size hypothesis with its evident prediction that Miranda, as the smallest moon with the highest surface-to-volume ratio, should be cratered and devoid of internal activity. I asked the students:

> As you take this exam, *Voyager 2* is sending back to Earth the first close-up pictures of Uranus and its moons. . . . On what basis might you predict that Miranda, although the smallest of these moons, is most likely to show some activity (volcanoes, for example) on its surface? We will probably know the answer before the exam ends!

(When I first wrote the exam in early January, I couldn't even provide my students with the moons' diameters, for they had not yet been measured precisely, though we knew that Miranda was smallest. Between writing and administering the exam, *Voyager* measured the diameters, and we rushed to the printers with an insert. Science, at its best, moves very quickly indeed.)

So I was ready for the final undoing of the size hypothesis, but not for the actual result of Miranda's countenance. The conjecture of my question turned out to be quite wrong. I had been thinking of Io and the gravity of a nearby giant planet. Since Miranda is closest to Uranus, I supposed that it might be lit with modern volcanoes. But no volcanoes are belching forth sulfur, or anything else, on Miranda. The actual observations, however, spoke even more strongly against the size hypothesis and its prediction of a cold, cratered world.

I had made one good prediction, probably for the wrong reason: Miranda is the most geologically active of Uranian moons, despite its small diameter of but 300 miles. (The moons of Uranus, outdoing even the mythic splendor of Jupiter's satellites, bear lovely Shakespearean names—in order from Uranus out: Miranda, Ariel, Umbriel, Titania, and Oberon. In addition, *Voyager* has discovered at least ten additional and much smaller moons between Miranda and the planet's surface.) The first photos of Miranda stunned and delighted the boys in Pasadena even more than her namesake had mesmerized Ferdinand on Prospero's island. Laurence Soderblom, speaking for the *Voyager* imaging team, exclaimed: "It's just mind boggling. . . . You name it, we have it. . . . Miranda is what you would get if you can imagine taking all the bizarre geological features in the solar system and putting them on one object." A brave new world, indeed. So much for the size hypothesis and its uniform blanket of craters.

Moreover, all the Uranian moons are surprisingly active (except for Umbriel, odd man out in more ways than one, as the only non-Shakespearean entry) in a gradient of increasing turmoil from the outermost king of *Midsummer Night* to the innermost daughter of the *Tempest*. "As you move closer to Uranus," Soderblom added, "we see an increasing ferocity, as though these bodies have been tectonically shuffled in a cataclysmic fashion."

I must save the details for another time, but for starters, the surface of Miranda is a jumble of frozen geological activity—long valleys, series of parallel grooves, and blocks of sunken crust. Most prominent, and also most notable for their lack of any clear counterpart on other worlds, are three structures that seem related in their formation. One has been dubbed a stack of pancakes, the second a chevron, and the third a racetrack. They are series of parallel grooves, or cracks, shaped to different forms according to their nicknames and full of evidence for massive slumping, rifting, and cliff making.

In short, Io failed the size hypothesis by its position too close to Jupiter. Venus may not conform by a particular history that left it moon free and cloud covered. Miranda has failed, we know not why, by showing signs of a frantic past when the hypothesis predicted a passive compendium of impacts. The physical principle invoked by the size hypothesis—the law of surfaces and volumes—is surely correct, but not potent enough to overwhelm other influences and lead to confident predictions by itself. As we learn more and more about the historical complexity of the heavens, we recognize that where you are (Io) and what you have been (Venus and Miranda) exert as much influence over a planet's surface as its size. After an initial success for our moon, Mercury, and Mars, the size hypothesis flunked all further tests.

The story of a theory's failure often strikes readers as sad and unsatisfying. Since science thrives on self-correction, we who practice this most challenging of human arts do not share such a feeling. We may be unhappy if a favored hypothesis loses or chagrined if theories that we proposed prove inadequate. But refutation almost always contains positive lessons that overwhelm disappointment, even when (as in this case) no new and comprehensive theory has yet filled the void. I chose this tale of failure for a particular reason, not only because Miranda excited me. I chose to confess my former errors because the replacement of a simple physical hypothesis with a recognition of history's greater complexity teaches an important lesson with great unifying power.

An unfortunate, but regrettably common, stereotype about science divides the profession into two domains of different status. We have, on the one hand, the "hard," or physical, sciences that deal in numerical precision, prediction, and experimentation. On the other hand, "soft" sciences that treat the complex objects of history in all their richness must trade these virtues for "mere" description without firm numbers in a confusing world where, at best, we can hope to explain what we cannot predict. The history

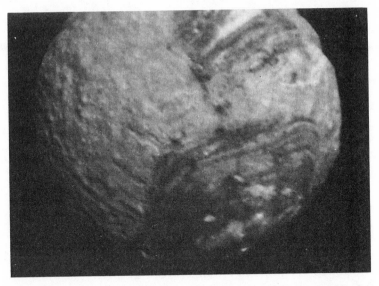

*Voyager* photograph of Miranda, showing fractured and reaggregated terrain. PHOTO COURTESY NASA/JPL.

of life embodies all the messiness of this second, and under-valued, style of science.

Throughout ten years of essays firmly rooted in this second style, I have tried to suggest by example that the sciences of history may be different from, but surely not worse than, the sciences of simpler physical objects. I have written about a hundred historical problems and their probable solutions, hoping to illustrate a methodology as powerful as any possessed by colleagues in other fields. I have tried to break down the barriers between these two styles of science by fostering mutual respect.

The story of planetary surfaces illustrates another path to the same goal of lowering barriers. The two styles are not divisible by discipline into the hard sciences of physical systems and the soft sciences of biological objects. All good scientists must use and appreciate both styles since large and adequate theories usually need to forage for insights in both physics and history. If we accepted the rigid dichotomy of hard and soft, we might argue that as physical bodies, planets should yield to predictive theories of the hard sciences. The size hypothesis represented this mode of explanation (and I was beguiled by it before I understood history better)—a simple law of physics to regulate a large class of complex objects. But we have learned, in its failure, that planets are more like organisms than billiard balls. They are intricate and singular bodies. Their individuality matters, and size alone will not explain planetary surfaces. We must know their particularities, their early histories, their present locations. Planets are physical bodies that require historical explanations. They break the false barrier between two styles of science by forcing the presumed methods of one upon the supposed objects of the other.

Finally, we should not lament that simple explanations have failed and that the "messy" uniqueness of each planet must be featured in any resolution. We might despair if the individuality of planets dashed all hope for general explanation. But the message of Io, Venus, and Miranda is not gridlock, but transcendence. We think that we understand Io, and we strive to fathom the moons of Uranus. Historical explanations are difficult, damned interesting, and eminently attainable by human cleverness. Whoever said that nature would be easy?

Prospero, after saving his foes from the tempest, asserts that he cannot relate the history of his life too simply, for " 'tis a chronicle of day by day, not a relation for a breakfast." The tale is long and intricate, but fascinating and resolvable. We can also know the richness of history in science. Proper explanation may require a tapestry of detail. Our stories may recall the subtle skills of Scheherazade rather than the crisp epitome of a segment in *Sixty Minutes,* but then who has ever been bored by Sinbad the Sailor or Aladdin's magic lamp?

# 35 | The Horn of Triton

THE ARGUMENTS of "iffy" history may range from the merely amusing to the horribly tragic. If Mickey Owen hadn't dropped that third strike, the Dodgers might have won the 1941 World Series. If Adolf Hitler had been killed in the Beer Hall Putsch, the alliances that led to World War II might not have formed, and we might not have lost our war fleet at Pearl Harbor just two months after Owen's miscue.

I don't think that we would be so fascinated by conjectures in this mode if we felt that anything could happen in history. Rather, we accept certain trends, certain predictabilities, even some near inevitabilities, particularly in war and technology, where numbers truly count. (I can't imagine any scenario leading to the victory of Grenada over the United States in our recent one-day conflict; nor can I conjecture how the citizens of Pompeii, without benefit of motorized transport, could have escaped a cloud of poisonous gases streaming down Mount Vesuvius at some forty miles per hour.) I suspect, in fact, that our fascination with iffy history arises largely from our awe at the ability of individuals to perturb, even greatly to alter, a process that seems to be moving in a definite direction for reasons above and beyond the power of mere mortals to deflect.

In opening *The Eighteenth Brumaire of Louis Bonaparte*, Karl Marx captured this essential property of history as a dynamic balance between the inexorability of forces and the power of individuals. He wrote, in one of the great one-liners of scholarship in the activist mode: "Men make their own history, but they do not make it just as they please." (Marx's title is, itself, a commentary

499

on the unique and the repetitive in history. The original Napoleon staged his *coup d'état* against the Directory on November 9–10, 1799, then called the eighteenth day of Brumaire, Year VIII, by the revolutionary calendar adopted in 1793 and used until Napoleon crowned himself emperor and returned to the old forms. But Marx's book traces the rise of Louis-Napoleon, nephew of the emperor, from the presidency of France following the revolution of 1848, through his own *coup d'état* of December 1851, to his crowning as Napoleon III. Marx seeks lessons from repetition, but continually stresses the individuality of each cycle, portraying the second in this case as a mockery of the first. His book begins with another great epigram, this time a two-liner, on the theme of repetition and individuality: "Hegel remarks somewhere that all facts and personages of great importance in world history occur, as it were, twice. He forgot to add the first time as tragedy, the second as farce.")

This essential tension between the influence of individuals and the power of predictable forces has been well appreciated by historians, but remains foreign to the thoughts and procedures of most scientists. We often define science (far too narrowly, I shall argue) as the study of nature's laws and their consequences. Individual objects have no power to shape general patterns in such a system. Walter the Water Molecule cannot freeze a pond, while Sarah the Silica Tetrahedron does not perturb the symmetry of quartz. Indeed, the very notion of Walter and Sarah only invites ridicule because laws of chemical behavior and crystal symmetry deny individuality to constituent units of larger structures. What else do we mean when we assert that hydrogen and oxygen make water or that silica tetrahedra sharing all their corner oxygen ions form quartz? (We could scarcely speak of a law if Ollie Oxygen willingly joined with Omar but refused to share with Oscar because they had a fight last Friday.) No actual quartz crystal has a perfect lattice of conjoined tetrahedra; all include additions and disruptions known as impurities or imperfections—but the very names given to these ingredients of individuality demonstrate that scientific content supposedly lies in the regularities, while uniquenesses of particular crystals fall into the domain of hobbyists and aestheticians.

(I don't mean to paint the world of science as a heartless place of perfect predictability under immutable laws. We permit a great deal of play and doubt under the guise of randomness. But randomness is equally hostile to the idea of individuality. In fact, classically random systems represent the ultimate denial of individuality. Coin-flipping and dice-throwing models rest upon the premise that each toss or each roll manifests the same probabilities: no special circumstances of time or place, no greater chance of a head if the last five tosses have been tails, or if you blow on the coin and say your mantra, or if Aunt Mary will die as a consequence of your failure to score—in other words, no individuality of particular trials. Individuality and randomness are opposing, not complementary, concepts. They both oppose the idea of clockwork determinism, but they do so in entirely different ways.)

Natural history does not share this consensus that individual units with particular legacies cannot shape the behavior and future state of entire systems. Our profession, although part of mainstream science since Aristotle, grants to individuals the potential for such a formative role. In this sense, we are truly historians by practice and we demonstrate the futility of disciplinary barriers between science and the humanities. We should be exploring our marked overlaps in explanatory procedures, not sniping at each other behind walls of definitional purity.

Natural history stands in the crossfire and should provoke a truce by reaching in both directions. Individual organisms can certainly set the local history of populations and may even shape the fate of species. Walter the Walrus and Sarah the Squirrel are friendly and congenial, rather than risible, concepts (and may be actual creatures at the municipal zoo). Two recent cases of extraordinary (although not particularly likable) individuals have led me to consider this theme and to grant more attention to the vagaries of one in my profession.

1. Jane Goodall's quarter century with the chimpanzees of Gombe will rank forever as one of the great achievements in scientific dedication combined with stunning results. With such unprecedented, long-term knowledge of daily history, Goodall can specify (and quantify) the major determinants of her population's fate. Contrary to our intuitions and expectations, the demography of the Gombe chimps has not been set primarily by daily rhythms of birth, feeding, sex, and death, but by three "rare events" (Goodall's words), all involving mayhem or misfortune: a

polio epidemic, a carnage of one sub-band by another, and the following tale of one peculiar individual.

With odd and unintended appropriateness as we shall see, for the word means "suffering," Goodall named one of the Gombe females Passion. Goodall met Passion in 1961 at the outset of her studies. In 1965, Passion gave birth to a daughter, Pom, and, as Goodall remarks (all quotes from *The Chimpanzees of Gombe,* Harvard University Press, 1986), "thereby gave us the opportunity to observe some extraordinarily inefficient and indifferent maternal behavior."

Nonetheless, Pom and Passion formed a "close, cooperative bond" as the daughter matured. In 1975, Passion began to kill and eat newborn babies of other females in her band. She could not easily wrest a baby from its mother and failed when acting solo, but Passion and Pom together formed an efficient killing duo. (Goodall observed three other "cannibalistic events" during nearly thirty years of work, all directed by males toward older chimps of other bands; Passion's depredations are the only recorded incidents of cannibalism within a band.) During a four-year period, Passion and Pom, in sight of observers, killed and ate three infants by seizing them from their mothers and biting through the skull bones (sorry, but nature isn't always pretty, and I hate euphemisms). They may have been responsible for the deaths of seven other infants. During this entire period, only one female successfully raised a baby. In studying Goodall's curves of Gombe demography, the depredations of Passion have as great an impact as any general force of climate or disease. Moreover, the effects are not confined to the short years of Passion's odd obsession (for reasons unknown, she stopped killing babies in 1977), but propagate well down the line. Since only one female was raising a baby in 1977, nearly all were in estrus, thus prompting a baby boomlet and sharp rise in population when Passion stopped her cannibalism.

Such observational work on the behavior of animals in their natural habitat requires a personal pledge to maximal noninterference. Passion taxed this principle to its absolute limit. Goodall told me that when Passion died "of an unknown wasting disease" in 1982, she (Jane, not Passion) watched with renewed faith in noninterference and some legitimate sense of moral retribution.

2. *Notornis,* the New Zealand ornithological journal, does not

show up in the scientific equivalent of the corner drug store; I was therefore delighted when Jared Diamond alerted me (via *Nature,* which does appear at our watering holes) to a fascinating article by Michael Taborsky entitled "Kiwis and Dog Predation: Observations in Waitangi State Forest" (see the bibliography). The Waitangi Forest houses the largest "known and counted" population of the brown kiwi *Apteryx australis*—some 800 to 1,000 birds. In June and July of 1987, Taborsky and colleagues tagged twenty-four birds with radio transmitters "so that their spacing and reproductive activities could be studied" (all quotations come from Taborsky's paper cited above).

On August 24, they found a dead female, evidently killed by a dog. Thus began a tale worthy of *The Hound of the Baskervilles.* By September 27, thirteen of the tagged birds had been killed. All showed extensive bruising, and most had defeathered areas; ten of the thirteen birds "were found partly covered or completely buried under leaf litter and soil." Scientists and forestry workers found ten more carcasses without transmitters, all killed and buried in the same way, and all dispatched during the same period.

It didn't take the sleuthing genius of Mr. Holmes to recognize that a single dog had wreaked this reign of terror. Distinctive footprints of the same form appeared by the carcasses, along with "dog droppings of one type and size." On September 30, a female German shepherd, wearing a collar but unregistered, was shot in the forest. Her "long claws suggested that she had not been on hard surfaces for some time, i.e., was probably living in the forest." The killings abruptly stopped. Taborsky tagged several more birds with transmitters, bringing the total to eighteen; all these birds survived to the end of the study on October 31.

This Rin Tin Tin of the Dark Side had killed more than half of the tagged birds in six weeks. As "there is no reason to believe that birds with transmitters were at greater risk than those without," the total killed may range to 500 of the 800 to 1,000 birds in the population. Lest this seem a staggering and unbelievable estimate, Taborsky provides the following eminently reasonable defenses. First, given the remote chance of finding a buried, untagged kiwi carcass, the ten actually located during the interval of killing must represent the tiny pinnacle of a large iceberg. Second, other evidence supports a dramatic fall in total population: Taborsky and colleagues noted a major drop in calling rates for these ordinarily noisy birds; a dog trained to find, but not to kill, kiwis could not locate a single live individual (although she found two carcasses) in a formerly well-inhabited section of the forest. Third, kiwis, having evolved without natural enemies and possessing no means of escape, could not be easier prey. Taborsky writes:

> Could a single dog really do so much damage? People working trained kiwi dogs at night know it is very easy indeed for a dog to spot and catch a kiwi. The birds are noisy when going through the bush and their smell is very strong and distinctive. When a kiwi calls, a dog can easily pick up the direction from more than 100 m away. With a kiwi density as high as it was in Waitangi Forest a dog could perhaps catch 10–15 kiwis a night, and the killing persisted for at least 6 weeks.

As to why a dog would kill so many animals "for sport," or at least not for food, who knows? We do, however, understand enough to brand as romantic twaddle the common litany that "man alone kills for sport, other animals only for food or in defense." The kiwi marauder of New Zealand may have set a new record for intensity of destruction, but she followed the killing pattern of many animals. In any case, she surely illustrated the power of individuals to alter the history of entire populations. Taborsky estimates that, given the extremely slow breeding of kiwis, "the population will probably need 10–20 years and a rigorous protection scheme to recover to previous densities."

These two stories may elicit both fascination and a *frisson,* but still strike some readers as unpersuasive regarding the role of individuals in science. To be sure, both Passion and the austral hell hound had a disturbing effect on their populations. But science is general pattern, not ephemeral perturbation. The Gombe chimps recovered in a few years, as a subsequent baby boom offset Passion's depredations. On the crucial issue of scale, individuals still don't set patterns in the fullness of time or the largeness of space. Predictability under nature's laws takes over at an amplitude of scale and a degree of generality meriting the name "science." I would offer three rebuttals to this argument.

First, scale is a relative concept. Who can set the boundary between perturbations in systems too small to matter and long-term patterns of appropriate generality? Human evolution is a tiny twig among millions on the tree of earthly evolution. But do all the generalities of anthropology therefore count only as details outside the more ample scope of true science? Earthly evolution may be only one story of life among unknown cosmic billions; are all the laws of biology therefore nothing but peculiarities of one insignificant example?

Second, small perturbations are not always reined in by laws of nature to bring systems back to a previous equilibrium. Perturbations, starting as tiny fluctuations wrought by individuals, can accumulate to profound and permanent alterations in much larger worlds. Much of the present fascination for chaos theory in mathematics stems from its attempt to model such agents of pattern, even in large systems operating under deterministic laws. The Gombe chimps may feel no long-term effect of Passion's cannibalism, but the Waitangi kiwis may never recover.

Third, some natural populations may be so small that individuality dominates over pattern even if a larger system might fall under predictable law. If I am flipping a coin 10,000 times in a row, with nothing staked on any particular toss, then an individual flip neither has much effect in itself nor influences the final outcome to any marked degree. But if I am flipping one coin once to start a football game, then a great deal rides on the unpredictability of an individual event. Some important populations in nature are closer in number to the single toss than to the long sequence. Yet we cannot deny them entry into the domain of science.

Consider the large, orbiting objects of our solar system—nine planets and a few score moons. The domain of celestial mechanics has long been viewed as the primal realm of lawfulness and predictability in science—the bailiwick of Newton and Kepler, Copernicus and Galileo, inverse square laws and eclipses charted to the second. We used to view the objects themselves, planets and moons, in much the same light—as regular bodies formed under a few determining conditions. Know composition, size, distance from the sun, and most of the rest follows.

As I write this essay, *Voyager 2* has just left the vicinity of Neptune and ventured beyond the planetary realm of our solar system—its "grand tour" complete after twelve years and the most colossal Baedecker in history: Jupiter, Saturn, Uranus, and Neptune. Scientists in charge have issued quite appropriate statements of humility, centered upon the surprises conveyed by stunning photographs of outer planets and their moons. I admit to the enormous mystery surrounding so many puzzling features—the whirling storms of Neptune, when the closer and larger Uranus appears so featureless; the diverse and complex terrain of Uranus's innermost moon, Miranda (see the preceding essay). But I do think that a generality has emerged from this confusing jumble of diverse results, so many still defying interpretation—a unifying principle usually missed in public reports because it falls outside the scope of stereotypical science.

I offer, as the most important lesson from *Voyager,* the principle of individuality for moons and planets. This contention should elicit no call for despair or surrender of science to the domain of narrative. We anticipated greater regularity, but have learned that the surfaces of planets and moons cannot be predicted from a few general rules. To understand planetary surfaces, we must learn the particular history of each body as an individual object— the story of collisions and catastrophes, more than steady accumulations; in other words, its unpredictable single jolts more than daily operations under nature's laws.

While *Voyager* recedes ever farther on its arc to the stars, we have made a conceptual full circle. When we launched *Voyager* in 1977, we mounted a copper disk on its side, with a stylus, cartridge, and instructions for playing. On this first celestial record, we placed the diversity of the earth's people. We sent greetings in fifty-five languages, and even added some whale song for ecumenical breadth. This babble of individuality was supposed to encounter, at least within our solar system, a regular set of worlds shaped by a few predictable forces. But the planets and moons have now spoken back to *Voyager* with all the riotous diversity of that unplayed record. The solar system is a domain of individuality by my third argument for small populations composed of distinctive objects. And science—that wonderfully diverse enterprise with methods attuned to resolve both the lawful millions and the unitary movers and shakers—has been made all the richer.

This is my third and last essay on *Voyager.* The first, which shall be my eternal incubus, praised the limited and lawful regularity that *Voyager* would presumably discover on planetary surfaces throughout the solar system. I argued, following the standard "line" at the time (but bowing to convention is never a good excuse), that simple rules of size and composition would set planetary surfaces. With sufficient density of rocky composition, size alone should reign. Small bodies, with their high ratios of surface to volume, are cold and dead—for they lose so much internal heat through their relatively large surface and are too small to hold an atmosphere. Hence, they experience no internal forces of volcanism and plate tectonics and no external forces of atmospheric erosion. In consequence, small planets and moons should be pristine worlds studded with ancient impact craters neither eroded nor recycled during billions of years. Large bodies, on the other hand, maintain atmospheres and internal heat machines. Their early craters should be obliterated, and their surfaces, like our earth's, should bear the marks of continuous, gentler action.

The first data from planetary probes followed these expectations splendidly. Small Mercury and smaller Phobos and Deimos (the moons of Mars) are intensely cratered, while Mars, at its intermediary size, showed a lovely mixture of ancient craters and regions more recently shaped by erosion and volcanic action. But then *Voyager* reached Jupiter and the story started to unravel in favor of individuality granted by distinctive histories for each object.

Io, Jupiter's innermost major moon, should have been dead and cratered at its size, but *Voyager* spotted large volcanoes spewing forth plumes of sulfur instead. Saturn's amazingly complex rings told a story of repeated collisions and dismemberments. Miranda, innermost moon of Uranus, delivered the *coup de grâce* to a dying theory. Miranda should be yet another placid body, taking its lumps in the form of craters and wearing the scars forever. Instead, *Voyager* photographed more signs of varied activity than any other body had displayed—a geological potpourri of features, suggesting that Miranda had been broken apart and reaggregated, perhaps more than once. Brave new world indeed.

I threw in the towel and wrote my second essay in the *mea culpa* mode (reprinted as the preceding essay). Now I cement my conversion. *Voyager* has just passed Neptune, last post of the grand tour, and fired a glorious parting shot for individuality. We knew that Triton, Neptune's largest moon, was odd in one important sense. All other bodies, planets around the sun and moons around the planets, revolve in the same direction—counterclockwise as you look down upon the plane of the solar system from above.* But Triton moves around Neptune in a clockwise direction. Still, at a size smaller than our moon, it should have been another of those dead and cratered worlds now mocked by the actual diversity of our solar system. Triton—and what a finale— is, if anything, even more diverse, active, and interesting than Miranda. *Voyager* photographed some craters, but also a complexly cracked and crumpled surface and, most unexpectedly of all, volcanoes, probably spewing forth nitrogen in streaks over the surface of Triton.

In short, too few bodies, too many possible histories. The planets and moons are not a repetitive suite, formed under a few simple laws of nature. They are individual bodies with complex histories. And their major features are set by unique events— mostly catastrophic—that shape their surfaces as Passion decimated the Gombe chimps or the austral hound wreaked havoc on the kiwis of Waitangi. Planets are like organisms, not water molecules; they have irreducible personalities built by history. They are objects in the domain of a grand enterprise—natural history—that unites both styles of science in its ancient and still felicitous name.

As *Voyager* has increased our knowledge, and at least for this paleontologist, integrated his two childhood loves of astronomy and fossils on the common ground of natural history, I cannot let this primary scientific triumph of our generation pass from our solar system (and these essays) without an additional comment in parting.

Knowledge and wonder are the dyad of our worthy lives as intellectual beings. *Voyager* did wonders for our knowledge, but

---

*Several Southern Hemisphere colleagues wrote to protest the indefensible parochialism of this image. The solar system has no natural "above" or "below." We think of the Northern Hemisphere as "above" only by cartographic convention in the Eurocentric tradition. I could have silently changed my restrictive image, but elected this correction by footnote since the illustration of such parochialism, however embarrassing, always serves to illustrate the power of unconscious bias and the need for continuous self-scrutiny.

performed just as mightily in the service of wonder—and the two elements are complementary, not independent or opposed. The thought fills me with awe—a mechanical contraption that could fit in the back of a pickup truck, traveling through space for twelve years, dodging around four giant bodies and their associated moons, and finally sending exquisite photos across more than four light-hours of space from the farthest planet in our solar system. (Pluto, although usually beyond Neptune, rides a highly eccentric orbit about the sun. It is now, and will be until 1999, within the orbit of Neptune and will not regain its status as outermost until the millennium. The point may seem a bit forced, but symbols matter and Neptune is now most distant. Moments and individualities count.)

The photos fill me with joy for their fierce beauty. To see the most distant moon with the detailed clarity of an object shot at ten palpable paces; the abstract swirling colors in Jupiter's great spot; the luminosity and order of Saturn's rings; the giant ripple-crater of Callisto, the cracks of Ganymede, the sulfur basins of Io, the craters of Mimas, the volcanoes of Triton. As *Voyager* passed Neptune, her programmers made a courtly and proper bow to aesthetics and took the most gorgeous picture of all, for beauty's sake—a photograph of Neptune as a large crescent, with Triton as a smaller crescent at its side. Two horns, proudly independent but locked in a common system. Future advertisers and poster makers may turn this exquisite object into a commercial cliché, but let it stand for now as a symbol for the fusion of knowledge and wonder.

*Voyager* has also served us well in literary allusions. Miranda is truly the "brave new world" of her most famous line. And Triton conjures all the glory and meaning of his most celebrated reference. You may have suppressed it, for forced memorization was a chore, but "The World Is Too Much with Us" remains a great poem (still assigned, I trust, by teachers). No one has ever matched Wordsworth in describing the wonder of childhood's enthusiasms—a wonder that we must strive to maintain through life's diminution of splendor in the grass and glory in the flower, for we are lost eternally when this light dies. So get to know Triton in his planetary form, but also remember him as Wordsworth's invocation to perpetual wonder:

        . . . Great God! I'd rather be
A Pagan suckled in a creed outworn;
So might I, standing on this pleasant lea,
Have glimpses that would make me less forlorn;
Have sight of Proteus rising from the sea;
Or hear old Triton blow his wreathed horn.

The most elegant photograph of the *Voyager* mission—a symbol of both knowledge and wonder. The horns (crescents) of Neptune and Triton.
PHOTO COURTESY NASA/J.P.L.

# Bibliography

Altholz, Josef L. 1980. The Huxley-Wilberforce debate revisited. *Journal of the History of Medicine and Allied Science,* 35(3):313–316.

Baker, Jeffrey J.W., and Garland E. Allen. 1982. *The study of biology.* Reading, MA: Addison Wesley.

Barbour, T., and J.C. Phillips. 1911. Concealing coloration again. *The Auk,* 28(2):179–187.

Bates, Henry Walter. 1863. Contributions to an insect fauna of the Amazon Valley. *Transactions of the Linnaean Society.*

Bates, Henry Walter. 1876. *The naturalist on the river Amazons: A record of adventures, habits of animals, sketches of Brazilian and Indian life, and aspects of Nature under the equator, during eleven years of travel.* London: J. Murray.

Berman, D.S., and J.S. McIntosh. 1978. Skull and relationships of the Upper Jurassic sauropod *Apatosaurus. Bulletin of the Carnegie Museum of Natural History,* No. 8.

Bierer, L.K., V.F. Liem, and E.P. Silberstein. 1987. *Life Science.* Heath.

Bligh, W. 1792. *A voyage to the South Sea undertaken by command of his Majesty for the purpose of conveying the bread-fruit tree to the West Indies in his Majesty's ship the* Bounty, *commanded by Lt. W. Bligh, including an account of the mutiny on board the said ship.* London.

Bohringer, R.C. 1981. Cutaneous receptors in the bill of the platypus (*Ornithorhyncus anatinus*). *Australian Mammalogy* 4:93–105.

Bohringer, R.C., and M.J. Rowe. 1977. The organization of the sensory and motor areas of cerebral cortex in the platypus *Ornithorhynchus anatinus. Journal of Comparative Neurology,* 174(1): 1–14.

Borges, Jorge Luis. 1979. *The book of sand.* London: A. Lane.

Brody, Samuel. 1945. *Bioenergetics and growth: with special reference to the efficiency complex in domestic animals.* New York: Hefner.

Browne, Janet. 1978. The Charles Darwin–Joseph Hooker correspondence: An analysis of manuscript resources and their use in biography. *Journal of the Society for the Bibliography of Natural History,* 8:351–366.

Brownlie, A.D., and M.F. Lloyd Prichard. 1963. Professor Fleeming Jenkin, 1833–1885 pioneer in engineering and political economy. *Oxford Economic Papers,* NS, 15(3):204–216.

Burnet, Thomas. 1691. *Sacred theory of the earth.* London: R. Norton.

Burrell, Harry. 1927. *The platypus: Its discovery, position, form and characteristics, habits and life history.* Sydney: Angus and Robertson.

Bury, J.B. 1920. *The idea of progress: An inquiry into its origin and growth.* London: Macmillan and Co., Ltd.

Busse, K. 1970. Care of the young by male *Rhinodrma darwini. Copeia,* No. 2, p. 395.

Calder, A., III. 1978. The kiwi. *Scientific American,* July.

Calder, Wm. A., III. 1979. The kiwi and egg design: Evolution as a package deal. *Bioscience,* 29(8):461–467.

Calder, A., III. 1984. *Size, function and life history.* Cambridge, MA: Harvard University Press.

Caldwell, W.H. 1888. The embryology of Monotremata and Marsupialia. Part I. *Philosophical Transactions of the Royal Society of London* (B), 178:463–483.

Camper, Petrus. 1791. *Physical dissertation on the real differences that men of different countries and ages display in their facial traits; on the beauty that characterizes the statues and engraved stones of antiquity; followed by the proposition of a new method for drawing human heads with the greatest accuracy.* Utrecht: B. Wild & J. Altheer.

Cavalli-Sforza, L.L., A. Piazza, P. Mendozzi, and J. Mountain. 1988. Reconstruction of human evolution: Bringing together genetic, archaeological, and linguistic data. *Proceedings of the National Academy of Sciences,* August, pp. 6002–6006.

Chamberlin, T.C., and R. Salisbury. 1909. *College Geology* (American Science Series). New York: Henry Holt & Co.

Chambers, Robert. 1844. *Vestiges of the natural history of creation.* London: J. Churchill.

Chapman, Frank M. 1933. *Autobiography of a bird lover.* New York: D. Appleton-Century Co., Inc.

Coletta, Paolo E. 1969. *William Jennings Bryan:* Vol. 3, *Political Puritan.* University of Nebraska Press.

Cope, E.D. 1874. *The origin of the fittest, and the primary factors of organic evolution.* New York: Arno Press.

Corben, C.J., G.J. Ingram, and M.J. Tyler. 1974. Gastric brooding: Unique form of parental care in an Australian frog. *Science,* 186:946.

Cott, Hugh B. 1940. *Adaptive coloration in animals.* London: Methuen.

Crosland, M.J., and Ross H. Crozier. 1986. *Myrmecia pilosula,* an ant with only one pair of chromosomes. *Science,* 231:1278.

Darnton, Robert. 1968. *Mesmerism and the end of the Enlightenment in France.* Cambridge, MA: Harvard University Press.

Darwin, Charles. 1859. *On the origin of species by means of natural selection.* London: John Murray.

Darwin, Charles. 1863. Review of Bates's paper. *Natural History Reviews,* pp. 219–224.

Darwin, Charles. 1868. *Variation of animals and plants under domestication.* London: John Murray.

Darwin, Erasmus. 1794. *Zoonomia, or the laws of organic life.* London: J. Johnson.

Darwin, Francis. 1887. *The life and letters of Charles Darwin.* London: John Murray.

David, Paul A. 1986. Understanding the economics of QWERTY: The necessity of history. In W.N. Parker, ed., *Economic history and the modern economist,* pp. 30–49. New York: Basil Blackwell, Inc.

Davis, D.D. 1964. Monograph on the giant panda. Chicago: Field Museum of Natural History.

De Quatrefages, A. 1864. *Metamorphoses of man and the lower animals.* London: Robert Hardwicke, 192 Piccadilly.

Disraeli, Benjamin. 1880. *Endymion.*

Douglas, Matthew M. 1981. Thermoregulatory significance of

thoracic lobes in the evolution of insect wings. *Science,* 211:84–86.

Dyson, Freeman. 1981. *Disturbing the universe.* New York: Harper & Row.

Eibl-Eibesfeldt, I. 1975. *Krieg und Frieden aus der Sicht der Verhaltensforschung.* Munchen, Zurich: Piper.

Eldredge, N., and S.J. Gould. 1972. Punctuated equilibria: An alternative to phyletic gradualism. In T.J.M. Schopf, ed., *Models in paleobiology,* pp. 82–115. San Francisco: Freeman, Cooper and Co.

Erasmus, Desiderius. 1508. *Adagia.*

Flower, J.W. 1964. On the origin of flight in insects. *Journal of Insect Physiology,* 10:81–88.

Force, James. 1985. *William Whiston, honest Newtonian.* Cambridge, England: Cambridge University Press.

Freud, Sigmund. 1905–1915. Three essays on the theory of sexuality.

Gates, Richard G. 1978. Vision in the monotreme Echidna (*Tachyglossus aculeatus*). *Australian Zoologist,* 20(1):147–169.

Gilley, Sheridan. 1981. The Huxley-Wilberforce debate: A reconsideration. In K. Robbins, ed., *Religion and humanism,* Vol. 17, pp. 325–340.

Gilovich, T., R. Vallone, and A. Tversky. 1985. The hot hand in basketball: On the misperception of random sequences. *Cognitive Psychology,* 17:295–314.

Gladstone, W.E. 1858. *Studies on Homer and the Homeric Age.*

Gladstone, W.E. 1885. Dawn of creation and of worship. *The Nineteenth Century,* November.

Goldschmidt, R. 1940 (reprinted 1982 with introduction by S.J. Gould). *The material basis of evolution.* New Haven: Yale University Press.

Goldschmidt, R. 1948. Glowworms and evolution. *Révue Scientifique,* 86:607–612.

Goodall, Jane. 1986. *The chimpanzees of Gombe.* Cambridge, MA: Harvard University Press.

Gould, S.J. 1975. On the scaling of tooth size in mammals. *American Zoologist,* 15:351–362.

Gould, S.J. 1977. *Ontogeny and phylogeny.* Cambridge, MA: Harvard University Press.

Gould, S.J. 1980. *The panda's thumb.* New York: W.W. Norton.

Gould, S.J. 1981. *The mismeasure of man.* New York: W.W. Norton.

Gould, S.J. 1985. *The flamingo's smile.* New York: W.W. Norton.

Gould, S.J. 1985. Geoffroy and the homeobox. *Natural History,* November, pp. 22–23.

Gould, S.J. 1987. Bushes all the way down. *Natural History,* June, pp. 12–19.

Gould, S.J. 1987. *Time's arrow, time's cycle.* Cambridge, MA: Harvard University Press.

Gould, S.J. 1989. Winning and losing: It's all in the game. *Rotunda,* Spring 1989, pp. 25–33.

Gould, S.J., and Elisabeth S. Vrba. 1982. Exaptation—a missing term in the science of form. *Paleobiology,* 8(1):4–15.

Grant, Tom. 1984. *The Platypus.* New South Wales University Press.

Grant, Verne. 1963. *The origin of adaptations.* New York: Columbia University Press.

Gregory, William King. 1927. *Hesperopithecus* apparently not an ape nor a man. *Science,* 66:579–587.

Gregory, William King, and Milo Hellman. 1923. Notes on the type of *Hesperopithecus haroldcookii* Osborn. American Museum of Natural History *Novitates,* 53:1–16.

Gregory, William King, and Milo Hellman. 1923. Further notes on the molars of *Hesperopithecus* and of *Pithecanthropus. Bulletin of the American Museum of Natural History,* Vol. 48, Art. 13, pp. 509–526.

Griffiths, M. 1968. *Echidnas.* New York: Pergamon Press.

Griffiths, M. 1978. *The biology of monotremes.* San Diego/Orlando: Academic Press.

Haller, John S. 1971. *Outcasts from evolution.* University of Illinois Press.

Hegner, Robert W. 1912. *College zoology.* New York: Macmillan Co.

Heinrich, B. 1981. *Insect thermoregulation.* San Diego/Orlando: Academic Press.

Hite, Shere. 1976. *The Hite Report: A nationwide report on sexuality.* New York: Macmillan.

Holway, John. 1987. A little help from his friends. *Sports Heritage,* November/December.

Home, Everard. 1802. A description of the anatomy of the *Ornithorhynchus paradoxus*. *Philosophical Transactions of the Royal Society,* Part 1, No. 4, pp. 67–84.

Howes, G.B. 1888. Notes on the gular brood-pouch of *Rhinoderma darwini*. *Proceedings of the Zoological Society of London,* pp. 231–237.

Hrdy, Sarah. 1981. *The women that never evolved.* Cambridge, MA: Harvard University Press.

Hunter, George William. 1914. *A civic biology.* New York: American Book Company.

Hutton, J. 1788. Theory of the Earth. *Transactions of the Royal Society of Edinburgh,* 1:209–305.

Hutton, J. 1795. *The theory of the Earth.* Edinburgh: William Creech.

Huxley, Julian. 1927. On the relation between egg-weight and body-weight in birds. *Journal of the Linnaean Society of London,* 36:457–466.

Huxley, Leonard. 1906. *Life and letters of Thomas H. Huxley.* New York: Appleton.

Huxley, T. H. 1880. On the application of the laws of evolution to the arrangement of the Vertebrata, and more particularly of the Mammalia. *Proceedings of the Zoological Society of London,* No. 43, pp. 649–661.

Huxley, T. H. 1885. The interpreters of Genesis and the interpreters of Nature. *The Nineteenth Century,* December.

Huxley, T. H. 1888. The struggle for existence in human society. *The Nineteenth Century,* February.

Huxley, T. H. 1896. *Science and education.* New York: Appleton & Co.

Huxley, T. H. 1896. Mr. Gladstone and Genesis. *The Nineteenth Century.*

Huxley, T.H., and Julian Huxley. 1947. *Evolution and ethics.* London: Pilot Press.

Ingersoll, Ernest. 1906. *The life of animals.* New York: Macmillan.

Ingram, G.J., M. Anstis, and C.J. Corben. 1975. Observations on the Australian leptodactylid frog, *Assa darlingtoni. Herpetologia,* 31:425–429.

Jackson, J.F. 1974. Goldschmidt's dilemma resolved: Notes on the larval behavior of a new neotropical web-spinning Mycetophilid (Diptera). *American Midland Naturalist,* 92(1):240–245.

James, William. 1909. Letter quoted in *The autobiography of Nathaniel Southgate Shaler.* Boston: Houghton Mifflin.

Jenkin, F. 1867. Darwin and the origin of species. *North British Review,* June.

Jerison, H.J. 1973. *Evolution of the brain and intelligence.* San Diego/Orlando: Academic Press.

Jordan, David Starr, and Vernon L. Kellogg. 1900. *Animal life: A first book of zoology.* New York: D. Appleton & Co.

Kellogg, Vernon L. 1907. *Darwinism to-day.* London: G. Bell & Sons.

Kellogg, Vernon L. 1917. *Headquarters nights.* Boston.

Keyes, Ralph. 1980. *The height of your life.* Boston: Little, Brown.

Kidd, Benjamin. 1894. *Social evolution.* New York/London: Macmillan & Co.

Kidd, Benjamin. 1918. *The science of power.* New York.

Kingsland, S. 1978. Abbott Thayer and the protective coloration debate. *Journal of the History of Biology,* 11(2):223–244.

Kingsolver, J.G., and M.A.R. Koehl. 1985. Aerodynamics, thermoregulation, and the evolution of insect wings: Differential scaling and evolutionary change. *Evolution,* 39:488–504.

Kinsey, A.C. 1929. The gall wasp genus *Cynips:* A study in the origin of species. *Indiana University Studies,* Vol. 16.

Kinsey, A.C. 1936. The origin of higher categories in *Cynips. Indiana University Publications Science Series,* No. 4.

Kinsey, A.C., W.B. Pomeroy, C.E. Martin, and P.H. Gebhard. 1953. *Sexual behavior in the human female.* Philadelphia: W.B. Saunders.

Kirby, W., and W. Spence. 1863. *An introduction to entomology,* 7th ed. London: Longman, Green, Longman, Roberts & Green.

Kirwan, Richard. 1809. *Metaphysical essays.* London.

Kropotkin, P.A. 1902. *Mutual aid: A factor of evolution.* New York: McClure Phillips.

Lacépède, B.G.E. 1801 (1'an IX de la République). *Discours d'ouverture et de clôture du cours de zoologie donné dans le Muséum d'Histoire naturelle.* Paris: Plassan.

Lamarck, J.-B. 1809. *Philosophie zoologique.* Paris.

Lavoisier, A., B. Franklin et al. 1784. *Rapport des commissaires chargés par le roi de l'examen du magnétisme animal.* Paris: Imprimerie royale.

Lende, R.A. 1964. Representation in the cerebral cortex of a primitive mammal: Sensorimotor, visual and auditory fields in the Echidna (*Tachyglossus aculeatus*). *Journal of Neurophysiology,* 27:37–48.

Levine, L.W. 1965. *Defender of the faith: William Jennings Bryan, the last decade.* New York: Oxford University Press.

Linnaeus, C. 1758 (facsimile reprint 1956). *Systema naturae. Regnum animale.* London: British Museum (Natural History).

Livingstone, David N. 1987. *Nathaniel Southgate Shaler and the culture of American science.* University of Alabama Press.

Lovejoy, A.O. 1936. *The great chain of being.* Cambridge, MA: Harvard University Press.

Lucas, J.R. 1979. Wilberforce and Huxley: A legendary encounter. *The Historical Journal,* 22:313–330.

Lyell, Charles. 1830–1833. *Principles of geology, being an attempt to explain the former changes of the earth's surface by reference to causes now in operation.* London: John Murray.

MacFadden, B.J. 1986. Fossil horses from "Eohippus" (*Hyracotherium*) to *Equus:* Scaling, Cope's law, and the evolution of body size. *Paleobiology,* 12(4):355–369.

Marsh, O.C. 1874. Notice of new equine mammals from the Tertiary Formation. *American Journal of Science,* Series 3, Vol. 7, pp. 247–258.

Marsh, O.C. 1877. Notice of new dinosaurian reptiles from the Jurassic Formation. *American Journal of Science,* 14:514–516.

Marsh, O.C. 1879. Principal characters of American Jurassic dinosaurs, Part II. *American Journal of Science,* 17:86–92.

Marsh, O.C. 1879. Notice of new Jurassic reptiles. *American Journal of Science,* 18:501–505.

Marsh, O.C. 1883. Principal characters of American Jurassic Dinosaurs. Part VI: Restoration of *Brontosaurus. American Journal of Science,* Series 3, Vol. 26, pp. 81–85.

Marsh, O.C. 1896. *The dinosaurs of North America.* 16th Annual Report of the U.S. Geological Survey, Washington, D.C., Part 1, pp. 133–414.

Marshack, Alexander. 1985. *Hierarchical evolution of the human capacity: The Paleolithic evidence.* James Arthur Lecture on the evolution of the human brain. New York: American Museum of Natural History.

Masters, William H., and Virginia E. Johnson. 1966. *Human sexual response.* Boston: Little, Brown.

Matthew, W.D. 1903. *The evolution of the horse.* American Museum of Natural History pamphlet.

Matthew, W.D. 1926. The evolution of the horse: A record and its interpretation. *Quarterly Review of Biology,* 1(2):139–185.

Mayr, Ernst. 1982. *The growth of biological thought.* Cambridge, MA: Harvard University Press.

McDiarmid, R.W. 1978. Evolution of parental care in frogs. In G.M. Burghardt and M. Bekoff, eds., *The development of behavior: Comparative and evolutionary aspects.* New York: Garland Press.

Mivart, St. George J. 1871. *On the genesis of species.* New York: D. Appleton & Co.

Moore, Ruth. 1957. *Charles Darwin.* London: Hutchinson.

Morris, Desmond. 1969. *The naked ape.* New York: Dell Publishing Co.

Nisbet, R. 1980. *History of the idea of progress.* New York: Basil Books.

Osborn, Henry Fairfield. 1904. The evolution of the horse in America. *Century Magazine,* November.

Osborn, Henry Fairfield. 1921. Preface to Madison Grant's *The passing of the great race.* New York: C. Scribner's Sons.

Osborn, Henry Fairfield. 1922. *Evolution and religion.* New York: Charles Scribner's Sons.

Osborn, Henry Fairfield. 1922. *Hesperopithecus,* the first anthropoid Primate found in America. *American Museum of Natural History Novitates,* No. 37, pp. 1–5.

Osborn, Henry Fairfield. 1925. *The Earth speaks to Bryan.* New York: Charles Scribner's Sons.

Othar, L.K., and J. Rhodes. 1978. Instrumental learning in the Echidna (*Tachyglossus aculeatus setosus*). *Australian Zoologist,* 20(1):131–145.

Otto, J.H., and A. Towle. 1977. *Modern biology.* New York: Holt, Rinehart & Winston.

Paul, Diane B. 1987. The nine lives of discredited data. *The Sciences,* May.

Pirlot, P., and J. Nelson. 1978. Volumetric analysis of monotreme brains. *Australian Zoologist,* 20(1):171–179.

Pirsson, L.V., and C. Schuchert. 1924. *Textbook of geology,* Part 2. New York: John Wiley & Sons, Inc.

Playfair, John. 1802. *Illustrations of the Huttonian theory of the Earth.* Edinburgh: William Creech.

Poulton, E.B. 1894. The structure of the bills and hairs of *Orni-thorhynchus* with a discussion of the homologies and origin of mammalian hair. *Quarterly Journal of Microscopical Science,* 36: 143–199.

Prothero, D.R., and Neil Shubin. 1989. The evolution of Oligo-cene horses. In D.R. Prothero and R.M. Schoch, eds., *The evolution of perissodactyls,* pp. 142–175. Oxford: Oxford University Press.

Pugh, George E. 1977. *The biological origin of human values.* New York: Basic Books.

Rahn, H., C.G. Paganelli, and A. Ar. 1975. Relation of avian egg weight to body weight. *The Auk,* 92(4):750–765.

Read, W. Maxwell. 1930. *The Earth for Sam.* New York: Harcourt, Brace & Co.

Reid, B., and G.R. Williams. 1975. The kiwi. In G. Kuschel, ed., *Biogeography and ecology in New Zealand,* pp. 301–330. The Hague: D.W. Junk.

Renfrew, Colin. 1987. *Archaeology and language.* Cambridge, En-gland: Cambridge University Press.

Réville, Alfred. 1881. *Prolegomenes de l'histoire des religions.* Paris: Fischbacher.

Riggs, Elmer. 1903. Structure and relationships of opisthocoel-ian dinosaurs. Part 1: *Apatosaurus. Geological Series of the Field Columbian Museum,* Publication 82, Vol. 2, No. 4, pp. 165–196.

Roger, J. 1976. "William Whiston," entry in the *Dictionary of Scientific Biography,* Vol. 14, pp. 295–296. New York: Charles Scribner's Sons.

Romer, A.S. 1966. *Vertebrate paleontology,* 3rd ed. Chicago: Univer-sity of Chicago Press.

Roosevelt, Theodore. 1911. Revealing and concealing coloration in birds and mammals. *Bulletin of the American Museum of Natural History,* 30(8):119–231.

Sagan, Carl, and A. Druyan. 1985. *Comet.* New York: Random House.

Saint-Hilaire, Etienne Geoffroy. 1803. Extrait des observations anatomiques de M. Home, sur l'échidné. *Bulletin des Sciences par la Société philomathique,* an 11 de la République.

Saint-Hilaire, Etienne Geoffroy. 1827. Sur les appareils sexuels et urinaires de l'Ornithorhynque: *Mémoires du Muséum d'Histoire Naturelle.*

Schuchert, Charles. 1928. Charles Doolittle Walcott (1850–1927). *Proceedings of the American Academy of Arts and Sciences,* 62: 276–285.

Schweber, S.S. 1977. The origin of the *Origin* revisited. *Journal of the History of Biology,* 10:229–316.

Scott, W.B. 1913. *A history of land mammals in the Western Hemisphere.* New York: Macmillan.

Shaler, N.S. 1861. Lateral symmetry in Brachiopoda. *Proceedings of the Boston Society of Natural History,* 8:274–279.

Shaler, N.S. 1901. *The individual: A study of life and death.* New York: D. Appleton & Co.

Shaw, George. 1799. *The naturalist miscellany,* Plate 385.

Simpson, G.G. 1951. *Horses.* Oxford, England: Oxford University Press.

Smith, Grafton Elliot. 1902. On a peculiarity of the cerebral commissures in certain Marsupialia, not hitherto recognized as a distinctive feature of the Diprotodonta. *Proceedings of the Royal Society of London,* Vol. 70, No. 462, pp. 226–230.

Smith, W.H. 1975. *The social and religious thought of William Jennings Bryan.* Coronado Press.

Solecki, Ralph S. 1971. *Shanidar, the first flower people.* New York: Knopf.

Stanton, William. 1960. *The leopard's spots.* Chicago: University of Chicago Press.

Stebbins, G. Ledyard. 1966. Chromosomal variation and evolution. *Science,* 152:1463–1469.

Stevenson, R.L. 1887. *Papers literary, scientific, etc. by the late Fleeming Jenkin.* London: Longmans, Green & Co.

Strahan, R., ed. 1983. *Complete book of Australian mammals.* Sydney: Angus & Robertson.

Symons, Donald. 1979. *The evolution of human sexuality.* New York: Oxford University Press.

Taborsky, Michael. 1988. Kiwis and dog predation: Observations in Waitangi State Forest. *Notornis,* 35:197–202.

Thayer, Abbott H. 1896. The law which underlies protective coloration. *The Auk,* 13:124–129.

Thayer, Abbott H. 1900. Arguments against the banner mark theory. *The Auk,* 17(2):108–113.

Thayer, Abbott H. 1903. Protective coloration in its relation to mimicry, common warning colours, and sexual selection.

*Transactions of the Entomological Society of London,* Article 26, pp. 553–569.

Thayer, Abbott H. 1909. An arraignment of the theories of mimicry and warning colors. *Popular Science,* December.

Thayer, Abbott H. 1909. *Concealing-coloration in the animal kingdom.* New York: Macmillan.

Thayer, Abbott H. 1911. Concealing coloration. *The Auk,* 28: 146–148.

Todes, Daniel P. 1988. Darwin's Malthusian metaphor and Russian evolutionary thought, 1859–1917. *Isis,* 78:537–551.

Tyler, M.J. 1983. *The gastric brooding frog.* London/Canberra: Croom Helm.

Ulinski, Philip S. 1984. Thalamic projections to the somatosensory cortex of the Echidna (*Tachyglossus aculeatus*). *Journal of Comparative Neurology,* 229(2):153–170.

Vonnegut, Kurt. 1963. *Cat's cradle.* New York: Delacorte Press.

Vorzimmer, Peter J. 1970. *Charles Darwin: The years of controversy.* Temple University Press.

Walcott, Sidney. 1971. How I found my own fossil. *Smithsonian.*

Whiston, William. 1696. *A new theory of the Earth from its original to the consummation of all things, wherein the creation of the world in six days, the universal deluge, and the general conflagration, as laid down in the Holy Scriptures, are shewn to be perfectly agreeable to reason and philosophy.* London.

Whiston, William. 1708. *New theory,* 2d ed. London: B. Tooke.

White, Nelson G. 1951. *Abbott H. Thayer: Painter and naturalist.* Connecticut Printers.

Wilberforce, W. 1860. Review of "On the Origin of Species." *Quarterly Review,* 108:225–264.

Wolf, John, and J.S. Mellett. 1985. The role of "Nebraska Man" in the creation-evolution debate. *Creation/Evolution,* 5: 31–43.

# Index

Abbie, A.A., 287–88
Abel, Othenio, 442
abortion, 57–58
Adam (biblical), 239–40, 372
Adams, Thomas Boylston, 22
adaptationism, 60, 128, 137
*Aepyornis*, 110
Agassiz, Elizabeth Cary, 314
Agassiz, Louis
  fossils collected by, 309
  William James and, 318–21
  Shaler and, 312–18
  Shaler's letter to, 15
agnosticism, 403, 406
Agricola, Georgius, 402
Allen, Francis H., 209, 210
Allen, Garland E., 165
allometry, 117–20
alloploidy, 484
Altholz, Joseph L., 397–98
*Alytes obstetricans* (midwife toad), 301
Amadon, Dean, 110
*Anchitherium*, 169
Anderson, Sir Charles, 397
Andrews, Roy Chapman, 95
Angell, Roger, 159–60, 163
animal magnetism, 184–87
  Royal Commission (France, 1784)
    to investigate, 188–96
Anstis, M., 301
ant eaters, *see* echidnas
anthropic principle, 115

Antioch College, 341
*Apatosaurus*, 86–93
*Apemon*, 263n
*Apteryx* (kiwis), 110
*Apteryx australis* (brown kiwi), 503–5
Ar, A., 118
*Arachnocampa luminosa* (glowworms),
  261–65
*Archaeopteryx*, 144–45
architecture, 61–62
Arian heresy, 368
Aristotle, 256, 458–59, 501
Artiodactyla, 181
asbestos, 474
*Assa darlingtoni* (frog), 301
association cortexes, 289
assortative mating, 346–47
*Athenaeum*, 390–92
atheism, accusations of, 448–49
Austen, Jane, 54
autoploidy, 483–84
Aztecs, 168–69

Bacall, Lauren, 479
Bach, Johann Sebastian, 366
Backus, David, 160
Bacon, Francis, 475
Baker, Jeffrey J.W., 165
*Baluchitherium*, 181
Barbour, T., 222–24
Barnum, P.T., 42, 45

baseball
DiMaggio's hitting streak in,
    463–65, 467, 468, 470–72
frequency of streaks of winning
    and losing games in, 466
origins of, 46–58
Baseball Hall of Fame (Cooperstown,
    New York), 46, 52
basketball, 465–66
Bates, Henry Walter, 294–96, 304
Batesian mimicry, 295–97
*Beagle*, HMS (ship), 25–28, 297
beauty, Camper's conception of,
    234–37
Beecher, Henry Ward, 400
Bell, Christopher, 352
Belloc, Hilaire, 474
Berman, D.S., 87
Bible
    Bryan on, 421
    Genesis in, 403–15, 455
Bierer, L.K., 155
biological determinism, 428
biology
    cultural evolution compared with,
        63–66
    taxonomy in, 80–85
    textbooks in, 155–67
Bird, Larry, 466
birds
    *Apteryx australis* (brown kiwi), 503–5
    *Archaeopteryx*, 144–45
    chromosomes of, 481
    flamingos, 210–11, 216–19, 226
    kiwis, 110–17, 121
    peacocks, 219–21
    relative size of eggs of, 117–18
blacks
    facial angle of skulls of, 237–39
    in Magi tales, 229–30, 233–34
    Shaler on, 316
blending inheritance, 343–44
Bligh, William, 281–82
boa constrictor, 84
Bogart, Humphrey, 479
Bohringer, R.C., 279, 280
Boveri, Theodor, 481–82
boxing, 432

brachiopods, 314–15
brains
    of echidnas, 285–92
    relative weights of, 118–19, 121
Brazil, 318
breasts, 127
British Association for the
    Advancement of Science, 81–82,
    385
Brody, Samuel, 118, 119
*Brontosaurus*, 80
    controversy over name of, 85–93
brooding, in frogs, 300–305
Brown, Barnum, 442
Brown, Dayton Reginald Evans, 227
Browne, Janet, 389, 398
brown kiwis *(Apteryx australis)*, 503–5
Brownlie, A.D., 352, 353
Bryan, William Jennings, 175, 415,
    417–19
    debate between Osborn and,
        432–36, 441, 443–44, 446
    evolution opposed by, 419–23,
        427–28
    at Scopes trial, 416
Buchmann, 290–91
Buffon, George Louis Leclerc de, 357,
    369
*Bufo marinus* (toad), 303
Burgess Shale fossils, 242–50, 252
Burnet, Thomas, 372
Burns, Robert, 380, 436
Burrell, Harry, 270, 278
Burt, Sir Cyril, 156–57, 226
Busse, K., 302
Butler, Samuel, 260
butterflies
    Batesian mimicry in, 295
    development of, 258, 259
    disruptive markings of, 215
    mimicry in, 304

Calder, William A., III, 115–17
Caldwell, W.H., 269–70, 272–75
calendars, French Revolutionary,
    355–56
*Camarosaurus*, 91–92

Cambrian explosion, 242
camouflage, 211
    disruptive markings in, 215
    naval, 227–28
Camper, Petrus, 230–40
Canada, 22
cancer, 474–75
cannibalism, 502
Canning, George, 16, 23–24, 27, 29, 31
Capra, F., 339
Cardiff Giant, 42–45
Carnegie, Andrew, 242
Carrick, Frank, 277
Carson, Johnny, 106
Carter, D.B., 298
Carter, Jimmy, 15, 321–24, 479
Cartwright, Alexander Joy, 56–57
Castlereagh, Viscount (Robert Stewart), 23–29, 449
catastrophism, Whiston's, 368, 379
caterpillars, 258, 259
Cavalli-Sforza, L.L., 35–37
Cavendish, Henry, 359
cerebral cortexes, 289
Cerion (land snail), 241
Chadwick, Henry, 51
Chambers, Robert, 280
chaos theory, 505
Chapman, Frank M., 225
Chapman, John Jay, 224
chemistry, Lavoisier as founder of, 358–60
Chen, 290
Chesterton, G.K., 460
chimpanzees, 501–2, 504, 505
chromosomes, 481–86
Churchill, Winston, 416
cicadas, 260
clams, 314–15
clitoris, 127–31
Cnidaria (corals), 259
Cobb, Ty, 467
Coffinhal, 354
Colbert, Ned, 95
Coletta, Paolo E., 419
Colias butterflies, 148

colors of animals
    Batesian mimicry in, 295–97
    Thayer's theories of, 210–25
comets, 369–70
    role in mass extinctions of, 378–79
    in Whiston's history of earth, 371, 373, 375
competitive behavior, 328, 332, 333
    Kropotkin on, 334–38
Condorcet, Marquis de Marie Jean Antonie Nicholas de Caritat, 124
consciousness, 31
contingent events, 69
Cook, Harold J., 434, 435, 437, 438, 443, 444
cooperative behavior, 331, 336–38
Cooperstown (New York)
    Baseball Hall of Fame at, 45, 46
    baseball "invented" in, 49–50
    Cardiff Giant located at, 42
Cope, Edwin D., 162
    Marsh and, 87, 170
    on natural selection, 139, 140
corals, 259
Corben, C.J., 298, 301
Corday, Charlotte, 362
Corliss, John, 83, 84
corpus callosum, 287–88
Cortés, Hernando, 168–69
cortexes, 289
cosmology
    anthropic principle in, 115
    in Genesis, 404, 413
Cott, Hugh B., 212
countershading, 210–21, 225
Crandall, L.S., 68
craniometry, 231
creationism
    Agassiz's belief in, 312
    court decisions on, 431
    Darwin on arguments against, 295–96
    Gladstone's, 403–15
    Hesperopithecus fossil and, 445
    in high-school textbooks, 165
    Paluxy Creek "human" footprints and, 447
    as "science," 455–56

creationism (*continued*)
Scopes trial of, 416–19
U.S. Supreme Court on, 456–59
in Whiston's history of earth, 372–73, 375
Cretaceous mass extinction, 17, 378–79
Crick, Francis, 137
Crosland, Michael W.J., 482, 483
Crozier, Ross H., 482, 483
cultural evolution, 63–66
Cuvier, Georges, 314

Danilevsky, N.I., 333–34
DaPonte, Lorenzo, 184*n*
Darnton, Robert, 184, 187–88
Darrow, Clarence, 416, 432–33
Darwin, Charles, 57, 60, 124
admired by Russian intellectuals, 332
argument between Jenkin and, 343–51
on Bates, 295–96
birthdate of, 21
Bryan on, 420, 433
Danilevsky on, 333–34
debate between Huxley and Wilberforce over, 385–401
discussed in high-school textbooks, 165–66
on evolutionary struggle for existence, 327–29
evolution defined by, 460
FitzRoy and, 25–29
Gladstone's attempt to reconcile Genesis with, 404–5
on incipient stages, 143–44, 305
inspired by Malthus, 333
on intermediate forms of vertebrate eye, 263*n*
Jenkin's argument with, 340–41
Kropotkin on, 331, 335, 338
limit of theorizing by, 455
on mimicry in animals, 296–97, 304
Mivart on, 139–43
on natural laws, 30

principle of functional shift of, 145–51
Shaler on, 314
social Darwinism and, 424–27
Tolstoy on, 327
*see also* evolution
Darwin, Erasmus, 124–26, 136, 137
Darwin, Francis, 386, 389, 391, 393, 395, 398
David (biblical), 294, 312
David, Paul A., 63, 68, 70
Davis, D. Dwight, 127, 128
Dawkins, W. Boyd, 392
Deimos (moon of Mars), 507
Del Valle, Edith, 203, 205
Dempsey, Jack, 432
Deslon, Charles, 191–94
Diamond, Jared, 503
Dickens, Charles, 124
DiMaggio, Joe, 463–65, 467, 468, 470–72
*Dinornis maximus*, 110
dinosaurs, 103
*Archaeopteryx*, 144–45
*Brontosaurus* name controversy, 85–93
Cretaceous mass extinction of, 378–79
recent popularity of, 94–99, 106
*Diplodocus*, 91–92
Diptera (flies), 481
Disraeli, Benjamin, 397, 402–3, 473
disruptive markings, 215–16
DNA (deoxyribonucleic acid), 480
dogs, 503–4
Doubleday, Abner, 46, 48–52, 58
Douglas, M.M., 148
Draper, 386, 390, 397, 398
*Drosophila* (fruit flies), 481
Dunbar, Carl, 243
Duplantier, 431
Dvorak, August, 63
Dvorak Simplified Keyboard (DSK), 63
dwarfism, 120–21
Dyson, Freeman, 115
Dyster, 394

Earth
  debate over age of, 448–54
  geological activity on, 490–91
  Venus compared with, 493–94
  Whiston's history of, 370–76
echidnas (anteaters)
  brains of, 285–92
  collected by Calswell, 274–75
  described by Bligh and Tobin, 282
  Geoffroy Saint-Hilaire on, 283–84
  as monotreme, 270, 271
  snouts of, 278–79
ecology, Darwin on, 328
economics, Jenkin's contributions to,
  351–53
Edison, Thomas, 68
education
  Bryan's attempt to ban teaching of
    evolution in, 419, 422–23,
    433–34
  science in, 99–100, 103
  state laws against teaching
    evolution in, 431
  status of teachers in, 101–2
  textbooks used in, 155–67
  U.S. Supreme Court on evolution
    taught in, 456–59
Edwards v. Aguillard (U.S., 1987), 403,
  416, 456–59
eggs
  of kiwis, 111–17, 121
  of platypuses, 270, 272–75
  relative weights of, 117–20
  of Rheobatrachus silus (frog), 303–4
  of Rhinoderma darwini (frog), 297
Eibl-Eibesfeldt, I., 131
Ekelund, Robert B., Jr., 351–52
embryology, 256
  Erasmus Darwin on, 125
  debate over differentiation in, 414
  of humans, 127
Emil, David, 104
encephalization quotient (EQ), 286
Enlightenment, 359
environment, struggle against, 334–38
environmentalism, 18
Eohippus (Hyracotherium), 175
  in evolutionary sequence, 178, 180

  fox terrier simile for, 158–65,
    173–74
  name of, 90
  weight of, 167
epidemiological statistics, 474–77
Epperson v. Arkansas (U.S., 1968), 431
Equus (horse), 169, 175, 181
Erasmus, Desiderius, 137–38
Eve (biblical), 239–40, 372
evolution
  Agassiz on, 312
  argument between Darwin and
    Jenkin over variations in, 343–50
  of baseball, 48, 52–58
  Batesian mimicry in, 295
  biological and cultural, 63–66
  of brooding in frogs, 300–301
  Bryan's campaign against, 417–23,
    419–20, 427–28
  Bryan's campaign against teaching
    of, 433–34
  Chambers's defense of, 280
  cooperation in, Kropotkin on,
    330–31
  Darwin's definition of, 460
  in Erasmus Darwin's Zoonomia, 125
  debate between Huxley and
    Wilberforce over, 385–401
  discussed in high-school textbooks,
    155–56, 165–67
  genetic measurement of distance in,
    35–39
  Gladstone's attempt to reconcile
    Genesis with, 404–5
  of horses, 158–65, 169–81
  human, Jimmy Carter on, 322–24
  of human sex organs, 129
  of human sexuality, 130–33
  imperfections in, 61
  incipient stages dilemma in, 143–44
  of kiwis, 112–13, 121
  of language, 34–36, 39–40
  of mimicry, 304
  Mivart on, 139–41
  not study of life's origins, 455
  passage of time in, 252
  platypus, place in, 275–78
  preadaptation in, 144n

evolution (*continued*)
    public acceptance of, 60
    Russian discussions of, 332–34
    Shaler on, 315–20
    social Darwinism and, 424–27
    state laws against teaching of, 431
    struggle for existence in, 327–28,
        334–38
    Tolstoy on, 327
    U.S. Supreme Court on, 456–59
    of vertebrate eye, 263*n*
    of wings, 141–43, 145–51
exaptation, 144*n*
Ext, Anna, 203–4
extinctions
    Cretaceous, 17
    Permian, 17–18
    of *Rheobatrachus silus* (frog), 306
    role of comets in, 378–79
eyes, evolution of, 263*n*

facial angle
    Camper's discovery of, 230–33
    in conceptions of human beauty,
        234–37
    racial differences in, 237–39
Farmer's Museum (Cooperstown, New
        York), 42
Farrar, Canon, 395–96
females
    in haplodiploidal reproduction, 483
    human, clitoris of and orgasms in,
        128–36
    kiwis, 111–12
    *Rheobatrachus silus* (frog), 298–99
Ferme Générale (Tax Farm; France),
        360–64
fertilization, 57
Fitzgerald, Edward, 467
FitzRoy, Robert, 25–29, 387, 391
flamingos, 226
    red color of, 210–11, 216–19
Flausen, von, 424
flies, chromosomes of, 481
Flower, J.W., 146, 147
Flower, William Henry, 287
flying animals, 407

Forbes, David, 389
Force, James, 378
Forestier, Amedee, 444–45
fossils
    from Burgess Shale, 243, 245–50,
        252
    discussed in high-school textbooks,
        155–56
    in Harvard's Museum of
        Comparative Zoology, 309–12
    of *Hesperopithecus*, 434–47
    of horses, 169
    of soft-bodied organisms, 242
fox terriers, 155, 159–65, 167, 173–74
Foxx, Jimmy, 464
France
    Napoleon's coup in, 500
    pre-revolutionary tax collection in,
        360–62
    Revolution in (1789), 355–58, 363
Francis (saint), 11, 12
Franklin, Benjamin, 184, 187, 363
    on Royal Commission (France,
        1784), 189–97
frauds (hoaxes)
    animal magnetism, 184–96
    Burt's twin studies, 156–57, 226
    Cardiff Giant, 42–45
    mermaids, 271
    Paluxy Creek "human" footprints,
        447
    Piltdown Man, 444, 445
    in *Weekly World News*, 182–83
French Revolution (1789), 355–58,
        360, 363
Freud, Sigmund, 129, 133–36
frigidity
    Freud on, 134
    Kinsey on, 136
frogs
    brooding in, 300–302
    extinctions of, 306
    *Rheobatrachus silus*, 294, 297–300,
        302–5
    *Rhinoderma darwini*, 297
fruit flies (*Drosophila*), 481
functionalism, 130, 136–37
functional shift principle, 145–47

Gaia theory, 339
Galileo Galilei, 11, 12, 354
Gandhi, Mohandas, 325
Garfield, James A., 310
*Gastrotheca riobambae* (frog), 301
Gates, Richard G., 291
Geikie, Archibald, 160
genes, 345
  carried on chromosomes, 482
Genesis (bible), 413–15
  in creationist arguments, 455–56
  Gladstone on, 403–6, 410
  T.H. Huxley on, 406–8, 411–12
genetics
  complexity in, 480–81
  evolutionary distances measured by,
    35–39
  number of chromosomes and,
    481–86
  textbooks discussions of, 156–57
Geoffroy Saint-Hilaire, Etienne,
    272–75, 283–84, 292–93
Germany, social Darwinism in, 424–26
Ghent, Treaty of (1814), 22, 28
Gilbert, Sir William Schwenck, 305
Gilbreth, Frank B., 63
Gilley, Sheridan, 389, 399
Gilman, Page, 105
giraffes, 165, 166
Gladstone, William Ewart, 402–15
glowworms, 260–65, 267
God, 412–15
  Bryan on, 433
  Jimmy Carter on, 323
  in creation science, 456
  Darwin on, 30
  in debate over age of earth, 450
  Gladstone on, 405–6
  Hutton on, 452
  Shaler on existence of, 317,
    319–20
  in Whiston's history of earth,
    371–73, 375–77
Goldschmidt, Richard, 263*n*
Goliath (biblical), 294
Goodall, Jane, 501–2
Gordon, Charles George ("Chinese"),
    343, 403

Grant, Eli, 310–12, 320–21
Grant, Madison, 162
Grant, Tom, 277
Grant, Vernon, 484
Graves, Abner, 48, 50
Gray, Asa, 30, 397
Great Britain
  in War of 1812, 22–23
Greeks, ancient, 234–37
Green, J.R., 392–94
Greenberg, Hank, 464
"green" movements, 17
Gregory, William King, 434, 436,
    438–42, 446
Grenada, 499
Griffiths, M., 285
Grimm, Jacob, 32, 33
Grimm, Wilhelm, 32
Guillotin, Joseph Ignace, 189, 356

Haller, John S., 232
Halley, Edmond, 369–70
Hancock, Nathaniel, 369
Handel, George Fredrick, 59–60
haplodiploidy, 483
Hardy, Oliver, 255
Hawking, Stephen, 103
Hawkins, Waterhouse, 97
Hegel, Georg Wilhelm Friedrich, 500
height (human), 479
Heitz, Tom, 46, 48, 56
Hellman, Milo, 440, 441
Henrich, Tommy, 470
Henslow, John Stevens, 391, 392,
    397
Hepburn, Katharine, 479
heredity
  chromosomes in, 481–82
  of IQs, textbook discussions of,
    156–57
  Mendelian, 345
  pre-Mendelian view of, 340,
    343–44, 346–47
  in stories of Jukes and Kallikak
    families, 429
  of variations in single individuals,
    341

*Hesperopithecus* ("Nebraska Man"),
    435–47
Hinton, E.H., 146
*Hipparion,* 169
hippopotamuses, 121
Hite, Shere, 128, 129
Hitler, Adolf, 499
hoaxes, *see* frauds
Hobbes, Thomas, 329, 334
Hogarth, William, 369
Holway, John, 470–71
Home, Everard, 272, 279, 282
Homer, 404
Honig, Donald, 51
Hooker, Joseph, 341, 391–93, 396–98,
    401
Hoover, Herbert, 402
Hornsby, Rogers, 467
horses
    evolution of, 169–81
    evolution of, discussed in textbooks,
        158–65
    introduced into Americas, 168–69
    racehorses, 348
Houdini, Harry, 183
Howes, G.B., 297
Hrdy, Sarah, 131–32, 136
Hull, George, 42, 44
humans
    in anthropic principle, 115
    beginning of personhood in, 57
    conceptions of beauty in, 234–37
    contemporaneousness of dinosaurs
        and, 95–96
    development of, 256, 259
    embryonic development of, 414
    evolution of, Jimmy Carter on,
        322–24
    evolution of, high-school textbooks
        on, 165
    evolution of, Shaler on, 317–20
    evolution of languages of, 34–36,
        39–40
    female, clitoris of and orgasms in,
        128–36
    genetic data on evolution of, 37–39
    in "green" analysis, 17
    height of, 479

*Hesperopithecus* fossil mistaken for,
    434–46
T.H. Huxley on struggle for
    existence among, 329
Kropotkin on cooperation among,
    331
male, nipples of, 125–28
measurement of skulls of, 230–33,
    237–40
Paluxy Creek "footprints" of, 447
polygenist view of racial groups of,
    316
reproduction of, Erasmus Darwin
    on, 125
social Darwinism and, 424–27
Hummer, P.J., 165
Hunter, George William, 174–75,
    428–29
Hutton, James, 448–56, 460
Huxley, Aldous, 489
Huxley, Julian, 117–18, 427
Huxley, Leonard, 170–71, 386, 393,
    395, 398
Huxley, Thomas Henry, 139, 168
    in debate with Wilberforce,
        385–401
    on evolution and ethics, 427
    on Gladstone, 403–15
    on horse evolution, 169–73, 175,
        177
    Kropotkin's response to, 330–31
    on natural selection, 328–29
hybirds, 484
Hymenoptera, 482–83
*Hyracotherium (Eohippus),* 175
    in evolutionary sequence, 178, 180
    fox terrier simile for, 158–65,
        173–74
    name of, 90
    weight of, 167
hyraxes, 160

imaginal disks, 255–56, 259
imagos, 255–57, 259, 260
    of glowworms, 261–64
Immigration Restriction League, 316
incipient stages dilemma, 140–44, 146

Indian caste system, 357–58
Ingersoll, Ernest, 164
Ingersoll, Robert G., 419
Ingram, Glen J., 298, 299, 301
insectivores, 286
insects
    development of, 255–60
    division of labor in, 265
    glowworms, 260–64, 267
    haplodiploidy in reproduction of,
        483
    wings of, 146–51
intelligence
    of echidnas, 290–92
    human, William James on, 320
intercourse, 129–36
*International Code of Zoological
    Nomenclature*, 80–81
International Commission on
    Zoological Nomenclature, 80, 83
International Union of Biological
    Sciences, 81
International Zoological Congress
    (1913), 79–80, 79–83
interspecific scaling, 118
intraspecific scaling, 118
Io (moon of Jupiter), 492–93, 495,
    496, 498, 507
IQs, heritability of, textbook
    discussions of, 156–57
*Isis* (journal), 331


Jackson, Andrew, 21–23, 28, 29
Jackson, J.F., 263*n*
Jackson, Shoeless Joe, 467
James, William, 14–16, 318–21, 324
Jefferson, Thomas, 188, 190, 402
Jenkin, Henry Charles Fleeming
    contributions to economics of,
        351–53
    Darwin's response to, 340–42
    debate between Darwin and,
        343–51
    Stevenson's biography of, 342–43
Jephthah (biblical), 59
Jesus, 155
Jevons, William Stanley, 353

Johnson, Jack, 432
Johnson, Philip, 62
Johnson, Virginia E., 128, 129
Jones, William, 32–33
Jordan, David Starr, 423, 430
Joyce, James, 251, 257
Jukes family, 429
Jupiter (planet), 489, 492, 507


Kael, Pauline, 105
Kahneman, Daniel, 468–69
Kallikak family, 429
Keeler, Wee Willie, 464
Keeton, W.T., 164–65
Kellogg, Vernon L., 423–25, 430
Kelvin, Lord (William Thomson),
    342, 349
Kennedy, John F., 402
*Keroplatus*, 263*n*
Keyes, Ralph, 479
Kidd, Benjamin, 423, 425–27
Kingsland, Sharon, 226
Kingsolver, Joel G., 145, 148–51
Kinsey, Alfred C., 128, 129,
    135–37
Kirby, William, 257–59
Kirkpatrick, Randolph, 225
Kirwan, Richard, 448–50, 454, 456
kiwis, 110–21, 503–5
Koehl, M.A.R., 145, 148–51
Kofoid, Charles, 219
Korea, 103
Kropotkin, Petr, 330–38, 427


lace crabs, 248–49
Lacépède, Bernard Germain Etienne
    de la Ville-sur-Illon, 356–58,
    364–66
Lafayette, Marquis de Marie Joseph
    Paul Yves Roch Gilbert du
    Motier, 188
Lagrange, Joseph Louis, 355, 358
Lamarck, Jean Baptiste, 125, 165–66,
    272–73
Lamarckian evolution, 125
    in cultural evolution, 65

Lamarckian evolution (*continued*)
  discussed in high-school textbooks, 165–66
  neo-Lamarckianism and, 315
land snails, 241
language(s)
  evolution of, 34–35
  genetic evolution of humans and, 39–40
  Grimm's work in, 32–33
  Latin, 80
  unified origin of, 33–34
Laplace, Pierre Simon de, 404
La Rosa, Julius, 202, 207–8
Larsen, Don, 464
larva, 255–60
  of glowworms, 260–64
Latin (language), 80
Laughton, Charles, 281
Laurel, Stanley, 255
Lavoisier, Madame, 364
Lavoisier, Antoine, 184, 356, 365–66
  execution of, 354–55
  founder of modern chemistry, 358–60
  on Royal Commission (France, 1784), 189–97
  as tax collector, 360–64
Lawrence, Thomas Edward, 74
Leather, Ted, 73
Le Boff, Gerald A., 184*n*
Le Conte, Joseph, 160
Lende, R.A., 289
*Leptomorphus*, 263*n*
Levine, L.W., 419
Liberty Bell, 122
Liem, K.F., 155
Lincoln, Abraham, 21
Linnaeus, Carolus, 81, 84, 255–56
Livingstone, David N., 314
Lloyd, Elisabeth, 130
Lloyd Prichard, M.F., 352, 353
Locke, John, 369
Longley, Ms., 70, 71
Louis-Napoleon (Napoleon III; emperor, France), 500
Louis XI (king, France), 356
Louis XVI (king, France), 188

Lufkin, Dan, 74
luminescence, of glowworms, 261–62
Lyell, Charles, 368, 380, 395, 448, 449
Lysippus, 235

McCarthy, Joe, 470
McDiarmid, R.W., 300
McDonald, K.R., 298
McGurrin, Frank E., 70, 71
McIntosh, J.S., 87
McKormick, Robert, 25
*McLean* v. *Arkansas* (U.S., 1982), 431
*Macroera*, 263*n*
"Madame Jeannette" (song), 201–7
*Maiasauria*, 97
males
  in haplodiploidal reproduction, 483
  kiwis, 111–12
  nipples of, 125–28
  platypuses, 278
  *Rhinoderma darwini* (frog), 297, 301–2
Malthus, Thomas Robert, 328, 332, 334, 335, 337
mammals
  breasts of, 127
  chimpanzee, 501–2, 504, 505
  echidnas, 282–92
  kiwi's similarity to, 117
  orgasms in, 131
  platypuses, 270–80
Maori people, 110
Marat, Jean-Paul, 362–63
Maris, Roger, 464
Marmorek, Ernest F., 184*n*
*Marrella*, 245–46, 248
Mars (planet), 491, 507
Marsh, Othniel C., 139
  *Brontosaurus* named by, 87–92
  on Cardiff Giant, 44
  *Eohippus* described by, 160–63
  T.H. Huxley and, 170–73, 175, 177
Marshall, Alfred, 352, 353
Marx, Karl, 340, 499–500
mass extinctions
  Cretaceous, 17

Permian, 17–18
role of comets in, 378–79
Masters, William H., 128, 129
masturbation, 129, 133
Mather, Bryant, 104–5
Matthew, W.D., 434
on evolution of horse, 162,
174–75, 177
*Hesperopithecus* fossil and, 444, 446
*Prosthennops* described by, 443
Mayr, Ernst, 82, 84
mean (statistical), 473
Meckel, Johann Friedrich, 272–75
Medawar, Sir Peter, 475
median (statistical), 473, 475–77
Mellett, James S., 443
Melson, Lewis R., 227–28
men, *see* males
Mencken, H.L., 416–17
Mendel, Gregor, 345, 481
Menotti, Gian-Carlo, 229
Menozzi, P., 35
Mercury (planet), 491, 507
mermaids, 271
Merriam, C. Hart, 214–15
Mesmer, Franz Anton, 184–94
mesmerism, 187–96
*Mesohippus*, 178–80
mesothelioma, 474–77
metamorphosis of insects, 255–60
metric system, 356
Mexico, 168–69
Micah (prophet), 408
Michener, Charles, 82
midwife toads, 301
Mill, John Stuart, 352
Mills, A.G., 48, 52
Mills Commission, 48–52
mimicry in animals
in color, 295–97
Cott on, 212
Darwin on, 304
Thayer on, 213
*Miohippus*, 160, 162, 178–80
Miranda (moon of Uranus), 489,
494–96, 507, 509
Mivart, St. George, 139–43, 146, 149
moas, 110, 116

Monaco, 79
monarch butterfly, 295
monkeys, 121
monogeny, 239
Monotremata, 271, 283
monotreme mammals
echidnas, 282–92
platypuses, 270–80
moon (earth's), 490–91
Moore, Ruth, 386–88, 399
morality
Darwinism and, 427
evolution and, 327
evolution and, Bryan on, 421
T.H. Huxley on, 408
Jenkin on, 351
Kropotkin on cooperation in, 331
science and, 429–30
Morris, Desmond, 131
Moses (biblical), 371, 375, 406
Mountain, J., 35
Mozart, Wolfgang Amadeus, 184
"Murphy's Law," 241
Museum of Comparative Zoology
(Harvard University), 309–13
mutations, 345
Mycetophilidae, 261
*Myrmecia pilosula* (ants), 482–83

Napier, Sir Charles, 269
Napoleon Bonaparte (emperor,
France), 356, 499–500
Napoleon III (Louis-Napoleon;
emperor, France), 500
*Naraoia*, 246
National Science Teachers
Association, 155
natural history, 501
natural selection, 60
animal coloration and, 221–22
Bryan on, 421, 433
current utility and historic origin in,
114–15
Danilevsky on, 334
discussed in high-school textbooks,
165
T.H. Huxley on, 328–29

natural selection (*continued*)
  among kiwis, 112–14, 119
  Kropotkin on cooperation in, 331
  mimicry in animals and, 304
  Mivart on, 139–43
  mutual aid in, 338
  social Darwinism and, 426–27
  variations in, 349
*Naturphilosophen,* 414
naval camouflage, 227–28
"Nebraska Man" *(Hesperopithecus),*
  435–47
Nelson, J., 288
nematodes (worms), 481
neocortexes, 286–89
neo-Lamarckianism, 315
Neptune (planet), 488, 507–9
Newell, Norman D., 243
New Orleans, Battle of (1815), 21–23,
  28
Newton, Sir Isaac, 368–71, 376–80,
  455
New York All-City High School
  Chorus, 201–8
*New York Times,* 380, 433
New Zealand, 110
nipples, 125–28
Nixon, Richard M., 385
Noah's flood, 372–75
Nostratic (hypothetical language), 33,
  40

O'Brien, P., 303
O'Casey, Sean, 251
Oland, Warner, 293
Omar Khayyám, 16, 467, 468
*Ophioglossum,* 484–85
*Ophioglossum reticulatum,* 485
Oram, R.F., 165
*Orfelia,* 263n
*Orfelia aeropiscator,* 263n
orgasms, 128–36
*Ornithorynchus anatinus* (platypus),
  271–80
Osborn, Henry Fairfield
  debate between Bryan and, 432–34

  on evolution of horse, 162–64
  *Hesperopithecus* fossil and, 434–46
Otto, J.H., 164
Overton, William R., 431
oviparous animals, 270
ovoviviparous animals, 270
Owen, Mickey, 499
Owen, Richard, 160, 272–74, 390

Paganelli, C.V., 118
paleontology, 98
  *see also* dinosaurs
Paley, William, 323
Paluxy Creek footprints, 447
pandas, 61, 66–67, 127
pangrams, 73
*Parascaris equorum univalens*
  (nematode), 481, 482
Paris, 362, 363
Pass, John, 122
Paul, Daine B., 156–58
peacocks, 219–21
penises, 127
Perceval, Spencer, 24
Perissodactyla, 181
Permian mass extinction, 17–18
Phillips, J.C., 222–24
Phobos (moon of Mars), 507
Phyllopods, 247–48
Piazza, A., 35
pigs, fossils of, 442
Piltdown Man, 42, 44, 287, 444,
  445
Pirlot, P., 288
planets
  geological activity of, 490–93
  predicting features of, 498
  Venus, 493–94
  *Voyager* probes of, 488, 489–90,
    494–96, 505–9
plants
  in Genesis, 413
  genetics of, 480
  T.H. Huxley on origins of, 407–8
  polyploidal reproduction in, 484–85
Plato, 125–26, 469

platypuses, 270–80, 282, 284
  brain of, 292
  Geoffroy Saint-Hilaire on, 284
  neocortex of, 287
Platyura, 263*n*
Playfair, John, 449–54
Pluto (planet), 488, 509
polygenism, 316
polygeny, 239
Polynesians, 109–10
polyploidy, 480, 483–85
Pompeii, 499
Pope, Alexander, 60, 468
Popper, Karl, 437
population, in Malthusian theories,
  333
Poulton, E.B., 214, 225, 279
preadaptation, 144
Prescott, William H., 168–69
Preservation Hall (New Orleans),
  380–81
Pridmore, 290
primates
  chimpanzees, 501–2, 504, 505
  *Hesperopithecus* fossil mistaken for,
    434–47
probability, 471–72
  in Jimmy Carter's theology, 323–24
  experiments in perceptions of,
    468–69
  of "hot hands" in basketball,
    465–66
  in human evolution, Shaler on, 317,
    319–20
  in positions of stars and
    glowworms, 268
  in report of Royal Commission on
    animal magnetism, 195
  in variations around central
    tendency (statistical), 476–77
prostaglandin, 303–4
*Prosthennops*, 436, 442, 443
Prothero, Don, 179–80
*Pteranodon*, 92
Pugh, George, 131
*Punch* (magazine), 390
Purcell, Ed, 267–68, 466–67, 471

Quatrefages, A.de, 256–57
QWERTY keyboards, 62–63
  origin of, 66–68
  survival of, 68–72


racehorses, 348
racial groups
  genetic data on evolution of,
    37–39
  Lacépède on, 357–58
  polygenist view of, 316
  skull measurements of, 232,
    237–38
racism
  Jenkin's, 344
  Shaler's, 316
Rahn, H., 118
Randi, James, 183
randomness
  individuality versus, 501
  in position of stars and
    glowworms, 260–61, 266–68
ratites, 110
Read, W. Maxwell, 164
Rehnquist, William H., 457
Reid, Brian, 113, 114
religion
  accusations of atheism and, 448–49
  creationism as, 456
  in debate over age of earth, 450, 452
  T.H. Huxley on relationship
    between science and, 408–9
Renfrew, Colin, 33
reproduction
  assortative mating in, 346–47
  in evolutionary struggle for
    existence, 327–28
  in frogs, 300–302
  in glowworms, 262–63
  haplodiploidy in, 483
  in humans, Erasmus Darwin on,
    125
  in humans, orgasm in, 130–36
  in kiwis, 111–20
  in platypuses, 278
  polyploidy in, 484–85

reproduction (*continued*)
  in *Rheobatrachus silus* (frog), 294,
    297–300, 302–6
  in *Rhinoderma darwini* (frog), 297
reptiles, 272, 276
  Gladstone on, 410
  T.H. Huxley on, 411
Réville, Alfred, 404
*Rheobatrachus silus* (frog), 294,
    297–300, 302–5
  extinction of, 306
*Rheobatrachus vitellinus* (frog), 300
*Rhinoderma darwini* (frog), 297, 301–2
*Rhinoderma rufum* (frog), 302
Rhodes, 290–91
Riggs, Elmer, 89–90
Robespierre, Maximilien, 356, 364
Robinson, Frank, 466n
rodents, 286
Roger, Jacques, 369
Rolfe, Red, 470
Romans, ancient, 236–37
Romer, A.S., 164
Roosevelt, Theodore, 402
  on animal coloration, 209–11, 217,
    219, 222–24
Rose, Pete, 464
Ross, E.A., 421
rote learning, 99
Rowe, M.J., 279, 280
Royal Commission (France, 1784),
    188–96
Russia, 332–33
Ruth, Babe (George Herman), 464

salamanders, 480
Saturn (planet), 489, 507
Saul (biblical), 294, 312
Saunders, 290
Scalia, Anthony, 450, 456–59
Schuchert, Charles, 243–44
Schumpter, J.A., 352
Schweber, Sam, 388–89
science
  biological determinism and, 428
  creationism as, 455–56
  education in, 99–103

grand theories in, 454
"hard" and "soft," 496–97
T.H. Huxley on relationship
  between religion and, 408–9
lack of moral and ethical import
  of, 429–30
limits of theorizing in, 455
self-correction in, 437
textbooks in, 155–67
traditional definition of, 500
in Whiston's history of earth, 376
Scopes, John, 175, 428, 434, 456
Scopes trial, 416–19, 431–433,
    456
Scott, W.B., 164
*segue*, 98–99, 103–6
Seidel, Michael, 463n
sex, haplodiploidy in specification of,
    483
sex organs, of humans, 127, 129–36
Shakespeare, William, 489
Shaler, Nathaniel Southgate, 14–16,
    312–21, 324
Shaw, George, 271, 282
Shearman, D., 303
Sheldrake, R., 339
Sholes, C.L., 68, 71–73
Shubin, Neil, 179–80
Siamese twins, 182
Siberia, 337–38
Sidgwick, Isabel, 395
*Sidneyia inexpectans*, 249
Silberstein, E.P., 155
Simpson, G.G., 177–79
Sisyphus (mythical), 406
size
  allometric studies of, 117–20
  dwarfism, 120–21
  of glowworms, 261, 262
  human height, 479
  of *Hyracotherium (Eohippus)*, 158–67
  of kiwis, 115–17
  of planets of moons, 491–98
  success not correlated with, 480
  thermoregulation and, 145, 149–51
skepticism, Bryan on, 422–23
skin color, 345
  *see also* racial groups

skulls, human
  Camper's facial angle of, 230–33
  Camper's measurements of, by
    race, 237–40
  in conceptions of beauty, 234–37
Smith, Adam, 264–65, 334, 352
Smith, Grafton Elliot, 287, 292, 440,
  444–46
Smith, W.H., 419
Smooth, R.C., 165
snails, 241, 333
social Darwinism, 424–27
Society for American Baseball
  Research, 464
Soderblom, Laurence, 495
"Sod's Law," 241, 243
space exploration, 488, 489–90,
  505–9
Spalding, A.G., 48, 50–52
species
  alloploidal hybrids of, 484
  of Burgess Shale fossils, 243
  of echidnas, 285
  of humans, 239
  mimicry among, 295–97
  names for, 81–85
  polygenist view of races as, 316
  wedge metaphor for struggle
    among, 328
Spence, William, 257–59
Spencer, Herbert, 139, 406
spinal cords, 286
sports (variations), 346, 349
stamps, 79
  Brontosaurus controversy and,
    85–86, 92
stars, randomness in positions of,
  260–61, 266–68
statistics
  epidemiological, 474–77
  measurements of central tendency
    in, 473
  see also probability
Stebbins, G. Ledyard, 485
Stevenson, Robert Louis, 342–43,
  346, 351
Stewart, Balfour, 389–90, 392,
  395
Stewart, Robert (Viscount
  Castlereagh), 23–29, 449
Stow, John, 122
Strahan, R., 290
Strickland Committee, 81–82
struggle for existence, 327–29
  Kropotkin on, 334–38
  Russian discussions on, 332–34
  social Darwinism and, 427
Sturm, Johnny, 470
Sullivan, Sir Arthur Seymour, 305
Surveyor (spacecraft), 491
Swammerdam, Jan, 258
Swift, Jonathan, 368
Symons, Donald, 130, 131

Taborsky, Michael, 503–4
Tachyglossus aculeatus (echidna), 285
Taft, William Howard, 209
talapoins, 121
Taub, Louis, 70, 71
taxonomy
  Brontosaurus name controversy and,
    85–93
  of Burgess Shale fossils, 243
  Latin used for, 80
  of monotreme mammals, 274
  problems in, 81–85
teeth, of Hesperopithecus, mistaken for
  human, 434–47
Teilhard de Chardin, Pierre, 339
telegrams, 269
teleosts, 175–77
Tennyson, Alfred, 106, 328
Tetrahymena pyriforme, 83–84
textbooks, 155–67
Thayer, Abbott Henderson
  naval camouflage based on theories
    of, 227–28
  theories of animal coloration of,
    210–26
Thayer, Gerald H., 210
Thayer Expedition (Brazil), 318
thermoregularion, 145, 148–51
Thomas, Dylan, 103
Thomson, Albert, 442
time, Jenkin on, 348

Tobin, George, 282
Todes, Daniel P., 331–34, 337
Tolstoy, Leo, 325–28, 332
Toscanini, Arturo, 229
Towle, A., 164
Triton (moon of Neptune), 508
Truman, Harry S., 310
Tunney, Gene, 432
Tversky, Amos, 465, 468–71
Twain, Mark, 473, 478
twins
    Burt's data on, 156
    Siamese, 182
Tyler, Michael J., 298, 299, 302–3,
    305, 306
typewriter keyboards, 62–72

Ulinski, P.S., 289
Union Station (Washington, D.C.),
    109
United States, in War of 1812, 21–23
Uranus (planet), 489, 494–96, 507

Vander Meer, Johnny, 464
variations
    argument between Darwin and
        Jenkin over, 343–50
    around central tendency
        (statistical), 476–77
Venus (planet), 493–94
Vernon-Harcourt, A.G., 393
viceroy butterfly, 295
Vietnam War, 22
Viking (spacecraft), 491
viviparous animals, 270
Voltaire, 114, 151, 184
von Baer, Karl Ernst, 414
Vonnegut, Kurt, 169
Vorzimmer, Peter J., 345
Voyager (spacecraft), 488, 489–90,
    492–94, 506–9
Voyager 2 (spacecraft), 494–96, 505–8
Vrba, Elizabeth, 144n
vulgarisation, 11

Wadlow, Robert, 479
Wagner, Richard, 340
Waitomo Cave (New Zealand), 260,
    267
Walcott, Charles Doolittle, 242–50
Walcott, Helena, 245–48
Walcott, Sidney, 244, 248–50
Wallace, Alfred Russel, 294
    Darwin's correspondence with,
        341, 346
    inspired by Malthus, 333
Waptia, 246
War of 1812, 21–23, 28
Weekly World News, 182–83
Whig interpretation of history, 343
Whiston, William, 368–81
White, Nelson G., 212
White, Shep, 96, 97
Wigglesworth, Sir Vincent, 146
Wilberforce, Samuel, 385–401
Wilhousky, Peter J., 201–8
Willard, Jess, 432
Williams, G.R., 113
Williams, Ted, 464
Wilson, Woodrow, 209, 402, 417
wings
    of Archaeopteryx, 144–45
    evolution of, 145–51
    T.H. Huxley on origins of, 407
    origins of, 141–43
Wolf, John, 443
women, see females
Woodward, J., 372
Woodward, Arthur Smith, 440
Wordsworth, William, 509–10
Wren, Christopher, 376

Yastrzemski, Carl, 46
Young, Ned, 389

Zaglossus bruijni (echidna), 285
Zaglossus hacketti (giant echidna), 285
zebras, 215